Calcium Carbonate

From the Cretaceous Period into the 21st Century

Edited by F. Wolfgang Tegethoff
in collaboration with Johannes Rohleder and Evelyn Kroker

Springer Basel AG

Cover illustration: Scanning electron microscope image of a coccolith.

A CIP catalogue record for this book is available from the Library of Congress, Washington D.C., USA

Die Deutsche Bibliothek – CIP-Einheitsaufnahme

Calcium carbonate : from the Cretaceous Period into the 21st century / ed. by F. Wolfgang Tegethoff. In collab. with Johannes Rohleder and Evelyn Kroker. - Basel ; Boston ; Berlin : Birkhäuser, 2001
Dt. Ausg. u.d.T.: Calciumcarbonat
ISBN 978-3-0348-9490-6 ISBN 978-3-0348-8245-3 (eBook)
DOI 10.1007/978-3-0348-8245-3

Printed on Biberist Furiso, calcium carbonate matt art paper (115 g/m^2)
Layout: Karina Schwunk, Essen

ISBN 978-3-0348-9490-6

Index

Foreword

Anyone who takes the trouble to look up all the entries that have been filed under "Calcium Carbonate" in a university library will be surprised – if not overwhelmed – by the enormous amount of literature that is concerned in one way or another with this subject. Most of the corresponding books and journals can be assigned to the realms of chemistry, but they also include technical, geological and mineralogical literature. A continued search will eventually unearth works on the history of the arts and architecture. This is because calcium carbonate minerals, be they in the form of chalk, limestone or marble, symbolize defining moments in the history of our culture – the first prehistoric cave drawings in simple coloured chalk, the gigantic limestone blocks used to build the Egyptian pyramids and the marble statues of Michelangelo and Bernini.

This great diversity of what, at first sight, may appear to be only a simple chemical compound makes it rather difficult to explain to a layman what our company is producing and how the product should be defined. This task is much easier if you produce cars or furniture since only experts from different disciplines can associate certain concepts with calcium carbonate. The geologist and mineralogist will envision limestone and chalk or the different forms of marble. The engineer concerned with processing and application technologies will be familiar with the mineral's complicated crystalline properties, while the artist or art historian will think of the flawless "statuario" from the Carrara marble quarries that is sculptured into products that beautify our world.

The different terms and definitions are as numerous as the multitude of "aesthetic" and "useful" items that can be produced from $CaCO_3$ – the chemical formula of calcium carbonate. Moreover, such a diversity and profusion invariably tends to be bewildering rather than clarifying.

This book sets out to overcome this confusing array by indicating the multifarious interdependencies that exist in the world of calcium carbonate – its geology and the history of the arts, its extraction and processing and, obviously, its use in modern industry. Thus, the geologist will find out how a superwhite calcium carbonate slurry is produced from marble for use in paper making. The paper manufacturer, in turn, will be on home ground when the difference is explained between the calcium carbonate coating pigments produced from chalk or marble.

This book is therefore not targeted at a specific professional group with clearly defined knowledge and closely outlined interests, but rather at a very broad spectrum of specialists – among them geologists, processing technicians, paper manufacturers and agricultural scientists – as well as interested laymen who are interested beyond their own particular fields.

The book will reveal how, over a period of many millions of years, coccoliths (Coccolithoporida) have formed mountains composed of a vital filler that has become irreplaceable in modern industry for many everyday products, be they paints, varnishes, plastics or high-quality art paper. This book sets out to fulfil the formidable task of elucidating all these varied aspects – a truly demanding aim that places the highest expectations on the authors and the editorial work for this publication.

Since technical museums are responsible for presenting the sciences, technology and history in a popular form, and since calcium carbonate is extracted by mining methods, it was only natural that the German Mining Museum in Bochum was approached to take over the responsibility of realising this book project. I found two very competent and committed partners in Dr. Evelyn Kroker and Johannes Rohleder who, after three years of intensive and inspiring work, transformed the initial ideas into this remarkable compendium.

At this point I would also like to thank Professor Jacques Geyssant (Paris), Dr. Peter Hess (Cologne), Dr. Eberhard Huwald (Gummern), Dr. Ralph Kuhlmann (Wuppertal), Dr. Christian Naydowski (Oftringen) and Dieter

Strauch (Oftringen) for their invaluable work. Obviously, I knew beforehand that they were all outstanding experts in their respective fields of "Calcium Carbonate". However, their contributions also proved that, as authors, they had mastered in a very impressive manner the need for scientific exactitude in conjunction with a lucid, generally understandable presentation.

There was also another direct reason for our company to have this book published under the scientific guidance of the German Mining Museum, namely the fact that the company looks upon itself as a modern mining company as it gains calcium carbonate as a raw material from marble, limestone and chalk extracted by quarrying as well as open-cast and partially underground mining of its own extensive deposits in Europe, America and Australia. Moreover, the company refines the raw material by sophisticated technological processes to produce high-quality fillers for industrial production. Some of these processing and refining operations are actually conducted in close co-ordination with the buyers of a products – a partnership that cannot be taken for granted but which is very much to our mutual advantage.

Three years of hard work for a book that is exclusively concerned with calcium carbonate – Is such considerable effort really worth while? The answer is a clear yes! Calcium carbonate has not only had a long history that started long before the Cretaceous Period, but it also has a great future that will extend far beyond the 21st century. And it has been the purpose of this book to unite the two – the past and the future. "Calcium Carbonate – From the Cretaceous Period right into the 21st century" – All the current knowledge has been compiled in this book so that it will be available for tomorrow.

If this is not reason enough, then allow me to draw attention to the French philosopher Denis Diderot who, when asked about the reason for his "Encyclopédie ou Dictionnaire raisonné des sciences, des arts et des métiers", gave a very simple reply: "If the only thing to be saved in the Great Flood had been a copy of the 'Encyclopédie', then not everything would have been lost."

And I hope that Denis Diderot would not mind if I, as a person of the 21st century, were to add: In the event that a CD-ROM version of the "Encyclopédie" were to drift in the Great Flood onto some remote piece of land, there would bound to be someone who would revive the knowledge stored in bits and bytes to produce a beautiful art-print binding in the old bibliophile form to satisfy Man's thirst for knowledge.

F. Wolfgang Tegethoff

I.

Geology of Calcium Carbonate

From mineral to rock – the deposits

by Jacques Geyssant

For the Romans, the term "calx" meant both the rock and the product which they obtained from it by burning. Consistently, the process of lime burning was called "calcination" in the Middle Ages, and, applying the same logic, in 1808 Humphry Davy called the metal which he obtained by electrolysis from calcium chloride, along with chlorine gas, calcium.

The effects of this lack of clarity and inaccuracy in designating limestones and all the products of limestone are still felt today. For most people all these products are still called lime, although scientifically accurate designations were introduced long ago. Burnt lime is calcium oxide (CaO) and, if slaked, calcium hydroxide is obtained [$Ca(OH)_2$]. The limestones, on the other hand, consist of calcium carbonate ($CaCO_3$).

1. Features and characteristics of calcium carbonate

Calcium carbonate is composed of three elements which are of particular importance for all organic and inorganic material on our planet: carbon, oxygen and calcium.

The elements carbon, oxygen and calcium have their origin in the interior of what are known as giant stars. Their synthesis began when all the original hydrogen which existed was used up and converted to the next highest element helium (see figure).

Three helium atoms, each with 2 protons and 2 neutrons, now fused at a temperature of 100 million degrees Celsius to form a carbon atom with 6 protons and 6 neutrons. For oxygen there were two different methods of synthesis: four individual helium atoms could fuse to one oxygen atom or first a carbon atom was created, which then fused with another helium atom to oxygen. Calcium only formed when the temperature had fallen to 500 million degrees. Now two carbon atoms and one oxygen atom fused to a new nucleus consisting of 20 protons and 20 neutrons, to calcium (see box).

When these giant stars finally exploded in a supernova, carbon, oxygen and calcium were scattered in space, to collect millions of years later as a component of interstellar dust around a star of modest size. Today this star is our sun and the dust gathered in its gravitational field to form the planets.

1.1 Calcium carbonate – a special compound

Calcium carbonate exists only on earth and possibly on Mars. In Shergotty in India a meteorite fell from the sky which is assumed to have broken off from the crust of Mars due to the impact of a huge meteorite and was catapulted into the cosmos – and this small meteorite contains a little calcium carbonate, as well as traces of gypsum.

Calcium carbonate: chemical composition

Calcium carbonate is a simple salt which results from the reaction of carbon dioxide with burnt (I) or slaked lime (II), in accordance with the following formula:

$$CaO + H_2CO_3 \rightarrow CaCO_3 + H_2O \quad (I)$$
$$Ca(OH)_2 + H_2CO_3 \rightarrow CaCO_3 + 2\,H_2O \quad (II)$$

The chemical formula for calcium carbonate corresponds to a mass ratio of 56.03 percent calcium oxide to 43.97 percent carbon dioxide, or 40.04 percent calcium to 59.96 percent carbonate.

Like all carbonates, calcium carbonate is sensitive to acids:

$$CaCO_3 + 2HCl \rightarrow CaCl_2 + CO_2 \nearrow + H_2O$$

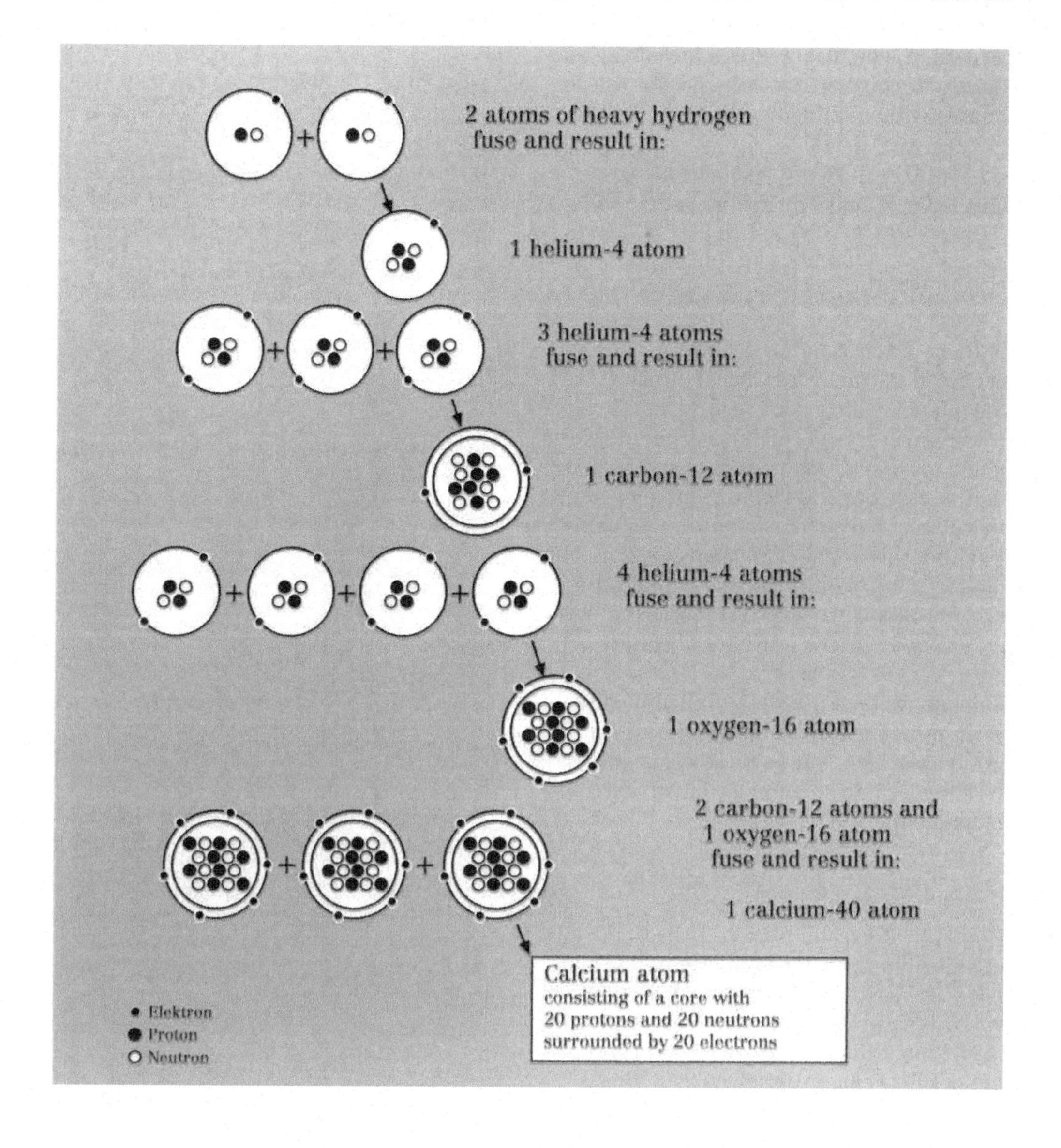

Synthesis of the three chemical elements carbon, oxygen and calcium, of which calcium carbonate is formed.

The most common method of recognition of carbonate rocks is based on this simple reaction: if a little hydrochloric acid is dropped onto a limestone, carbon dioxide is released and the fluid effervesces. If the rock is as porous as chalk, then the violent effects of the gas can even be heard, and the rock hisses and bubbles.

According to legend, this sensitivity to acid was also put to good use by the Carthaginian General Hannibal, when he crossed the Alps with his elephants in 218 B.C. In order to make it easier for the animals to climb among the rocks he ordered his soldiers to pour large quantities of vinegar on the compact limestone, which caused the rock to "weather" chemically and allowed the soldiers to break steps into the rock for the elephants without difficulty:

$$CaCO_3 + 2CH_3COOH \text{ (acetic acid)} \rightarrow Ca[CH_3COO]_2 \text{ (calcium acetate)} + CO_2 \nearrow + H_2O$$

Calcium carbonate and water

Is there water on Mars? This question has occupied astronomers since they discovered the planet which is closest to earth. And when tiny traces of calcium carbonate were found on Mars, many believed that proof had at last been found that at some time, there must have been water on this planet. Whether this is true is an open question, but on earth the formation, sedimentation and erosion of calcium carbonate rocks is always associated with the presence of water.

Calcium carbonate is very difficult to dissolve in pure water. Solubility is just 13 milligrams per litre, and it is the carbonate ion which goes into solution as a hydrogen carbonate ion:

$$CaCO_3 + CO_2 + H_2O \rightleftharpoons Ca(HCO_3)_2$$

However, if carbon dioxide is present, this equilibrium shifts far to the right, and solubility increases by more than 100 times. If water is saturated with dissolved carbon dioxide, more than 1 gram of calcium carbonate per litre can be dissolved.

Calcium: characteristics and occurrence

With a density of 1.55 grams per cubic centimetre, calcium is classified as a light metal. Calcium is very soft, melting at 842 degrees Celsius and boiling at 1484 degrees Celsius, it is silvery white in colour and burns to calcium oxide if it is heated in air. Calcium reacts violently with water under the influence of heat and the formation of hydrogen, and the resulting calcium hydroxide dissolves.

In the universe, the element calcium, at a concentration of 1.1 parts per million (ppm) occupies the 13th place in the list of the most common elements. On earth, on the other hand, the metal belongs to the 7 most common elements and, as a component of over 700 minerals, it even takes the 5th place in the earth's crust, at a proportion of 3.63 percent.

The high concentration of calcium in the earth's crust is a result of hydrothermalism (see section "$CaCO_3$ cycle"), which leads to an increased concentration of calcium carbonate, particularly on the oceanic ridge. The extent of this concentration is shown by a comparison with magnesium, which is chemically related. Whilst the occurrence of magnesium (mainly as a component of magnesium silicates) is twenty times greater in the earth's mantle rock, in the earth's crust approximately twice as much calcium as magnesium is found.

In daily life the measuring unit which has become established for the content of dissolved calcium carbonate is water hardness, measured in degrees; however, water hardness is defined differently in different countries. In France, for example, 1 degree of hardness corresponds to 10.3 milligrams of $CaCO_3$ per litre; in Great Britain, on the other hand, 10 milligrams per 0.7 litre and in Germany, 10 milligrams CaO per litre of water. This makes comparisons a complicated calculating task[1], because in Germany water is considered to be "hard" at hardness grade 14; in France, it is considered hard only as from hardness grade 20, whilst under this level, it is soft.

The differentiation between hard and soft relates to the behaviour of soap in water. If the water is hard, insoluble calcium oleates or lime soaps form which prevent the formation of lather, reduce cleaning power and also form deposits as white marks on clothes. But limestone can also deposit on immersion heater and flow heater elements, and when cooking with hard water there is an undesirable formation of calcium pect-

[1] A conversion of the varying hardness grades is possible by using the following formula: 1° dH (degree of German hardness) = 1.25° English hardness = 1.734° French hardness.

ates in the cell membranes of vegetables. While cooking is supposed to soften the cell membranes and hence the vegetables, calcium pectates make the vegetables hard and unpalatable.

As the solubility of calcium carbonate is proportional to the volume of dissolved carbon dioxide, it is determined by the same factors which also determine the concentration of the carbon dioxide (measured as partial pressure): pressure and temperature.

The influence of pressure

When water flows through the fine channels (diaclases) of a limestone or a volcanic rock containing calcium, it is under pressure. This also increases the partial pressure of the carbon dioxide, and the water can dissolve much more calcium carbonate. If this water then reaches the outside atmosphere, the water pressure adjusts to the atmospheric pressure, carbon dioxide escapes, calcium carbonate is precipitated and forms the typical limestone deposits. Travertines, more or less porous limestones, are formed in this way.

However, crust-like deposits of calcium carbonate also form at the outflows of underground springs in limestone areas and at springs in volcanic areas if the original rock has a high content of calcium silicates. For example, at Clermont-Ferrand there are the petrified springs of St. Alyre, which issue from the foot of the Chaîne des Puys, a chain of volcanoes which became extinct several thousand years ago.

The water there is rich in carbon dioxide, and the concentration of dissolved calcium carbonate is correspondingly high and can be used for "petrification". First the coloured iron carbonates in the water are removed by allowing the water to flow over wood chips and pebble stones. Then the pre-purified water falls in small cascades over a wooden stairway, on the steps of which the objects for petrification are placed: within two to three months they are completely covered with a thin, white, shiny layer of plimestone.

Just how much calcium carbonate can dissolve under natural conditions is shown at Blumau in Austria, south of Vienna. The underground thermal spring has been tapped by means of a bore, and now water which is almost boiling spurts from this bore at a pressure several times atmospheric pressure, and several tonnes of very white calcium carbonate are deposited in this way daily.

Limestone deposits in water courses

When a stream or a river flows over rapids or plunges over a waterfall, a drop in pressure occurs. The partial pressure of the dissolved carbon dioxide falls, limestone deposits form in the river bed and cause an obstruction to the flow, so that the turbulence of the water at this point even increases. In the water which accumulates in front of the obstacle a considerable amount of vegetation develops, consisting of algae, mosses and other plants, which activate the limestone deposits by drawing off more carbon dioxide from the water by assimilation. This forms travertines, which are rich in very finely formed plant impressions. As the plant remains rot away over time without leaving any residues, the porosity of these travertines is even greater than usual.

Because of the way they evolve, travertines often contain rich fossil finds, but the fossils found at Sézanne in the Marne Departement are particularly striking. About 50 million years ago in the Thanetian (Paleocene), large amounts of calcium carbonate must have been deposited over a short period in a small lake with lime-rich water, for the calcium carbonate petrified the stems and leaves of all the plants as well as the skeletons of the insects and vertebrates which once lived in the pond. In this way a small natural museum arose, which today allows us a glimpse of life 50 million years ago.

Dissolving of carbonates under pressure

Just as a loss of pressure leads to the precipitation of calcium carbonate, a rise in pressure can also increase the dissolving of calcium carbonate. This phenomenon is met in nature almost as frequently.

Irregular, ribbed planes can be observed in limestone beds, their cross section is reminiscent of skull sutures. The short projections or small columns of these planes, a few millimetres to centimetres long, are called stylolites and all lie in a parallel direction. Their surface is covered with a blackish or brownish skin of materials containing carbon and clay, which are more or less rich in iron oxides. These are mineral components of the rock which do not dissolve in water, even under pressure, and remain behind as solid residue.

As the direction of the stylolites corresponds to the direction of the maximum pressure which dissolved the limestone, styolites give the geologist important information on the natural stresses to which the limestone was subject in the course of geological times.

If the stylolites are vertical, they were formed by the pressure of the rock formations lying above them – the pressure was of lithostatic origin. However, if the stylolites are in an almost horizontal direction, tectonic dislocations were responsible for their formation. Measuring the direction of the horizontal stylolites in the limestone layers in southern Germany thus allows to reconstruct the different stages of the collision between Africa and Europe, which led to the formation of the eastern Alps 60 to 20 million years ago.

Some of the limestone dissolved on the stylolite surfaces is found in the secondary calcite which fills the tension cracks. In limestones such tension cracks are frequent, and they can be observed on polished rock surfaces as more or less straight or S-shaped white figures (see figure).

Influence of temperature

If limestone-rich water is heated, the dissolved carbon dioxide escapes and calcium carbonate is precipitated. "Scale" in kettles and on water heater elements is due to this reaction.

But in nature, too, cold water can dissolve large quantities of calcium carbonate and

Cross section through a limestone plate where stylolites and calcite bands have formed due to a horizontal dislocation (a). The connection between both phenomena is clearly discernible: whilst the stylolites form vertically to the fault pressure (black arrows) (b), the white bands are a consequence of the expansion of the rock vertically to the fault pressure (white arrows). This expansion leads to cracks which are then filled with calcite (c).

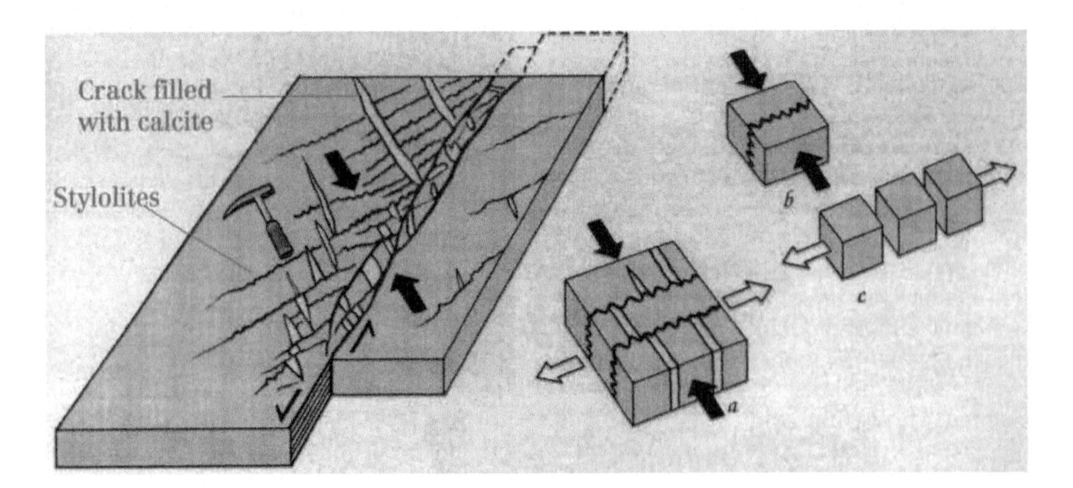

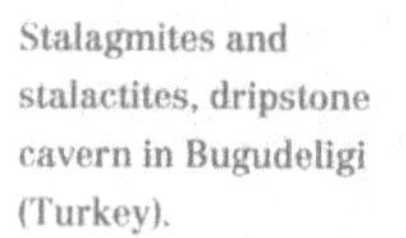

Stalagmites and stalactites, dripstone cavern in Bugudeligi (Turkey).

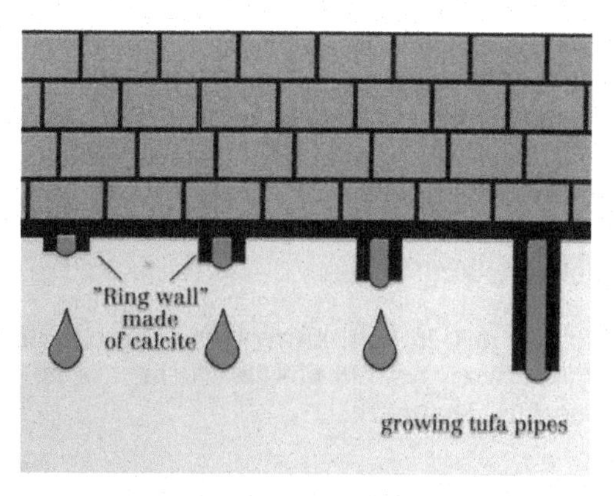

Formation of tufa pipes (from Lieber, p.161).

precipitate it as soon as it is heated. The cold water of mountain streams transports calcium carbonate from the rocks into the lower lying lakes. These lakes become warmer during the short, hot summer, calcium carbonate is precipitated and deposited on the bottom of the lakes. As materials containing clay are washed down to the lakes and deposited there during the following winter, layered sedimentation often forms at the bottom of mountain lakes. A dark, very thin layer of clayey material is deposited on top of a light-coloured layer of material containing carbonates deposited in summer. A light and a dark layer together form a varve, an annual layer approximately 1 centimetre thick. By counting the varves in the lakes of northern Europe, it was possible to draw up an absolute chronology for the retreat of the glaciers in the Quaternary period.

Precipitation by water loss

Just as the carbon dioxide content, the water volume too determines the balance between dissolved and undissolved calcium carbonate. If the proportion of water drops, a supersaturated solution results, from which the calcium carbonate precipitates. In nature there are two main ways in which the water content can fall: by evaporation or by freezing.

The best known example of the precipitation of calcium carbonates by evaporation is the formation of dripstones in the karst caves of limestone rocks: the stalactites and stalagmites (see figure).

Dripstones

Stalactites form on the roofs of caves. The water flows through a channel and deposits the calcium carbonate as shining crystals around this channel. Stalactites are therefore hollow and form real pipes or "organs" (see figure).

Stalagmites, on the other hand, are formed on the floor of a cave from the calcium carbonate which is dissolved in the drips which fall from the roof. For this reason stalagmites are not hollow and are structured in layers which cover one another from the bottom upwards.

Dripstones, in fact, always occur in one of these two ways, but possible forms are numerous. There are:

- monocrystalline stalactites in the form of fine, fragile tubes called “macaronis”, which can be several tens of centimetres long;
- very fine hanging structures only a few millimetres to centimetres thick, which can grow to several metres long;
- pearls with a diameter of a few millimetres, which consist of concentric layers of shiny calcite or aragonite crystals;
- and there is also “moonmilk” or “rock milk”.

This loose structure made of fine calcite needles of a length of several micrometers absorbs water very easily. Hence the observer easily gains the impression that he is looking at a fluid similar to milk. As X-ray analysis shows, rock milk contains, besides calcite, hydromagnesite [$Mg_5(CO_3)_4(OH)_2 \cdot 4H_2O$] and huntite [$CaMg_3(CO_3)_4$].

But water evaporation is a daily occurrence even in dry areas. If the water previously covered a plant, a stone or some other object, this is covered with a lime encrustation (calcrete) which conceals the underlying formations.

If evaporation describes the transition of water from a fluid to a gaseous condition, the freezing process leads to a solid material, in this case ice. Regardless of whether water vapour or ice, both aggregative conditions of water have in common that they contain practically no dissolved calcium. In contrast to everyday evaporation, precipitation of calcium carbonate by freezing is, however, rare in nature; it is significant above all in connection with global climatic shifts such as the ice ages. During the numerous ice ages in the Quaternary period,

Petrographic characteristics of rock-forming carbonate minerals. Dolomite as calcium magnesium carbonate has very similar characteristics to calcite in particular.

	Calcite	Aragonite	Vaterite	Dolomite
Crystal system	Trigonal	Rhomboid	Hexagonal	Trigonal - rhombohedral
Crystallographic forms	Hexagonal prisms Rhombohedron Scalenohedron	Hexagonal prisms	Hexagonal prisms	Rombohedron
Double refraction index	0.172	0.156	0.172	0.177
Character of double refraction	Optically negative	Biaxially negative	optically negative	optically negative
Specific density [g/cm^3]	2.72	2.94	2.72	2.8-2.9
Hardness	3	3.5-4	3	3.5-4

countless streams, rivers and lakes froze, the dissolved calcium carbonate was precipitated and cemented the loose rock of the alluvions and screes, as can still be seen today.

1.2 The crystal forms of calcium carbonate – mineralogy

Calcium carbonate is a polymorphous, variform compound, which occurs in three different crystal modifications: as vaterite, as aragonite and as calcite (see figure). It is true that in the shells of many snails all three varieties occur alongside one another, but the dominant modification in nature is clearly calcite. It is not only the predominant crystal in the massive limestone rocks; in combination with quartz, barite and fluorite it also forms the parent rock of many mineral veins. It can even be the only component of veins, the thickness of which can vary from a few centimetres to tens of metres.

In the elementary crystal lattice of calcite, calcium, carbon and oxygen atoms are arranged as follows: In a carbonate group $(CO_3)^{2-}$ the carbon atom is in the centre of an equilateral triangle, the corners of which are occupied by the three oxygen atoms. The individual carbonate groups are again arranged vertically to the C axis (symmetry axis 3rd sequence).

Calcite

Calcite crystallises in the rhombohedral system, where the elementary cell of the crystal lattice is a rhombohedral prism (see figure). This prism can be regarded as a cube which is pressed together or pulled apart in the direction of a diagonal. All the surfaces are rhombic and identical in width and length.

Calcite is one of the most common minerals in the Earth's crust. With its several hundred shapes, it is at all events the one with the largest number of shapes (see figure). The individual surfaces of the crystals can show

Calcite crystals with rhombohedral habit.

Calcite crystals with scalenohedron habit.

Idiomorphic calcite scalenohedron

very different characteristics, but always form the same angles to one another of 105 and 75 degrees. Often the crystals are very large, sometimes as much as several metres long, and there are quarries in which calcite is only excavated in the form of such "single crystals".

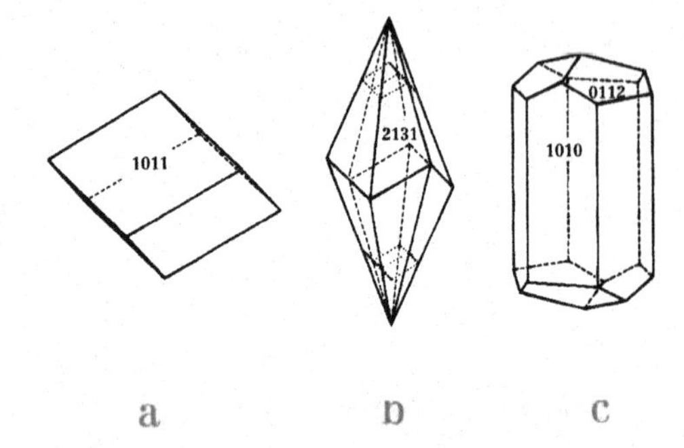

Some calcite crystal forms:
a) Rhombohedron,
b) Scalenohedron,
c) Prism.

Crystals and twin crystals

The rhombohedron in its basic form is surprisingly rare in natural calcite, but it occurs as a split form of most calcite crystals. Hexagonal (six sided) prisms and scalenohedrons are more common (see figure).

Twin crystals also occur in large numbers. These are intergrowths of several crystals of the same type forming according to certain laws. Examples are the heart or butterfly shaped crystal complexes consisting of individual scalenohedrons, which grow into one another pseudosymmetrically. Distortion of the individual crystals by 60 degrees towards one another can also occur.

The role of calcite in the discovery of crystallography

Calcite can be crystallised in all forms and combinations of the rhombohedral system, and the importance of this characteristic for the derivation of the crystallography laws should not be underestimated, as history shows. As early as 1778 the English doctor William Pryce anticipated the basic principles of crystallography when he established, in "Mineralogia Cornubiensis", that all forms of calcite arise from simple splitting of the basic form of the rhombohedron – but he did not consider his observation to be particularly important, and so it remained for the French mineralogist René Juste Haüy (1743-1822) to develop the first crystallography useable in practice.

As so often there is a legend centred around Haüy's significant discovery. If, for Newton, it was the apple which fell from the tree and made the laws of gravity clear to him, in Haüy's case it was a large calcite crystal which fell from the table on to the floor and shattered into a thousand pieces. When he was picking up the countless fragments Haüy noticed that, in fact, they all had a shape different from that of the original crystal, but all resembled the rhombohedral Iceland spar. Haüy repeated the process with the various crystal forms of calcite, and each time obtained a rhombohedron. If he broke up this rhombohedron again, he obtained another rhombohedron. He concluded from this observation that the crystals arise from the repetition of the elementary lattice or the elementary cell in the three spatial directions. He set down his observations in 1781 and 1782 in his "Mémoire sur la structure des crystaux" (Study on crystal structure). This formulated the basic laws of crystallography and illustrated them using the example of calcite.

A calcite rhombohedron shows its double refraction.

A thin plate (0.03 mm) cut from a marble (Austria) in polarised light (16 x magnification), showing calcite crystals with numerous twin lamella and ochre yellow tremolite crystals.

Of course, Haüy's crystallography was not yet perfect, and the German mineralogist Ludwig Seeber developed it in his "Versuch einer Erklärung des inneren Baus der festen Körper" (Attempt to explain the internal structure of solid bodies) in 1824 – but that is another story, in which calcite has no role to play.

Double refraction and other optical characteristics

All non-cuboid minerals pervious to light have two main refraction indices and therefore demonstrate the phenomenon of double refraction: they break each ray of light into two parts when it passes through the crystal. With most of these minerals the difference in the refractive indices is only slight, and the division of the ray of light is not visible with the naked eye. In the basalt rocks of Iceland, a calcite is found in which the double refraction counts among the strongest of all minerals: Iceland or double spar calcite. The difference between the refraction index of the ordinary part ray ($n_o = 1.658$) and that of the extraordinary part ray ($n_e = 1.486$) is 0.172 for this mineral.

When Iceland spar is examined with the naked eye, a dual image is seen (see figure). However, this is only true, if it is looked through from the right position, as a calcite crystal is not only uniaxially negative, it is also anisotropic. This means that certain characteristics of the crystal depend on the direction in which they are determined, including double refraction. This does not occur if the light ray passes through the crystal in the direction of the optical axis. If the ray passes through the crystal vertically to this axis, on the other hand, the double refraction is particularly strong.

Calcite not only has double refraction, but also polarises the light, so that after it has passed through the crystal it can only oscillate linearly in a single direction vertically to the direction of transmission. This phenomenon was exploited in 1828 by the English physicist William Nicol. He bisected an Iceland double spar, cemented the two parts together with Canada balsam and hence obtained a simple polarisation filter. His simple device is still used today as a "Nicol prism" in polarisation microscopy. Large, transparent calcite crystals, such as those found in many veins of pure calcite, can also be used for its manufacture.

If a thin cut calcite plate, at the very maximum 30 micrometers thick, is examined with a polarisation microscope, very bright polarisation colours can be seen (see figure).

The various cross sections of the calcite crystals shimmer in grey, pink and white colour tones, which are reminiscent of fire in fine pearls. If the crystals are split instead of being cut, very fine lines result which intersect at an angle of 120 degrees. With polysynthetic twin crystals, on the other hand, very fine small plates result, which are discernible due to the different polarisation colours.

Physical characteristics

Each crystal lattice is based on a fixed spatial arrangement of its atoms. If it is changed by an external force, stresses arise in the crystal lattice which sometimes lead to surprising phenomena. For example, simply pressing a calcite crystal together with the fingers creates a positive electrical charge.

But in normal condition, too, free of all outside influences, the arrangement of the atoms to a crystal is responsible for individual physical characteristics such as hardness. And as only very few crystal forms are completely symmetrical, the individual surface of crystals often have different characteristics. This also applies to calcite, as is shown in particular by the relative hardness. This is measured on the main surfaces at hardness 3 in accordance with Mohs' hardness scale (see figure), whereas on the base of the crystal it is only 2.5; here calcite can be scratched with the finger nail.

However, it is not possible to draw conclusions about the hardness of a rock from the hardness of a mineral, as the former is characterised to a considerable extent by the cohesion of the individual crystals to one another. This means that a limestone can be as soft as chalk if it consists of calcite grains which are not adequately cemented together. A compact, hard limestone, on the other hand, contains the same calcite crystals of the same hardness, but here they are consolidated by a calcitic cement.

Mohs' hardness scale: the average relative hardness of a mineral is determined on the basis of a series of 10 minerals with increasing hardness. Each mineral abrades the previous one and is abraded by the next one. Calcite (3) is thus abraded by fluorite (4), but abrades gypsum (2). This hardness scale was drawn up in 1822 by the German mineralogist Friedrich Mohs (1773 - 1839).

Absolute hardness	Relative hardness	Reference mineral	Chemical formula	Comparisons
800 – 1100	10	Diamond	C	
400	9	Corundum (sapphire, ruby)	Al_2O_3	
200	8	Topaz	$Al_2[SiO_4\|(OH,F)_2]$	
	7.5 - 8	Beryl (emerald)	$Al_2Be_3[Si_6O_{18}]$	
100	7	Quartz	SiO_2	←Steel file 6.5
50	6	Orthoclase	$K[AlSi_3O_8]$	← Window glass 5.5 ← Knife blade 5+
25	5	Apatite	$Ca_5\ (F,Cl,OH)\ [PO_4]_3$	
	4	Fluorite	CaF_2	← Copper 3+
8.3	3	Calcite	$CaCO_3$	← Finger nail 2+
	2	Gypsum	$CaSO_4 \cdot 2H_2O$	
1	1	Talc	$Mg_3(OH)_2[Si_4O_{10}]$	Graphite

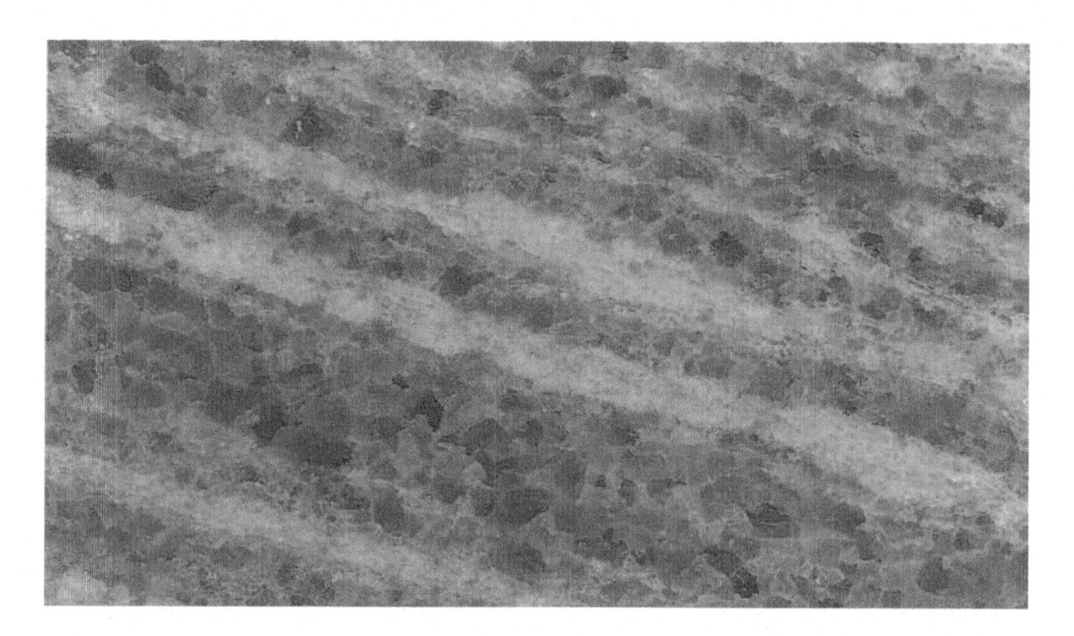

Polished section through a marble from Kenya with blue calcite crystals (original size).

The theoretical density of the calcite is 2.71 grams per cubic centimetre [g/cm^3]; the actual density varies, though, between 2.6 and 2.8 g/cm^3, depending on the extent to which calcium ions in the crystal lattice are replaced by other metal ions such as iron, manganese or zinc.

The colour of calcite

Pure calcite is transparent and colourless. In nature, however, it is found only rarely. Apart from Iceland spar, natural calcite is usually coloured honey yellow to yellowish brown, whilst massive varieties are milky white.

The varying coloration of calcite results when ions from other metals such as iron, zinc, cobalt or manganese replace calcium ions in the crystal lattice. Iron, for instance, results in a yellowish brown coloration, similar to that of siderite ($FeCO_3$); zinc leads to a grey-white coloration, comparable to that of smithsonite ($ZnCO_3$); cobalt gives rise to pink tones rather like those caused by spherocobaltine ($CoCO_3$), and manganese, finally, gives mauve or violet coloured shades which are in contrast to the red of rhodocrosite ($MnCO_3$) and the matt pink of kutnahorite [$CaMn(CO_3)_2$]. In addition, varieties containing manganese are often fluorescent carmine red.

If a small quantity of the emerald green to blackish green malachite [$CuCO_3\ Cu(OH)_2$] is mixed in with the calcite, it can even take on a green coloration, such as may be observed in the secondary calcite veins of the limestone massif of Vizarron in Central Mexico. All these colorations often underline the individual zones of calcite crystal growth.

More extraordinary is the colour of the sky to lavender blue calcite (see figure). This coloration is due to defects in the crystal lattice of the calcite caused by radiation from radioactive minerals. These defects lead to a differential absorption of light, where only the light waves corresponding to blue light are reflected.

The fact that the blue coloration has purely physical causes is shown when a blue calcite crystal is ground up. Its beauty disappears very rapidly, and the resultant powder is as white as any other calcite. Even in the crystal or rock the blue coloration does not continue forever; it diminishes over time, and if the crystals are exposed to sunlight, it disappears completely after a few months. It is even possible to accelerate this process. If the crystals are heated to 275 degrees Celsius, the discoloration takes only twenty minutes.

But today, physical processes can be reversed, provided suitable tools and methods are available. Any ordinary, colourless calcite crystal, can be turned into one of the rare and beautiful blue ones by artificial radiation.

A blue colour is extremely rare in ornamental stones, and the interest in such stone was, and is, correspondingly great. As artificial manufacture would be too complex for commercial use and, above all, too expensive, people stuck to the classical methods and for a long time quarried blue calcite marbles on an industrial scale. There were large quarries in Tatlock in Canada and in Crestmore in California, where "sky blue marble" was extracted. But blue marble can be found in Scandinavia, too, in particular in marble deposits with granite intrusions, as these are usually rich in radioactive minerals.

Aragonite

Considerably rarer than calcite is aragonite, where calcium carbonate is crystallised in the orthorhombic form. The name aragonite is derived from one of the most significant occurrences of the mineral, the gypsum and salt containing marls of Molina in Aragon in northern Spain.

The most characteristic form of the mineral aragonite is a twin crystal: three individual, prism shaped crystals are grouped together here to a pseudohexagonal prism with deep vertical ribs. The base of this twin crystal is ribbed in three directions, whereby each rib shows where the originally individual crystals have grown together. The crystal can easily be split along this dividing line (see figure).

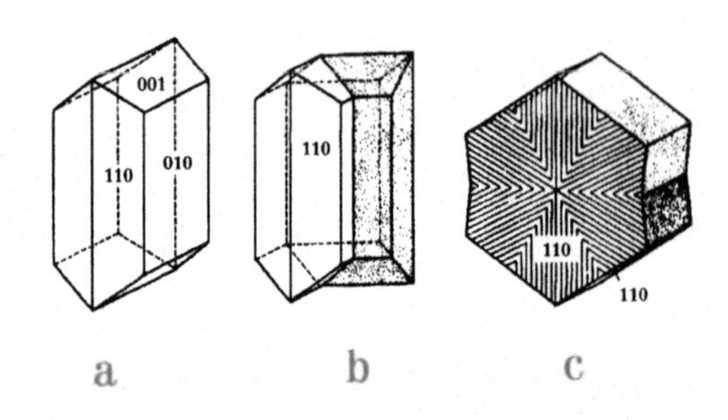

Characteristic aragonite crystals. The single crystal (a) is comparatively rare, twins (b) and the pseudohexagonal triplets (c), on the other hand, are very common.

The double refraction, at a value of 0.155, is lower than the value for calcite and the aragonite crystal is biaxially negative. Usually, aragonite is fairly pure, but it can contain up to four percent strontium. Sometimes the mineral is colourless, often honey yellow, but mostly white and non pervious to light with a glassy to resinous shine at the point of fracture. Hardness and density are greater than for calcite (see page 8).

The weakest of the split surfaces is the one parallel to the longitudinal direction; it allows a differentiation from calcite using the polarisation microscope. But a warm solution of cobalt nitrate [$Co(NO_3)_2$] can also serve as a test reagent for aragonite. If a calcium carbonate crystal takes on a pinkish lilac hue after adding a few drops of this solution, the substance is aragonite; if it is calcite, no reaction occurs, or only a very slight and delayed reaction.

At a geological scale, aragonite is not stable and is thus found much more rarely in nature than calcite. But aragonite is frequently the dominant component in the calcareous shells of numerous organisms, such as the shells of the giant tridacnas, as it was originally formed there and transformed into calcite only gradually. Pearls and mother-of-pearl are also a mixture of aragonite lamella and organic material in changing composition (see page 16). And even in the mineral kingdom there is aragonite. In the iron ore mines of the Erzberg in Styria one comes across "needle spar", a coral-form variety of aragonite, and in northern Germany "Schaumspat" or slate spar is commonly found. This mother-of-pearl type Triassic formation is a pseudomorphosis of gypsum into aragonite.

Little aragonite is deposited in fresh water and individual occurrences are characteristic of marine environments. In contrast to most other mineral substances, calcium occurs more rarely there than in fresh water, as numerous organisms draw the mineral from the water, in order to build their calcareous shells and skeletons. This is why the ratio of magnesium to calcium has also shifted further and further to the side of magnesium in the course of geological time, and as magnesium favours the formation of aragonite compared to that of calcite, it is rather aragonite which tends to form today in the marine environment.

Besides magnesium, traces of other metals such as strontium, lead, barium and calcium sulphate, and temperatures over 50 degrees Celsius, also favour its crystallisation. However, it is transformed over long periods into the more stable calcite, and therefore aragonite is rare in old carbonate rocks.

In the Bahamas in particular, as well as on the Bermuda Islands, aragonite sands have been forming for several thousand years in the tidal area, which are suitable for industrial use. The deposits are excavated by dredger and used for cement production.

Vaterite

Vaterite, named after the German chemist and mineralogist Heinrich Vater, is the hexagonal crystallising form of calcium carbonate. It is extremely unstable and only occurs exceptionally in nature. For artificial precipitation of calcium carbonate the conditions can, however, be selected to favour the formation of vaterite. Vaterite has also been found in snail shells along with calcite and aragonite, it is formed at the beginning of the creature's life and is then transformed into the more stable calcite in the course of time.

The physical and optical characteristics of vaterite are similar to those of calcite. The crystals are always small, and vaterite generally occurs in fibrous form; sometimes, though, it forms fine, microscopically small plates.

2. The limestones – development and classification

Calcium carbonate is the main component of chalk, limestone and marble. These rocks have a specific behaviour with respect to erosion. In contrast to magmatic rocks, they are subject to very little alteration, but readily soluble in water containing carbon dioxide.

In the course of the Earth's history even the most mighty limestone massifs stretching over an area of several thousand square kilometres and several thousand metres in height have thus disappeared due to dissolution. Over millions of years their calcium carbonate is washed into the sea by rivers, the waters of which remain beautifully blue and clear thanks to the high concentration of calcium ions.

During their slow disappearance the limestone massifs retain their relief, only their dimensions continually reduce. This behaviour distinguishes limestone from the harder, magmatic rocks such as basalt and granite. These rocks are broken up by weathering and erosion until only round fragments of rock and coarse grained sands remain.

2.1 Sedimentation

Sedimentation is the rock forming process from which all limestones originate, with the exception of the very rare carbonatites; marble, too, is ultimately only a sedimentary limestone which has undergone metamorphism through the influence of pressure and/or temperature.

Sedimentary rocks generally form in two stages: initially loose materials are deposited in layers, and then, in the stage of diagenesis, they are consolidated to rock by pressure or cementation.

Calcium carbonate is deposited either by chemical precipitation, by biochemical processes and by organogenic sedimentation. Whilst chemical precipitation is mainly associated with fresh water, the last two processes take place in salty seawater. Which crystal form occurs during the process is dependent both on the type of organisms (see figure) and also on the temperature of the

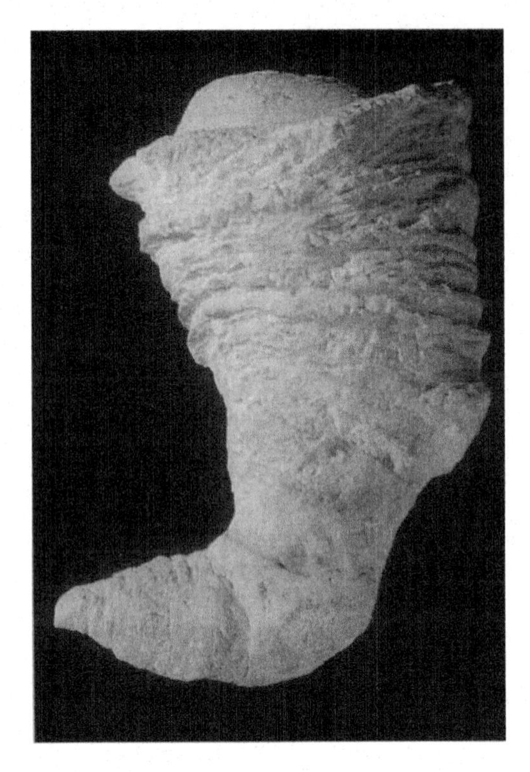

Biradiolites cornupastoris, Upper Cretaceous (Turonian), Dordogne, France (height 13.5 cm).

Mineralogical composition of the carbonate shells of major organisms.

Organisms	Aragonite	Calcite with little magnesium	Calcite with much magnesium	Calcite und Aragonite
Molluscs	+		o	o
Corals	+	+	+	
Sponges	+	+	+	
Polyzoons	+		+	+
Echinoderms			+	
Foraminifers	o	+		
Algae	+	+	+	

+ = Usual mineral
o = Occasional mineral

seawater. Warm water, for instance, favours the formation of aragonite and calcites, which contain between 4 and 15 percent magnesium carbonate.

Organogenic sedimentation

In the case of organogenic or bioclastic sedimentation, the original material forming the rock is of biological origin. This mainly concerns the inorganic remains of invertebrates, which deposit on the sea bed and become consolidated in the course of time. The size of the component elements fluctuates considerably, ranging from whole and broken shells of mussels and other mollusks, to coccoliths measuring a few thousands of a millimetre, discoid skeletal parts of single-celled marine algae.

If the deposits remain loose and are not cemented, but rather have the appearance of shelly sand, they are referred to as shelly limestone. This type of limestone is found in the Tours region in central France, having developed in the Miocene period ten to fifteen million years ago, and containing more than a thousand types of fossils.

Ammonite, Lower Cretaceous (about 100 million years old).

If carbonate sediments of organic origin are consolidated by a calcite cement, they are called lumachelle. Certain facies of the German "Muschelkalk" (a stage of Middle Triassic) are oyster lumachelle, which developed 220 million years ago.

The mollusks of the mussel class have been forming flat marine sediments particularly since the upper Mesozoic period, 120 million years ago. In the Cretaceous period, for example, the rudistes – mussels with a thick-walled, towerlike, rigid shell (see figure, page 16) – formed veritable reefs in the Mediterranean area. From these, 115 million years ago, arose the massive, white limestones of the Urgon facies, which today are being quarried in Orgon in southern France.

In contrast to the mussels, mollusks of the cephalopods class such as the cuttlefish or nautilus do not play a large part in organogenic sedimentation today. This was different in the Mesozoic period. At that time cephalopods were widespread and, above all, ammonites and belemnites, today extinct, were involved in the formation of limestones. Both ammonites and belemnites possessed an unstable aragonite shell, which transformed during progressive sedimentation into crystalline calcite.

Due to their wide geographical spread and their rapid evolution, ammonites in particular are excellent stratigraphic fossils for the different periods from the Paleozoic through to the Mesozoic (see figure, page 17). Thousands of types, classified into more than 1800 genera, make it possible to divide this period of around 400 million years into numerous epochs and spatial zones.

Corals are organisms which form colonies. From their calcareous skeleton, the polyps (see figure) form reefs in warm seas near the water surface. Reef limestones can form masses up to several hundred metres thick, which are usually not layered. Their high calcite content and the low impurity rate make reef limestones a rock of great economic significance. This is all the more true in that reefs are often the bedrock for major mineral oil deposits. When the reef limestones developed they contained large

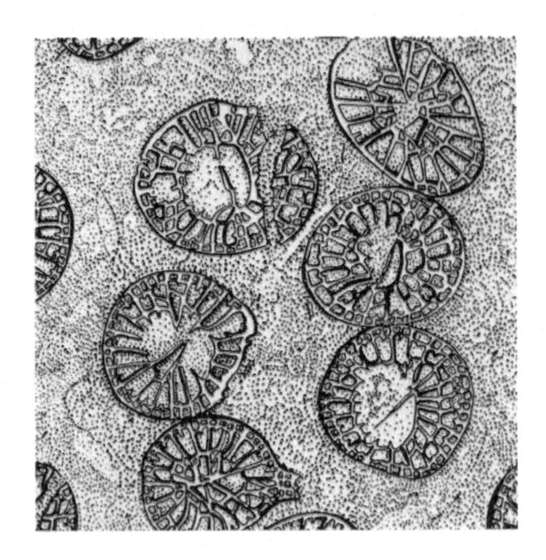

Drawing of a thin plate cut from a polyp limestone of Carboniferous age. The polyp tubes with their ray-form dividing walls can be seen, which are embedded in a calcite cement (micrite).

quantities of organic material which rotted over time, forming mineral oil or other bituminous substances. Viscous mineral oil collected in the course of the geological time in a vast reservoir, contained within the porous, fissured reservoir rocks.

Warm and clear water are necessary conditions for corals to thrive, which restricts its geographic development to the tropical zones. This gives the geologist the opportunity of reconstructing the position of the continental plates during the geological eras.

For example, during Upper Jurassic, 150 million years ago, the reef limestone belt which stretches from the Paris basin to southern Germany lay between 15 and 30 degrees latitude north; today it lies between the 48th and 49th parallel. An even more astonishing example is Spitzbergen. The group of islands off the Norwegian coast consists partly of reef limestones, which means that, when the corals colonised Spitzbergen during the Permian and Carboniferous periods 250 to 300 million years ago, a warm, tropical sea broke on its coastline.

Today Spitzbergen lies on 80 degrees latitude north – on the edge of the polar sea.

Sea urchins and, in particular, the crinoids and sea lilies belong to the echinoderms. These creatures were involved in the bioclastic sedimentation of limestones through the whole period from Paleozoic to Mesozoic. Limestones developed from the stems (trochites) of the sea lilies or the skeletal remains of other echinoderms, which are easy to recognise by their large calcite monocrystals with shiny (spathic) fracture planes. Due to their crystalline appearance these crinoidal limestones are also called "microgranite" (see figure).

Foraminifers are single-celled protozoa 0.05 to more than 10 millimetres in size, the calcareous shells of which frequently remain preserved in carbonate sediments. Thanks to their rapid evolution and wide spread, they too are very good stratigraphic fossils in carbonate sediments, which is of particular economic benefit in the search for mineral oil.

Drawing of a thin plate cut from a nummulitic limestone of Eocene age. Note the numerous thick-walled nummulitic cross sections, which are composed of fibres running vertically to the shell wall. The cement is calcareous clay with sand grains.

Drawing of a thin plate cut from a crinoidal limestone of Jurassic age. Note the accumulation of numerous broken and worn sea lily stems, the net-like structure and axial channels of which are still easy to see.

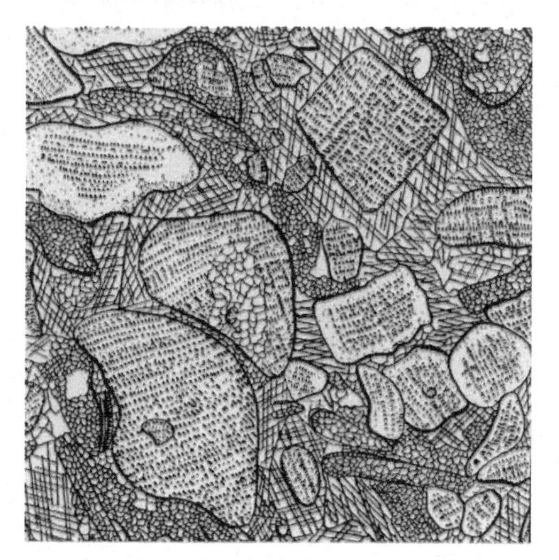

Nummulites, large foraminifers with discoid shells, are characteristic of carbonate sedimentation in the Paris Basin at the beginning of the Cenozoic era, 50 million years ago (see figure). The nummulitic limestones were a sought-after building stone in Paris and the surrounding area during the Middle Ages and in the early modern age. In order to meet the great demand, underground quarries were excavated in the southern parts of the city. As the shells of the nummulites, the size of coins, were still recognisable even in the sedimentary limestone, the citizens of Paris called the nummulitic limestone "pierre à liard", as in size and shape the shells resembled the "liard" or farthing, a low value bronze or copper coin then in circulation.

Coccolithophores are planktonic, single-celled marine algae. Their skeleton, the coccosphere, resembles a spherical capsule and consists of numerous discoid calcite plates, which are arranged radially (see figure, next page). These small plates, the

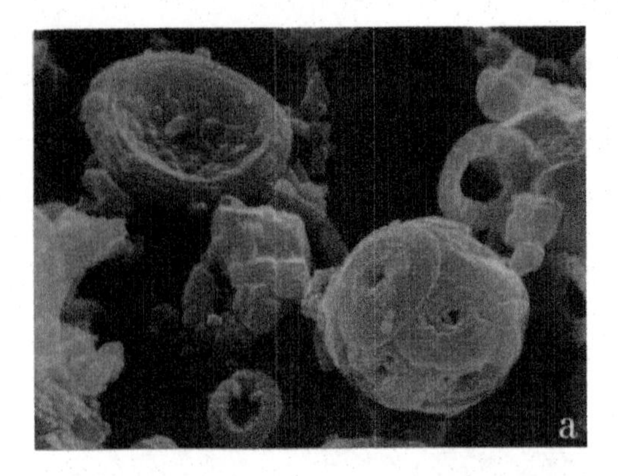

Scanning electron microscope (SEM) images of coccoliths, increasing magnification.
a) Coccolithophorides, Upper Cretaceous, from Omey/France (2 200 times magnification),
b) Coccosphere of a coccolithophoride (2 200 times magnification),
c) Ray-form calcite parts, which form a discoid plate of a coccosphere (6 500 times magnification),
d) Picture of a thin section of a coccolith (6 500 times magnification).

coccoliths, are only about ten micrometers in diameter, but they are the main component of the oceanic limestone sediments across several latitudes. Today a large "flowering of algae" still takes place each spring in the North Atlantic, when the sea is coloured white from the hundreds of millions of coccospheres contained in each litre of water. If this figure is extrapolated to the entire region covered by the flowering algae, several million tonnes of calcium carbonate originate in only a few weeks – as skeletal material of tiny limestone-depositing algae "Emiliana huxleyi" which belong to the family of coccolithophores.

In the Upper Cretaceous, 80 million years ago, a white, porous, soft and crumbly limestone developed in many places due to the build-up of these tiny calcareous plates. The formation of chalk took place in two stages: firstly a calcareous ooze developed from the sedimentation of billions of coccoliths on the sea bed, which then gradually consolidated by pressure, squeezing out the water at the same time. The resultant white rock is characterised by numerous cavities approximately 0.5 micrometers in size, which can constitute up to 40 percent of the total volume of the rock. The fine pores in chalk result in a large inner surface of up to 5 square metres per cubic centimetre and this leads to a high capillary ascension speed of the order of 5 centimetres in 15 minutes. This explains the high absorption capacity of chalk, which sticks to the tongue like clay.

Chalk is not only porous; the cohesion between the individual particles of the rock is also very low; chalk may be described as an "unfinished" limestone with little or no cementation. The soft rock can easily be scratched with the fingernails and is also very friable, that is why chalk has been a popular writing material for slates and similar bases for centuries – at every stroke millions of coccoliths remain behind on the rough surface.

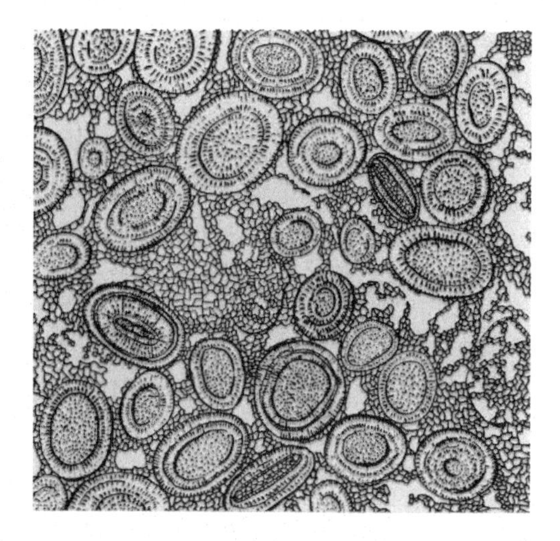

Drawing of thin cut plate from an oolitic limestone of Jurassic age. The small oolites are embedded in a mosaic-type calcite cement (sparite) with numerous interstices.

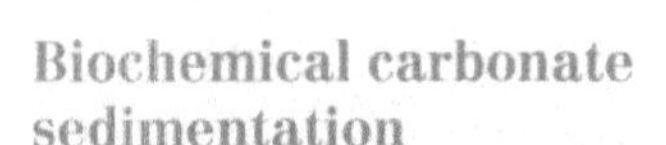

Biochemical carbonate sedimentation

The physical and chemical precipitation of calcium carbonate is often closely connected with biological or biophysical mechanisms, such as those which also occur during the development of blue-green algae. In the tidal area of the seas, but also in swampland and on the lake shores of temperate zones, discus-shaped or mammillary structures form, the stromatoliths. These beds of algae are structured in layers, and each layer corresponds to a growth phase of the algae colony. Petrification is caused by the precipitation of calcium carbonate, and a stratified limestone forms. Fine sediments are also intercalated in the upper section of the algae mat.

Ooliths are small calcium carbonate spheres which look like fish roe. Their diameter is from 0.5 to 2 millimetres and they consist of calcite layers which have been deposited concentrically around a core (see figure). The core or nucleus may be a grain of sand or a fragment of shell. Ooliths form especially in marine environment in shallow, warm turbulent waters rich in dissolved carbonates. The accumulation of the small individual spheres leads to the formation of oolithic limestones as those widely spread during Middle and Upper Jurassic (180 to 150 million years), from the Paris Basin to Bavaria.

Limestones of chemical origin

Certain natural conditions must prevail for dissolved calcium carbonate to be precipitated and deposited as limestone. Travertines, for instance, form at the outflows of underground springs and dripstones occur in karst caves in the limestone mountains. There is also secondary calcite, which fills the tension cracks in rocks, and finally the geodes, concretions of calcite crystals which form in the cavities of various rocks. These are mainly limestone rocks, but geodes are also found in volcanic rocks such as Iceland basalt or in sandstones.

Onyxes are zonal or banded white limestones which are differently coloured by red, yellow or green metal oxides. They are mostly deposited on the shores of calciferoues springs, as in Pamukkale in Turkey (cotton castle, see figure), but can also occur if calciferous water flows out of cracks at high pressure. Onyxes are often used for de-

Limestone terraces at Pamukkale.

corative purposes on and in buildings, and are therefore described as onyx marbles, like all polishable limestones. But they have just as little to do with marble as real onyx, a semiprecious stone made of coloured chalcedony.

Under very special conditions such as those in temporary lakes, calcium carbonate is precipitated in form rhombohedral calcite crystals just 0.1 millimetre in size. These carbonate sediments never consolidated and form a loose, fine powdery rock which is similar to an artificially precipitated calcium carbonate. Such a limestone is quarried in Villeau, south of Chartres.

2.2 Diagenesis – from sediment to rock

When a carbonate sediment forms, it is first saturated with water. The porosity of this sludgy mass can be up to 90 percent, whilst a solid rock has a porosity of a only few percent. In order to compress a loose sediment to a hard limestone, diagenesis with numerous physical and chemical processes must take place.

Compression

When an offshore drilling is carried out from a ship or a platform in the carbonate sediment which has been deposited homogeneously on the sea bed, it is evident that the hardness of the sediment increases with the depth of the drilling. Within a few hundred metres one first comes across a sediment saturated with water, then a chalk and finally a limestone with a density of 2.4 to 2.6 grams per cubic centimetre and a water content of only 10 to 20 percent.

On the Bahamas, but also on numerous other coasts between the latitudes of 35 degrees north and 35 degrees south, the diagenesis of the carbonate sediments takes place at very low depths from one to fifty metres. Fossils, first and foremost the remains of plants and animals which lived in the sea, may be embedded in the limestone. But the waste from our civilisation is also found in these seas, and occasionally "fossilised" coca cola bottles are found in perfectly hardened limestone rocks.

Cementing

A sedimentary rock contains numerous pores which are only gradually closed. Water which is saturated with calcium carbonate collects in the pores. It is then forced out under the great weight of the overlying rock layers, the calcium carbonate remains behind and is deposited as calcite, which closes the pores like a cement. How porous the original carbonate sediments can be is clear

Classification of limestones by their cement and components.

Components \ Matrix	Grain size > 4 μm = ...sparitic	Grain size < 4 μm = ... micritic
Bioclast	Biosparitic	Biomicritic
Oolite	Oosparitic	Oomicritic
Peloid limestone	Pelsparitic	Pelmicritic
Intraclasts (angular, re-deposited fragments)	Intrasparitic	Intramicritic
Reef limestones	Biolithic	Dismicritic

from the fact that the percentage proportion of cement in the solid rock is often just as great or even greater than the proportion of original sediments.

If the cement in a limestone rock is finely crystalline, it is called micrite or micro-crystalline calcite. If the cement consists of coarsely crystalline calcite, it is referred to as a sparite (see figure).

Dolomitisation

During the diagenesis of limestone rocks dolomitisation may occur, where the calcium ions in the crystal lattice are replaced by magnesium ions. The exchange takes places either directly during sedimentation, when the seawater is richer in magnesium than calcium, or considerably later during secondary dolomitisation, when magnesium-rich water circulates through the finished rock.

Dedolomitisation occurs much more seldom. This is always the case when water poor in magnesium and rich in sulphate flows through a dolomite rock. As dolomite has a greater density than calcite, a rise in porosity can be observed.

But not only the components have an influence on the porosity of the limestone rocks; the spatial arrangement of the components and the surrounding matrix also have a large part to play. For this reason a measurement of the porosity can shed light on the composition of the rock in question. These correlations are useful in the search for mineral oil, for example, as the porosity of a limestone rock is an indicator for potential of a mineral oil deposit and the theoretical output.

2.3 Classification of the limestones

Limestone is not simply limestone. The different limestone types can be classified according to certain specific criteria. The most important of these criteria are the structure, the texture, the type of components and the carbonate content.

Structure

As regards the structure, it is possible to distinguish under the microscope three different components of a limestone rock:

- the grain, which can be of biological or chemical origin;
- the matrix, which consists of a finely crystalline calcite ooze with particle sizes smaller than 4 micrometers – micrite;
- the cement or sparite, which is constituted by calcite crystals which are larger than 10 micrometers, often even reaching sizes of 20 to 100 micrometers.

Limestones are classified according to the importance of these 3 components. A limestone is called :

- mudstone, a finely grained limestone which consists almost exclusively of matrix and very few grains;
- wackestone, if the proportion of the matrix is over 50 percent and the proportion of the grains is about 10 percent;
- packstone, if the grains predominate in the calcite ooze;
- grainstone, if the grains are cemented by well crystallised sparite;
- boundstone, if it is a constructional limestone of reef type.

Texture

Even with the naked eye or with a magnifying glass it is possible to make classifications according to texture. Limestones are:

- compact, when the grains are too fine to distinguish them with the magnifying glass;
- crystalline, when the grains are larger than a few millimetres. If they are also poorly cemented to one another, the rock is referred to as a saccharoidal limestone;
- conglomeratic, if its components, several centimetres in size, are round like pebble stones;

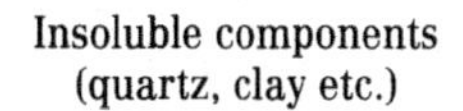

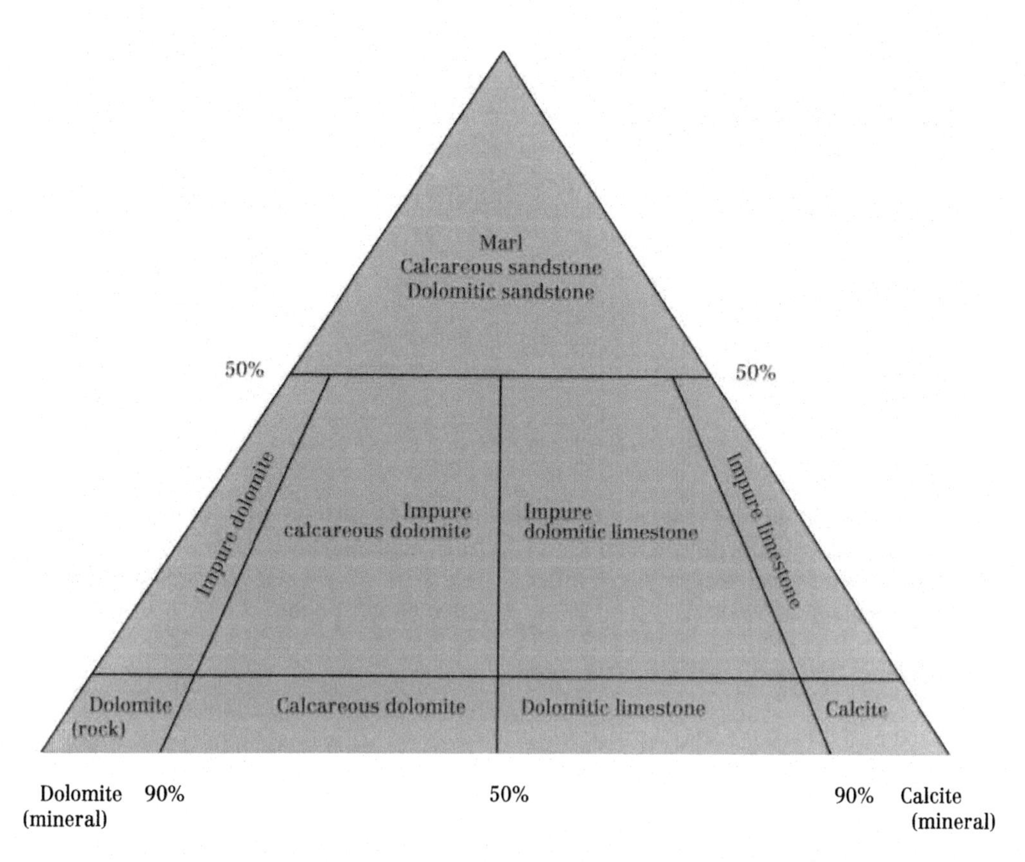

Classification of carbonate rocks on the basis of their content of calcite, dolomite and insoluble components.

- brecciated, if the components are angular;
- soft and porous, if the calcite is only slightly cemented;
- flinty, if nodules of flint a few centimetres to one metre in size are visible in the limestone.

Chemical analysis

The chemical composition of carbonate rocks can also be used for classification, taking into account the ratio of calcite to dolomite. There is:

- magnesium limestone with 5 to 20 percent magnesium carbonate;
- dolomitic limestone with 20 to 40 percent magnesium carbonate;
- pure dolomite with 40 to 46 percent magnesium carbonate.

The clay content allows a differentiation between:

- marly limestone or marl limestone with 5 to 36 percent clay;
- calcareous marl or marl, in which clay predominates at 35 to 65 percent.

These two types of classification can be combined (see figure).

Category / classification		Content [%]	
		$CaCO_3$	Equivalent CaO
1	Very high purity	> 98.5	> 55.2
2	High purity	97.0-98.5	54.3-55.2
3	Medium purity	93.5-97.0	52.4-54.3
4	Low purity	85.0-93.5	47.6-52.4
5	Impure	< 85.0	< 47.6

Classification of limestones by their purity.

Impurities

Apart from dolomite and clay, a limestone can contain other impurities, which are taken into account for classification. A differentiation is made between:

- sandstone limestone with quartz sand grains;
- ferruginous limestone, the iron-(III)-oxide (Fe_2O_3) content of which is greater than 3-4 percent;
- quartz limestone, where the quartz is invisible in the cement;
- phosphate limestone, which contains calcium phosphate in the form of apatite [$Ca_5(PO_4)_3 \cdot (F, Cl)$];
- pyrite limestone, which has traces of pyrite (FeS_2);
- graphite limestone, if the organic remains have been transformed into graphite during metamorphism.

Or limestone can be classified according to the degree of purity (see figure).

2.4 Metamorphism – from limestone to marble

The word marble comes from the Greek word *marmáreos*, which means shimmering or shining. The original meaning easily explains why, in general parlance, and above all in the building industry, all limestones are referred to as marble if they can be polished to a high shine. In geology, however, the concept of marble is reserved for limestones which have undergone re-crystallisation during metamorphism.

Conditions for metamorphism

Metamorphism is a transformation of the rock which always occurs when rock masses are subjected to high pressure of over 1000 bar and, at the same time, high temperatures from 200 to 500 degrees Celsius. These conditions differ radically from the conditions of diagenesis and force an adaptation of the minerals forming the rock, which can, in principle, take place in two different ways: by re-crystallisation or by chemical reactions between the individual components. All these transformations always take place in solid state, regardless of whether it is a regional metamorphism or a contact metamorphism.

Regional metamorphism

General or regional metamorphism is always the result of tectonics. Due to movements of the Earth's crust all rocks within a wide area are pushed down to great depths and the original rock is deformed. A typical characteristic of regional metamorphism is the foliated structure of the rocks, which forms due to the extensive unilateral pressure effects. The newly formed minerals are generally arranged in this foliation.

The marbles excavated in Norway belong to series which were metamorphised and deformed 420 million years ago during the formation of the Caledonian mountain range. Carrara marble, on the other hand, bears witness of the deformations in the depths of the Alpine range.

Originally, Carrara marble was a limestone which was deposited approximately 220 million years ago in a warm sea. When Africa and Europe approached one another, the formations of the Tuscan overthrust nappe, several kilometres thick, were pushed over the sedimented limestones which then came to a depth of 5 to 10 km. There the rock underwent extreme deformation at temperatures of nearly 300 degrees Celsius. About 15 million years ago the re-crystallisation ended and finally a last compression led to the uplifting of the marble. Gradually the overlying formations, which were still several kilometres thick, were worn away by erosion and the tectonic window of the Apuanian Alps, as geologists call it, was opened up and probably the most beautiful marble rock in the world was exposed.

Contact metamorphism

Contact metamorphism takes place when rocks come into contact with magmatic material and are thus exposed to extreme temperatures, which cause radical transformations. As this process takes place above all in the immediate area of contact between rock and magma, the contact metamorphism is thus limited to a small area: the greater the distance from the contact area, the lower the degree of metamorphism. It rarely extends over two kilometres or three at most,

Coarse crystalline marble block.

and the contact aureole (or zone) often is only a few tens to a hundred metres thick.

Even when the original limestone is very pure, contact metamorphosis never gives rise to such pure white marbles as in Carrara. The contact with the glowing hot, magmatic intrusion rock (mostly granite) causes complex reactions, in the course of which a special rock, skarn, is formed. Skarn is an old Swedish mining term and means light snuffer, and these coarse crystalline calcium silicate rocks do indeed contain numerous ore minerals which give them a characteristic lustre. But there are not only ore minerals present in these rocks; skarns are really "natural mineralogical museums": pink or brown calcium garnets, idocrase (also called vesuvianite), wollastonite, diopside and blue calcite – the multitude of minerals in all sizes and shapes is enormous.

In Thailand a greyish black limestone was transformed into a white marble during contact metamorphism. 150 million years ago, magmatic granites of Jurassic age shot through thick limestone layers which originated in the Carboniferous and Permian periods. Due to their high content of organic materials these limestones were originally dark in colour, but due to the intense heat in the metamorphism zone, the organic remains completely oxidised and a white marble was the result.

Characteristics of marbles

Colour, composition, grain size – apart from the conditions of the metamorphism, it is above all the original rocks which determine the characteristics of marbles. Due to re-crystallisation, a very pure limestone with a calcium carbonate content of over 98 percent changes to a white marble, consisting of over 99 percent calcite crystals which may reach a size of several centimetres. As marbles are almost completely crystalline, their porosity is markedly lower than that of limestones. In Carrara marble, for instance, it is just 0.01 to 0.22 percent.

If, however, the limestone contains impurities such as quartz, clay, iron oxide, iron sulphide or organic materials, new minerals form as well as the crystalline calcite. A simple, frequently occurring example is the formation of the calcium silicate wollastonite, which always forms when quartz is present in limestone:

$$CaCO_3 + SiO_2 \rightarrow CaSiO_3 + CO_2\nearrow$$

If the rock contains more than 5 percent of components which are insoluble in hydrochloric acid, it is referred to as cipolin or cipollino. The name is derived from the Italian 'cipolla' (onion), as the newly formed, silicate minerals encourage skin-like spalling of the marble during weathering.

Besides quartz and the deriving silicates, numerous sulphides and oxides form as components of crystalline marble. The most frequent metamorphism minerals are quartz, white mica (muscovite) and golden brown mica (phlogopite), green amphibole (actinolite) and white amphibole (tremolite), the green minerals diopside and serpentine, black graphite, the sulphides pyrite, marcasite and chalcopyrite and the feldspars, which are generally white. If iron oxides are present, they result in cream or pink coloured marbles, such as the famous pink marble from Portugal.

Synthesis of marble

The crystalline beauty of marble has fascinated mankind almost as much as the brilliancy of a diamond and it is not surprising, therefore, that attempts at re-crystallisation have been carried out repeatedly since the beginning of the 20th century, where the natural conditions for a metamorphism were recreated – successfully – in the laboratory. By heating limestone powder in an annealing box to temperatures of over 1100 degrees Celsius, a perfectly crystallised marble with a regular grain is achieved. But numerous experiments have been and are being carried out at high pressures. If a cylindrical piece of chalk is exposed to a very high pressure of 6000 to 7000 bar for a brief moment, the chalk is compressed; but if this pressure is maintained over a period of years – the record stands at 17 years – the chalk is also

The carbonatite lava flows, about half a metre in width, solidify after a few dozen metres to form solid rock. The solidification temperature is about 480°C. After a few days the apparently solid stone disintegrates to a white powder, as the air humidity dissolves the considerable salt quantities and destroys the structure of the rock.

metamorphised into a compact marble. And finally, artificial marble can also be created if limestone is melted in an electrical furnace at a temperature of 1050 degrees Celsius and at a pressure of 100 bar in a carbon dioxide atmosphere.

2.5 Carbonatites – extraordinary limestones

Besides the sediments chalk and limestone and the metamorphic marble, there are also magmatic calcium carbonate rocks – the carbonatites (see figure).

In the volcanic region of Kaiserstuhl on the Rhine a brown carbonate rock which was combined with volcanic intrusion rock puzzled geologists for a long time. A magmatic calcium carbonate rock was at that time still considered an impossibility, as calcium carbonate is not stable at the high temperatures of a lava. Not until 1989 a blackish, comparatively cold lava at a temperature of 500 degrees Celsius was observed in an active volcano in northern Tanzania, mainly consisting of calcite and dolomite (see figure). This was the proof of the magmatic origin of the carbonatites.

Carbonatites occur in two modifications, which differ by their content of dolomite. If the calcite content is between 50 and 100

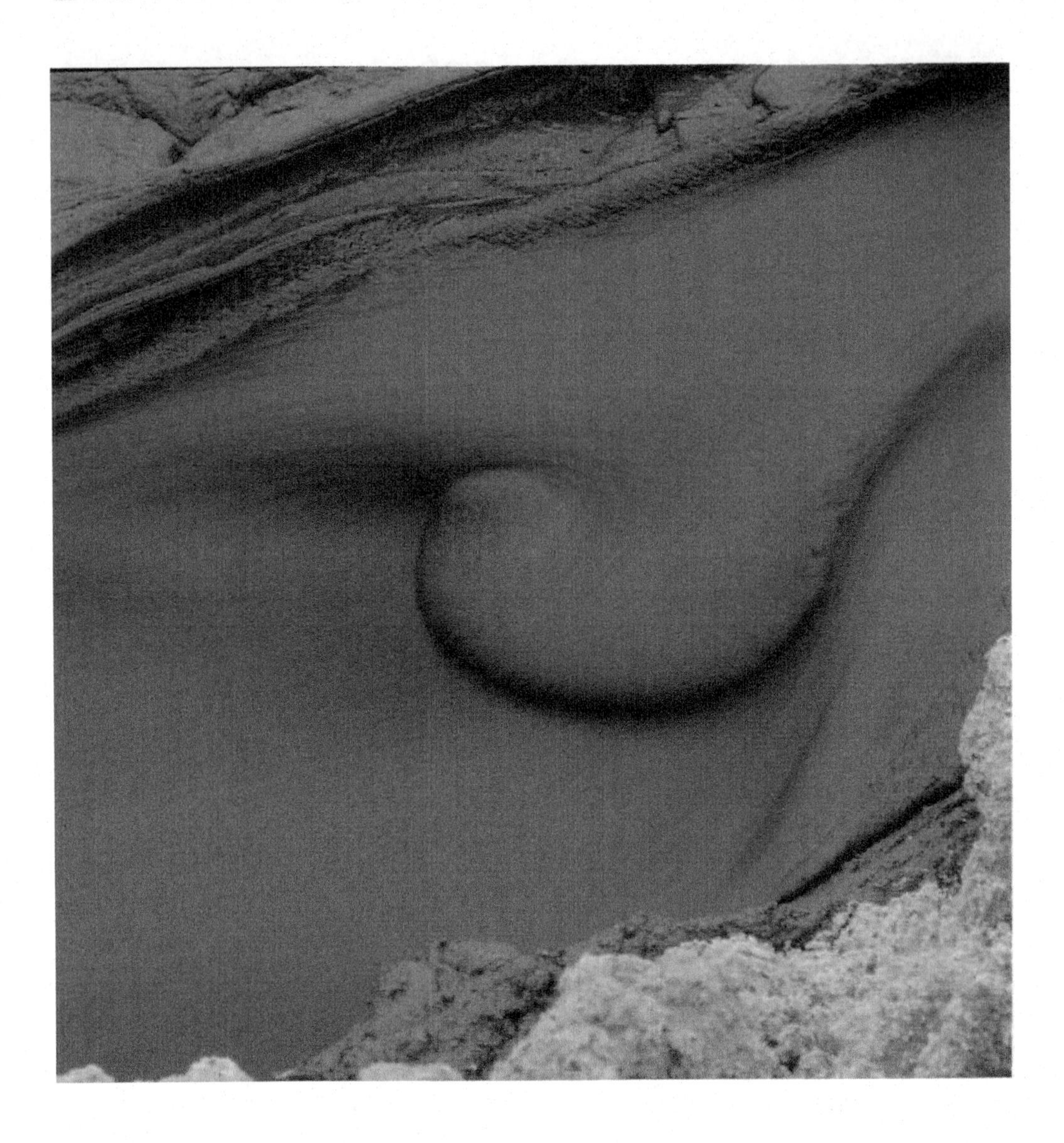

The thin liquid lava flows like hot asphalt through small channels. Bizarre flow patterns are continually being formed. During the eruption the carbonatite lava reaches maximum temperatures of 540°C – about 100°C less than scientists had originally expected.

percent, this is sövite. If dolomite predominates, it is called rauhaugite.

Both rocks are quarried locally, for instance, in the Tororo complex in Uganda, where they are used mainly to obtain calcium carbonate for lime and cement production. Carbonatites are generally exploited because of their content of phosphate, rare earth minerals and titanium compounds. In Siilinjärvi in Finland, about 20 kilometres north of the town of Kuopio, a large carbonatite deposit is exploited to obtain apatite. Calcite and the mica phlogopite result only as by-products of flotation.

3. Limestone deposits

Limestones are found almost throughout the whole world, but only a few of them are as easily and clearly identifiable as the soft Champagne chalk or the gleaming white Carrara marble. In many cases detailed tests are required to determine the type of carbonate rock. Some of these test methods are so simple that they can readily be carried out in the field.

3.1 Recognition of limestones

For tests in the field it is naturally preferable to use as few tools and chemicals as possible; one of the simplest methods, therefore, is to assess the rock by visual criteria such as colour, form and texture. No tools are needed for this except the eye, but with the necessary experience, the rock appearance is frequently sufficient to enable a number of statements to be made about it.

The colour of limestones

If the colour of a limestone is to be determined in the field, it should be noted that a wet and a dry limestone differ in colour, as a patina forms during drying, the colour of which is usually lighter than the original: a limestone which is usually greyish black often has a greyish white patina. For this reason the exact colour must always be determined on a fresh, dampened fracture of the rock.

A pure limestone is normally white. As with the mineral calcite, the rock may be coloured as a result of impurities, which have to be present only in small quantities. In many cases a concentration of 1 ppm is sufficient to cause discoloration, or, in other words, a few grams of a foreign substance per tonne of limestone are sufficient to change the colour completely.

For this reason the determination of the colour alone sheds light on the presence of a limestone only in very rare cases, although the colour often provides a good indication of the conditions which prevailed at the place of origin of the rock.

- Red tones correspond to an oxidising environment in which the iron is in "ferric" condition and is present as an iron (III) ion. The resultant oxides are more or less hydrated: stilpnosiderite [$Fe_2O_3 \cdot 2H_2O$) contains two molecules of water of crystallisation and is ochre yellow in colour; goethite ($Fe_2O_3 \cdot H_2O$) with one molecule of water of crystallisation is brownish red and finally, hematite (Fe_2O_3), which is anhydrous, is blood red in colour.
- Dark, green to bluish green colours, on the other hand, correspond to a reductive environment. Here the iron is present in "ferrous" condition as an iron (II) ion and is a component of complex hydroxides of the general formula $Fe(OH)_2$ or one of the numerous iron silicates such as greenish chamosite and dark green glauconite.
- Dark grey to black colours also indicate a reductive environment. In most cases microscopically small grains of the mineral pyrite (FeS_2) cause the blackish colour. Occasionally, though, the coloration occurs through organic remains which were more or less metamorphised during petrogenesis. In the metamorphic marbles, for example, pure black graphite often occurs.

The transition from "ferrous" to "ferric" condition, the passage through various oxidation stages, can be observed in many quarries. The upper part often has a cream, pink or reddish hue, whilst the unoxidised lower part still shows the original bluish coloration of the reductive environment in which the sedimentation took place.

The red and brown spotted limestones of Devonian age in the Pyrenees consist of nodules originating from the fossil shell of a cephalopod. As these nodules are surrounded by a slaty cement which is coloured green or red depending on the degree of oxidation of the iron oxides, two types of "marble" can be obtained by polishing: the red Griotte marble and the green Estours marble. Both types were used extensively in Versailles in numerous buildings dating from the 17th and 18th centuries.

Often moss-like or fern-like shape are discernible on the surface or on the fracture plane of the limestone strata, which consist of manganese oxides and manganese hydroxides of the formula $MnO_2 \cdot nH_2O$. These dendrites, as they are called, have nothing to do with organic formations despite their branch-like appearance. They are usually iron – manganese precipitations, which come from the water rich in manganese, flowing through the rock (see figure).

The smell of limestones

If a limestone is broken, it sometimes smells like bad eggs. This smell arises if the organic remains in the sedimentary rock disintegrate and form hydrogen sulphide gas and volatile organophosphate compounds. The gases remain embedded in the crystal lattice of the calcite or dolomite as tiny bubbles and even survive the re-crystallisation of limestone to form marble. If such inclusions are present in a marble in large numbers, this causes a diffraction of the light in the calcite crystals and the marble takes on a light grey colour.

Scratch resistance

The surface hardness of limestones can be determined by measuring the scratch resistance. For this a steel tip is loaded with a mass of three kilograms and is guided vertically over the limestone (see figure). The width of the resultant scratch informs about the hardness of the rock.

Dendrites.

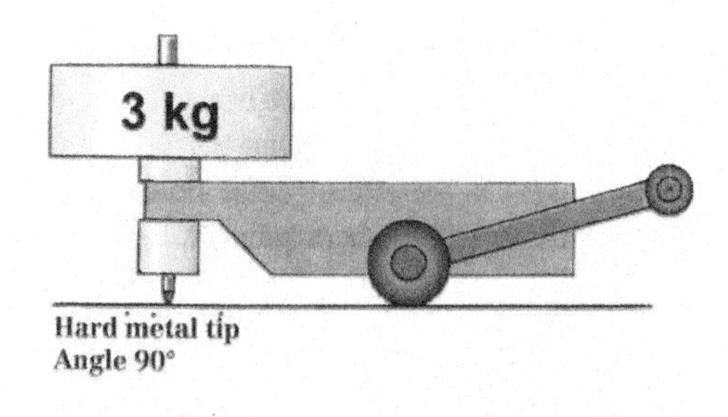

Measurement of crack width in limestone.

Soft limestones give values of the order of 2 millimetres. If the rock is covered by a newly formed calcite layer, the width of the scratch falls to 1.5 millimetres, and for hard limestones, such as plate-type limestones, scratch widths of 0.6 to 1 millimetre are obtained.

Calcite or dolomite

A considerable problem in identifying carbonate rocks is the differentiation between limestone rocks and dolomite rocks, as both often look very similar. The reaction with a dilute acid can give a first hint. It is true that the mineral dolomite, a double carbonate of calcium – magnesium of the formula $CaMg(CO_3)_2$ reacts with dilute hydrochloric acid only if it is simultaneously heated, but a slighter or delayed reaction can also occur with calcites under certain conditions.

However, there are several staining reactions by which the individual carbonate rocks can neatly be differentiated from one another. These reactions can be tested either on thin sections, on fractured surfaces or on individual grains of the rock. Particularly good results are achieved with the following two test reactions:

- If a 1 % solution of alizarin red S in 0.1 molar hydrochloric acid is dropped on to a calcite rock, the sodium salt of the alizarinsulfonic acid (alizarin red S) reacts to become an intensively red coloured calcium salt which is difficult to dissolve. Aragonite shows a bright purple colour.
- The direct dolomite test is successful with magneson (p-nitrobenzene azo resorcin). For this the rock sample is cauterised with 10 % hydrochloric acid and wetted with a drop of a solution of 0.002 grams of magneson in 100 millilitres of sodium hydroxide solution. A pale blue to violet film forms depending on the magnesium content of the rock.

3.2 Distribution on the Earth's surface

Marine carbonate sedimentation depends on numerous factors; the temperature and the salt content are particularly important. But the depth of the water, the depth to which the light penetrates, the current conditions and the amount of dissolved substances contained in the water, especially carbon dioxide, also play a part (see figure, next page).

A major aspect of carbonate sedimentation is the dispersion of organisms with calcareous shells. In warm, tropical waters foraminifers, mollusks and corals are abundant, and correspondingly a large amount of calcium carbonate is produced there daily. If the average seawater temperature is under 18 degrees Celsius, on the other hand, the sedimentation based on foraminifers and mollusks is markedly lower (see figure, next page).

Carbonate compensation depth

The depth of seawater also has a direct influence on carbonate sedimentation. Calcium carbonate is not easily soluble in the water near the surface, as this is supersaturated with hydrogen carbonate. In deeper water the situation changes drastically, as with increasing depth the average water temperature sinks to 2 degrees Celsius and pressure rises at the same time. The relative carbon dioxide content in the cold deeper waters is correspondingly six to seven times higher than in the warmer water near the surface. On one hand, this causes the pH

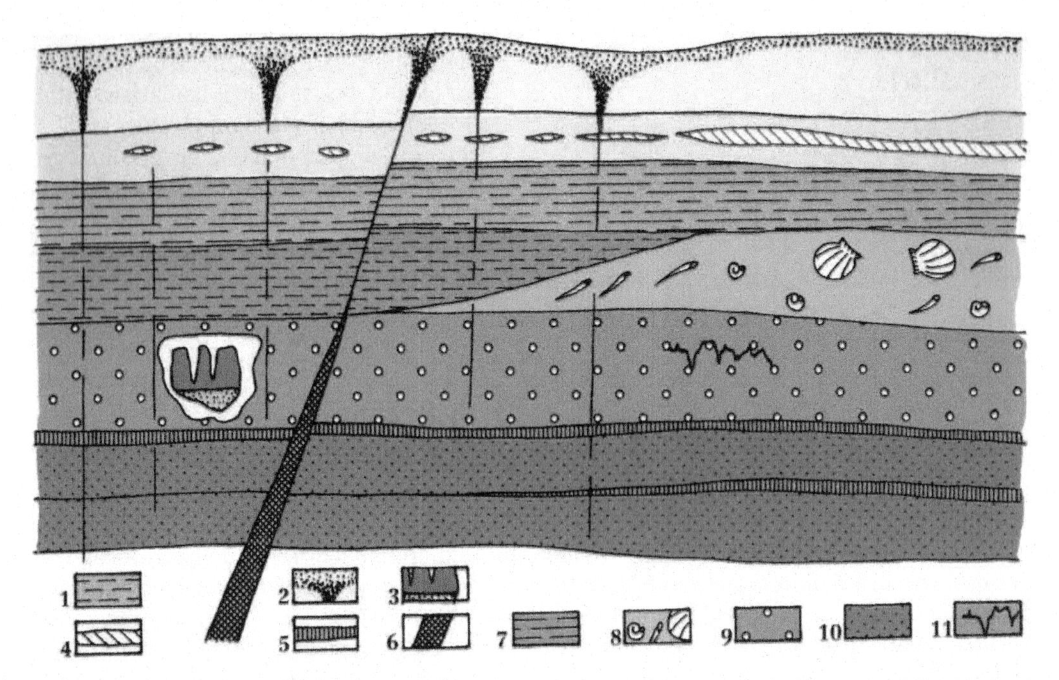

Geological problems which occur in limestone quarries (from Scott and Dunham):
1. Layers of varying thickness.
2. Surface weathering and leaching, mostly starting from fissures.
3. Solution cavities, partly filled with new limestone and clay/sand layers.
4. Nodules or bands of flint or flinty slates.
5. Clay or marl layers.
6. Fault with breccia on the fracture planes, also with mineral veins (e.g. fluorspar).
7. Fissures which break up individual layers.
8. Reef with varying structures, fossil types and porosity. The fossils may be empty, partly or entirely filled with cement.
9. Dolomitisation, silicification, fluorisation or sideritisation of limestone in the area of the mineral vein.
10. Almost complete dolomitisation.
11. Stylolite levels with clay.

value to drop significantly – deep waters are very corrosive – and on the other hand, the solubility of the carbonate increases greatly. Aragonite is completely dissolved from a depth of 3000 metres, whilst calcites can remain intact up to a depth of 4000 to 5000 metres. From this depth calcium carbonate is always dissolved: the carbonate compensation depth (CCD) is reached.

The CCD for the different oceans varies markedly. In the Atlantic, for example, it lies at 5000 metres, but in the Pacific at only 4200 to 4500 metres, as there the deep waters contain less oxygen and are therefore more acidic.

More or less extensive plankton productivity also has an influence on the CCD. Foraminifer shells have a vertical descent speed of about 2 centimetres per second and reach the bottom of the sea within a few weeks, whereas the smaller coccoliths sink more slowly and would only reach the sea bed after a few years had they not passed into solution long before. In fact, coccoliths can be found in the deep oozes, because they are contained in the excrement of plankton-eating crustaceans, deposited on the sea bed.

The fact that calcium carbonate is not found below the CCD supplies valuable information about the depth at which earlier sedi-

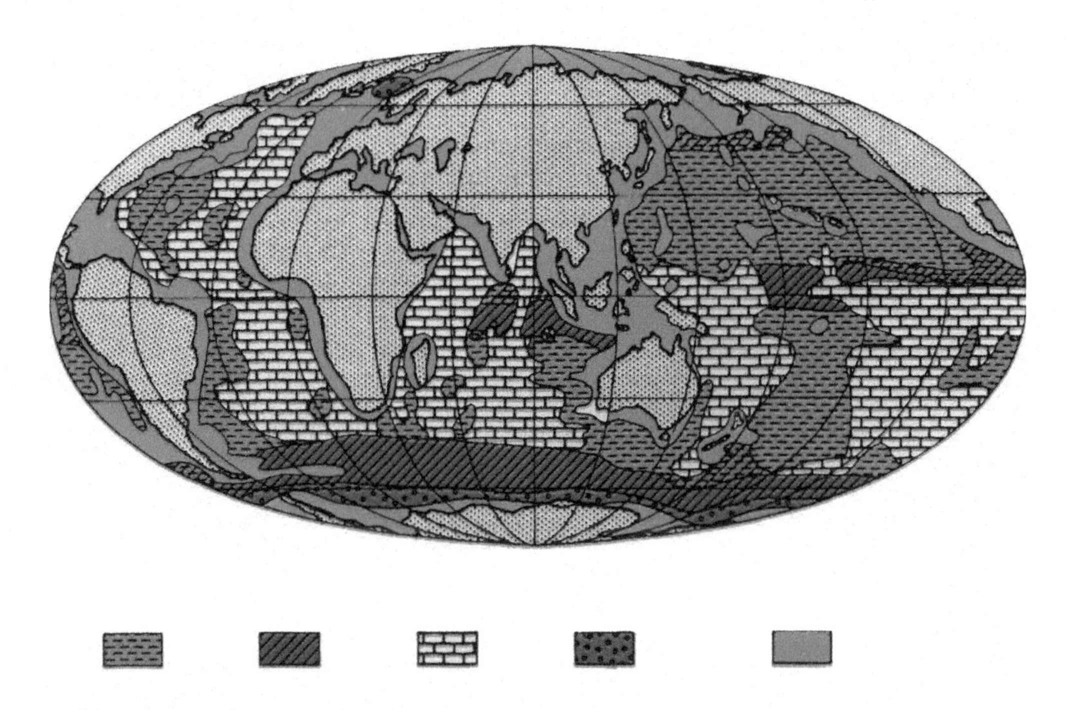

Map showing current
oceanic sedimentation.

ments were deposited. Just as the radiolarian ooze from plankton-type single-celled organisms with silicic acid skeletons is deposited today below the CCD, clayey siliceous rocks were also below the CCD at the time they developed. If a clayey siliceous rock is discovered within a sedimentary series, this layer thus indicates the position of the CCD at the time it was formed.

Composition of the sea bed

The carbonate facies are arranged in sequential zones from the continental platform to the ocean bed. The first deposits form on sea beds which still belong to the continental granite crust. These carbonate platforms are covered by sands and bioclastic oozes; the occurrence of oolites indicates turbulent zones in shallow water.

At the edge of the platforms, in warm regions, reefs and other limestone structures develop, behind which fine limestone sediments can be deposited. In the course of diagenesis the plate limestones are formed: these are limestones which, due to their fine grain, are used in lithography. But limestone reefs also erode. The products of their erosion are caught by the currents and distributed around the sea bed, where they form new carbonate sediments.

Algae and foraminifers feed those sediments which are deposited above the CCD on the oceanic basalt crust. The Mid-Atlantic ridge, for example, is mostly covered with carbonate oozes at a depth of 2000 metres, particularly as these areas are protected from terrigenous materials from the land, which large rivers as the Congo or the Amazon wash far into the sea. The thickness of these sediments grows from the ridge, where the oceanic crust develops, in direction of the margins.

As the oceanic crust cools down, its density increases and the rock, which becomes heavier, sinks down slowly, and plunges into the Earth's mantle rock through an oceanic trench sometime after about 150 million years. Due to this thermally conditioned sinking of the sea bed the carbonate sedimentations fall below the CCD and disap-

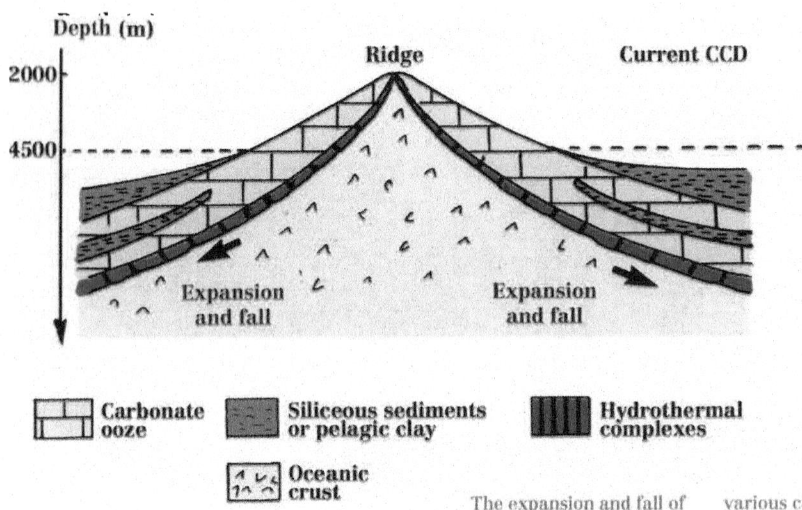

The expansion and fall of the oceans determine the sequence of sedimentation. This means that recessive carbonate deposits may be the consequence of a change in the CCD, which for its part can have various causes, for example, a change in biological productivity, fluctuations in climate, a fall in sea level, an increase in oxygen carrying deep currents or the passage of a moving plate through a more productive zone.

for its part can have
ductive zone.

pear. Red clays and biogenic siliceous oozes, which are less affected by dissolution, remain behind and cover the carbonate oozes (see figure).

If all this is taken into account, the whole history of an ocean can be constructed since its creation, simply by studying the sequence of the different sedimentation layers in a today's ocean and interpreting the results with the above correlations. Oceans which have long disappeared tell us their story, too, because their fossilised sea beds have survived through the ages in mountain ranges such as the Alps.

3.3 Limestone deposits in the geological ages

Carbonate deposits are nearly always associated with a biological activity, and just as life on earth underwent continuous change over millions of years, the limestones which formed from the carbonate deposits also changed (see figure).

The first limestone deposits

At the beginning of its history, 4 to 4.5 billion years ago, only a weak flow of heat reached the earth. The still young sun had just 70 percent of the power it has today and the earth escaped complete glaciation only due to the fact that its atmosphere was rich in carbon dioxide. If the sun today was still as feeble as it was at that time, the mean annual temperature on earth would be a frosty 0 degrees Celsius.

The fluctuations in the temperature and the gradually changing composition of the atmosphere naturally had an influence on all life on earth. For a long time single-celled blue algae were the only form of life on earth, and only they could convert the mineral carbon in carbon dioxide into the organic carbon of carbohydrates by means of photosynthesis. During this biological process oxygen arose as a waste product which was given off into the environment.

Era	Age [Mio years]	Period	Epoch	Significant orogeny phases	Carbonate sedimentation glaciation	O_2-content Living organisms
Cenozoic		Quarternary	Holocene			
	1.6		Pleistocene			
		Tertiary – Neocene	Pliocene	Alpine range	Limestone	Hominids
	2.3		Miocene		Molasse	
		Tertiary – Paleogene	Oligocene			Rapid development of mammals
			Eocene	Pyrenean range	Antarctic Nummulitic	
	65		Paleocene		limestone	Extinction of many
Mesozoic		Cretaceous	Upper Cretaceous	Laramide range (North America)		reptiles und ammonites 21 % O_2
			Lower Cretaceous		Chalk	Flowering plants
	130	Jurassic	Upper Jurassic (Malm)		Reef chalk	Birds
			Middle Jurassic (Dogger)	Andes range	and oolitic	Development of ammonites
	205		Lower Jurassic (Lias)	(South America)	limestones	Diversification of reptiles
		Triassic	Upper Triassic (Keuper)			First mammals
			Middle Triassic (Muschelkalk)	Cimmmerian phase	Dolomite	
	250		Lower Triassic (Buntsandstein)	(Asia)		First dinosaurs
Paleozoic		Permian	Upper Permian (Zechstein)			
	290		Lower Permian (Rotliegendes)			
		Carboniferous	Upper Carboniferous	Hercynian		Reptiles, polyzoons
	360		Lower Carboniferous	range (Europe)	Dark	insects
		Devonian	Upper Devonian		limestones	Amphibians; Ferns
			Middle Devonian			
	400		Lower Devonian			10 % O_2
		Silurian		Caledonian		Land plants
	420	Ordovician		range		Placoderms
	500			(Europe,		
		Cambrian		North America)	First limestones	Invertebrates
	530					with skeleton
cambrian	2600	Proterozoic		Several phases		2 % O_2
				in old shields	Stromatoliths	1 billion years ago
			(Africa, Brazil,			
	4600	Archaean		Canada etc.)		No O_2

Ice ages

Simplified geological time scale.

There are numerous proofs that there was still no oxygen in the Earth's atmosphere up to three billion years ago: for example, the occurrence of non-oxidised uranium minerals or of sedimentary pyrite, in which iron is present in its lowest oxidation level as an Fe^{2+} ion.

The absorption of carbon allowed the oxygen content on the earth to rise continually, firstly in the seas, but from about 1.7 billion years ago, in the atmosphere as well. But the fact that the ratio of oxygen to carbon dioxide changed so rapidly was also due to the formation of carbonate rocks, in which large volumes of carbon dioxide were permanently bound.

The first limestones were stromatoliths, which mainly developed in the Precambrian formations 3 to 0.5 billion years ago. Today it is often difficult to recognise their original, foliated structure, as the rocks have undergone extensive changes during numerous metamorphisms. The Scandinavian marble deposits, for example, developed from stromatoliths which formed 2 billion years ago.

The first limestone layers

During the Cambrian period, 540 million years ago, the first multi-cell, invertebrate organisms with calcareous shells developed in the seas: the mollusks, echinoderms and crustaceans. As these organisms, for the first time in the Earth's history, left behind numerous fossils, the fauna of the Cambrian period were formerly also known as original fauna. This is incorrect, however, as previously there were also what are called medusa, but these were not only invertebrates, but also shell-less, and left behind scarcely any traces. On the other hand, the numerous fossils in the limestone layers and slates of Cambrian age have allowed a relative stratigraphic chronology of these formations to be drawn up.

These Cambrian limestone formations are known in southern France, but they also occur on the margins of the large shields which are more than 600 million years old, including the Canadian, Siberian and Ethiopian shields.

In Upper Devonian and Lower Carboniferous, rank vegetation covered the continents and protected them against the intensive erosion to which they had been exposed in previous periods. This situation favoured the development of biochemical limestone sediments, which are characterised by a high calcium carbonate content and a low rate of impurities, such as clay and sand. In Europe thick series of these limestones were deposited from the Ardennes to the Urals, which are dark in colour due to their high proportion of organic material. At that time the first massive reef formations developed, merging laterally into normal limestone beds.

The Mesozoic period – the height of calcareous sedimentation

Looking at a geological map of Europe, the extent of the blue and green colours is immediately striking. Blue colours usually indicate a Jurassic formation, whilst green points to chalk formations. The characteristic rocks of these geological periods are mainly limestones. In these epochs the carbonate oozes were deposited on the continental crusts which had emerged from old mountain ranges of the Paleozoic period. These Caledonian and Hercynian mountain ranges were eroded and levelled off during the Permian period. Surrounded by limestone formations of Mesozoic age, these old rocks are still visible today in the French Massif Central, in the Ardennes, in the Rhenish Schiefergebirge and in Bohemia.

One must imagine that 200 to 70 million years ago these massifs were largely covered by a shallow, warm sea, at most 300 metres deep. At that time only a little land projected from the water, hardly any erosion took place and the carbonate sediments were thus very little contaminated by weathered rocks such as sand and clay.

When attempting to find an explanation for the flooding of the greater part of Europe, a general rise in sea level may be assumed, even though locally lowering of the sea bed (subsidence) may very well have taken place.

There are two factors which can cause such a significant change in the seawater level: a global climatic change and the deformation of the sea bed. For example, in the Quaternary period, the climate cooled down considerably, the glaciers extended, the water on the land froze and the sea level dropped by nearly 200 metres. In the Mesozoic period, though, there was no significant glaciation, nor any inland ice which could have melted. The average temperature of the seawater was 25 to 30 degrees at the surface and even the deep ocean waters still had a tempera-

Fossil shell of a Harpoceras. This ammonite was widespread 185 million years ago in the Lias. The black coloured lines correspond to the line of junction between the internal dividing walls; the clearly discernible ribs however are merely decorative.

ture of 10 to 15 degrees Celsius (against only 2 to 4 degrees today).

So the deformation of the sea bed was the cause of the flooding. An accelerated expansion of the ocean ridge released large volumes of heat, the sea bed expanded upwards and the capacity of the oceans fell. The depth of the newly formed oceans such as the Atlantic and parts of the Indian Ocean was just 2000 to 3000 metres; after cooling of the oceanic crusts, it is 4000 to 5000 metres today.

As the volume of water remained the same over the entire period, the oceans flooded onto the continental platforms. It is estimated that the water rose about 300 metres. This transgression, as it is called, reached its maximum in the Cretaceous period, as may be observed at many places in Normandy, in the Ardennes and in southern Russia. Here the chalk sedimented directly on the old base.

At that time only a few islands projected out of the water. These were the remains of the Hercynian mountain range which were covered by a humid tropical forest, which favoured a significant change in the granite rocks. The calcium minerals were dissolved as well as the silicon dioxide, and both flowed with the rivers into the seas, where the enriching of the warm seawater with dissolved calcium created ideal conditions for the development of a calcareous phytoplankton. Coccolithophores, in particular, multiplied rapidly, and as they died off, chalk was formed from the skeletal remains. As the solubility of silicon dioxide in salty seawater was markedly lower than in fresh water, the mineral precipitated and formed flint chalcedony, a siliceous rock of pure chemical origin.

It was seldom that shallow seas in a tropical climate could become so widespread and the conditions for the precipitation and sedimentation of calcium carbonate oozes were rarely as favourable as in the Mesozoic era, which can therefore be called, without exaggeration, the age of limestones.

Fossils in limestones

Fossils are the remains of living creatures of past epochs, which after death were subjected to secondary mineralisation (petrification) in a sedimentary rock (see figure, page 39). The occurrence of the first mineralised exoskeletons in the carbonate deposits of Cambrian age, 540 million years ago, marked the beginning of the classic fossilisation periods.

During mineralisation the calcium phosphate in the skeletons of organisms is generally replaced by cryptocrystalline or crystalline calcite. This means that environments in which calcareous sedimentation takes place, i.e., where calcium carbonate is abundant, are particularly favourable for the preservation of fossils. For this reason most fossils today are found in limestone layers, starting with the shells of invertebrates through to complete reptile skeletons. One of the most famous fossil deposits is enclosed in the limestones at Solenhofen in Bavaria.

A thinly stratified limestone has been quarried in the region around the small community of Solenhofen in the Altmühltal for several centuries, and is mainly used for sculpture and lithography (see chapter II, "The Cultural History of Limestone"). The deposits there were formed during Upper Jurassic,140 million years ago, in a tranquil lagoon, which was separated to the south from the ocean at that time, Thetys, by a reef (see figure). These were ideal conditions for the sedimentation of extraordinarily fine calcareous oozes, which, over the course of time, consolidated to form limestones. The numerous fossils in this ooze have been preserved to this day.

The best known fossil from Solenhofen is the archaeopteryx. The first example was found in 1855, and since then seven other archaeopteryxes have been exposed; seven of these eight examples come from the Solenhofen stratum. For a long time the archaeopteryx was considered to be a direct ancestor of today's birds. Its tail, its three-fingered claws and its teeth showed it to be a reptile, but its furcula and its feathers were unmistakable characteristics of birds – so this "original bird" could have been the long sought-for connecting link between the reptiles and the birds. This was a mistake, as was proved in 1992: the archaeopteryx is just another proof of the numerous paths of evolution.

The oil shales of the Swabian Jura are in reality a greatly hardened, bituminous marl. They also contain a series of impressive fossils, which have been collected since the 16th century. In particular the fossils of marine reptiles have appealed to collectors, so it is no wonder that since the 19th century more than 300 complete skeletons of ichthyosaurus have been discovered. These saurians were excellent swimmers, and their appearance was reminiscent of a dolphin or a shark. And they mainly fed on belemnites, as the numerous chitin hooklets of these mollusks in their stomachs show.

But limestones not only enclose traces of past life, they can also testify to the climatic changes which have taken place so often in the history of the earth.

In 1878, in Rüdersdorf near Berlin, the Swedish geologist Otto Torell discovered deep scores in the surface of the shelly limestone layers, for which he initially had no explanation (see figure, next page). After detailed investigations he interpreted these scores as

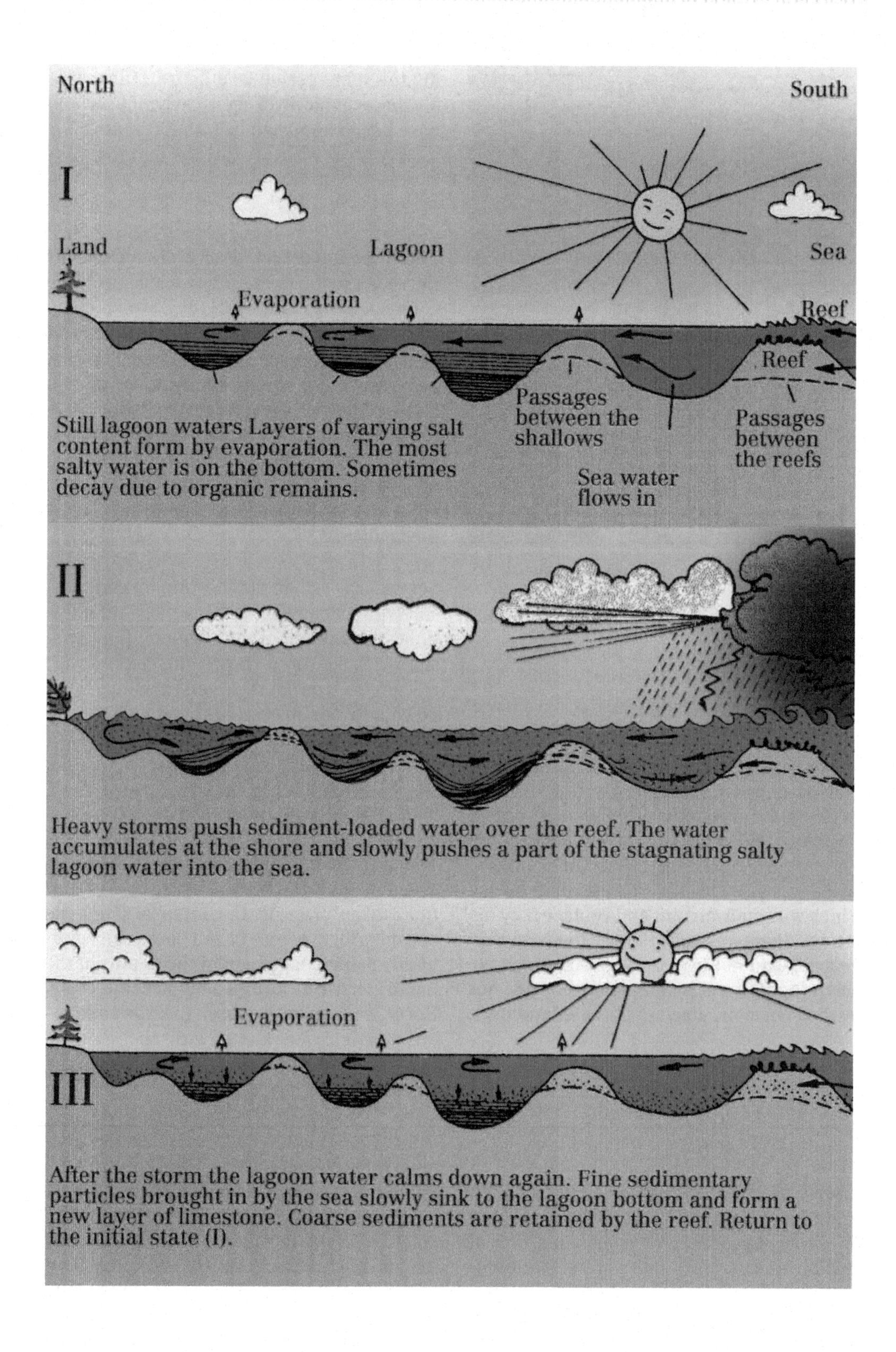

Sketch diagram showing the formation of litho-graphic limestones (from Barthel, p.199).

Glacial abrasions on a limestone, Rüdersdorf near Berlin.

traces of huge rocks which must have been carried along by a gigantic glacier. This was not only the scientific proof that northern Germany had been completely covered by ice during the Ice Age 20,000 years ago – Torell was also able to refute the drift theory which was then current and replace it by his glacial theory. Accordingly, the ice age deposits had not been brought from the far north to the south by massive icebergs, but by the enormous glaciers of the inland ice.

3.4 $CaCO_3$ cycle

The total mass of calcium carbonate on earth has constantly increased throughout the geological ages. Enormous quantities of calcium and carbon dioxide were continually drawn from the seawater by carbonate sedimentation. As the chemical composition of the seawater has not changed for at least 500 million years, however, the oceans must have been supplied with corresponding amounts of calcium and carbon dioxide during the same period (see figure). Otherwise all the oceanic hydrogen carbonate would have been exhausted in 150,000 years.

Supply of calcium

The most important source of calcium for the seas are the rivers. They transport large volumes of calcium every day, which come from the weathering of carbonate rocks and calcium and magnesium silicate rocks. However, these volumes only make up two thirds of the calcium which are being continually extracted from the sea by organisms and direct precipitation. The second source of calcium in the oceans was only discovered in the seventies: the submarine hydrothermalism.

Seawater is rich in magnesium. In comparison with calcium it contains more than five times the amount of this element. If magnesium-rich seawater comes into contact with calcium-rich basalt rocks, an exchange takes place between magnesium and calcium according to the following equation:

$$CaSiO_3 + Mg^{2+} + 2\ HCO_3^- \rightarrow$$
$$MgSiO_3 + CaCO_3 + CO_2 + H_2O$$

This transformation of calcium silicates into green magnesium silicate hydrates (serpentine) takes place in the regions along the oceanic ridge, where basalt rocks are continually being pushed out of the depths of the Earth's crust to form a new oceanic crust. Here Ca^{2+} ions are released in large volumes and are carried along with the thermal waters flowing out near the ridge, which are more than 300 degrees Celsius.

But carbon dioxide, too, is released by geological processes. In the collision and subduction zones, where tectonic plates collide, carbonate sedimentary rocks are transformed into metamorphic rocks: calcium and magnesium silicates develop from carbonate rocks containing clay or silica, and at the same time large volumes of carbon dioxide are released, which escape by primary volcanic emissions and by secondary thermal and gas sources.

In addition to the estimated 5 to 10 billion tonnes of CO_2 which are emitted year on year from active volcanoes alone, another significant source of carbon dioxide is to be mentioned : it is due to human activities in the last hundred years and reaches comparable levels.

Carbonate cycle in geological times

Global tectonics, i.e. the relation of the major plates, have a direct effect on the sedimentation rate of carbonates through submarine hydrothermalism and the continuously changing surface of the continents.

The extension rate of the oceans has changed continually over the geological ages. New sea beds formed at the oceanic ridge; the magnesium / calcium exchange between the oceanic crust and the seawater increased. Due to tectonic movements metamorphism and volcanism became more intensive; increasing quantities of carbon dioxide entered the atmosphere and favoured the greenhouse effect. The atmosphere heated up, which was favourable to the continental weathering of carbonates; here, too, large amounts of calcium were released.

At the same time tectonic activity again and again caused the sea beds to rise; the water flooded the shores and shallow seas formed on the continents, providing extraordinarily favourable conditions for carbonate sedimentation.

When looking at the different geological periods, it is striking that the conditions for the development of carbonate rocks must have been particularly favourable in the Cambrian period, from Upper Devonian to Lower Carboniferous, in Permian and Triassic periods and also in Jurassic and Cretaceous, because in these eras the deposited carbonate formations reached huge dimensions.

However, it should not be overlooked that limestones which are deposited on the oceanic basalt crust never reach the continental limestone masses. Dragged away with the oceanic crust in subduction areas, they disappear in the Earth's mantle rock, never to be seen again.

Cycle of calcium carbonate.

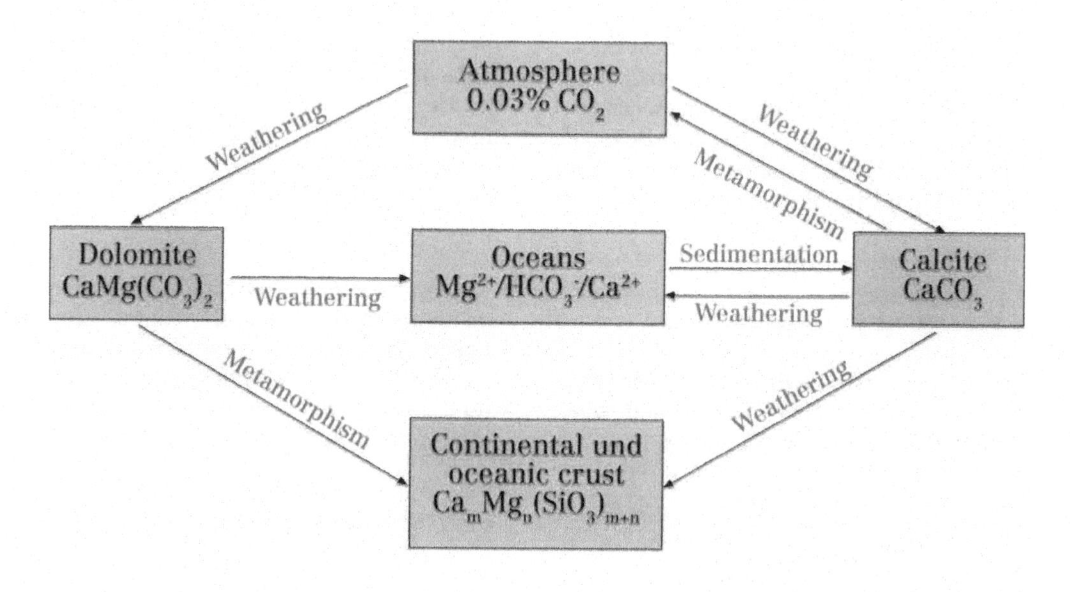

3.5 Industrially exploitable $CaCO_3$ deposits

Calcium carbonate rocks are widespread over practically the whole world. But not all deposits are worth quarrying, because the requirements in respect of calcium carbonate as a filler are high and can only be fulfilled if the raw material in the quarries meets certain criteria.

Selection criteria for carbonate deposits

The first criterion is purity. The calcium carbonate content of a rock must be at least 97-98 percent, which means that the proportion of insolubles in hydrochloric acid should not exceed 2 percent. Only certain marble deposits are an exception: when the metamorphism has transformed the original impurities into minerals large enough to be easily removed by flotation.

The brightness of the rock is also of great significance, although a very high calcium carbonate content is not necessarily synonymous with a high brightness. Even slight traces of brown or black organic material or a contamination by very finely distributed sulphides or iron oxides of some ppm may affect the brightness of the product when ground to powder.

Not least, the dolomite content must be under 5 percent for most applications, as this mineral, due to its higher hardness, can lead to problems during processing. However, finely ground fillers made of very white, pure dolomite are used in some areas of the paint and varnish industry.

Limestone deposits

Limestones were the first rocks to be quarried in human history. Due to their widespread occurrence they were easy to find, and due to their softness and layered structure they were easy to work (see chapter II, "Cultural History").

The limestones quarried today are classified according to their hardness, or rather their compactness, because they consist of an agglomeration of calcite grains of hardness 3 – the differences arise from the degree of cementation.

Soft limestones

Soft limestones seem very white, as long as they are dry. However, in order to establish their true level of brightness when prospecting in the field, the fracture plane must be moistened. Depending on the yellow index, it will take on a more or less cream coloured tone.

- **Chalk** was the first rock used as a filler (see section, "The Beginnings"). It forms very large deposits, which are to be found exclusively in northern Europe along a belt from England in the west to Russia in the east. These deposits formed in Upper Cretaceous, starting in the Cenomanian, 96 million years ago, and ending in the Maastrichtian, 65 million years ago. But only the chalk of the Senonian (80 to 70 million years ago) is sufficiently white and pure for the production of fillers, and here the deposits of the Paris Basin are especially noteworthy, although they also present differences.

  Whilst inclusions of flint are widespread in the deposits at Précy-sur-Oise, west of Paris, these nodules and bands of black silicon dioxide are completely absent from the Champagne chalks on the eastern edge of the Paris Basin. Besides these deposits near the towns of Châlons-en-Champagne and Troyes, there are other chalk deposits in France worth quarrying near Lille and Saint-Omer, and in the west of the Ile de France.

  In England the deposits are concentrated on the eastern part of the island. From London and Cambridge in the south up to Hull in the county of Humberside there are numerous chalk quarries in operation for the production of fillers.

  Crossing the channel, the next deposits are in Belgium near the town of Mons.

In Denmark the central part of the Jutland peninsula consists of a chalk which is covered by glacial deposits. A large quarry is sited near the town of Sterns on the island of Sjaeland. In Fakse, south of Copenhagen, a chalk is extracted which originated in the Danian stage at the limit between Cretaceous and Tertiary.

From Fakse it is not far from the Swedish deposits south of Malmö. In the opposite direction, in Lägerdorf in Schleswig-Holstein, one comes upon chalk strata which have been raised several hundred metres by "salt domes" (diapirs) lying beneath them. Two hundred kilometres east, on the island of Rügen, the chalk has been folded and fractured by glaciers.

Polish chalk in many cases contains marl and is silicified, and for this reason it is only suitable as a basic material for fillers to a limited extent. However, it is still used today as a building stone under the name "opokas".

The white chalk strata in south-western Russia are directly deposited on top of very old formations. This applies both to the deposits in the region of Voronež in the Don valley and to the chalk in the region of Staryj Oskol. Here, in the area of the magnetic anomaly of Kursk, the chalk covers a huge iron ore deposit.

- The **"young" reef or sub-reef limestones** from Tertiary or Quaternary periods also count among the soft limestones.

In Europe these include, in particular, the white limestones of Miocene age, which are found in northern Spain, south of Barcelona. They form irregular lenses which are inserted in a coarse, coloured limestone rock mixed with quartz grains. But soft limestones are found in Greece, too, on the Ionic islands of Cephalonia and Zakynthos.

In Indonesia limestones of Miocene age are widespread in the north and south of the eastern part of the island of Java. They have been cemented at the surface to a depth of some metres by calcite, which precipitated due to evaporation of the trapped water. These limestones are often associated to a soft, white dolomite rock, which can be sawn into blocks by hand. These blocks harden by drying and can be used as building material.

On Jamaica, reef limestones are quarried and ground to fine powders. White aragonite sands are extracted in the reef zones of the Bermuda Islands, they are due to calcium carbonate precipitation from supersaturated seawater.

An extraordinary calcium carbonate deposit exists in Villeau in France, where a fossil $CaCO_3$ deposit has never been consolidated over 30 million years, but has remained loose – in its composition and structure this deposit is similar to artificially precipitated calcium carbonate, PCC.

- **Urgonian limestones** in southern France show a similar method of formation. They correspond to a special facies of Lower Cretaceous and formed 110 to 120 million years ago in a marine sub-reef environment. The soft Urgonian limestones, which are scarcely consolidated, form large lenses within hard, well-stratified limestones, more or less coloured by iron oxides. The deposit at Orgon, south-west of Avignon, was chosen by the French geologist Alcide d'Orbigny in 1847 to define the Urgon facies.

- **Mylonitised white limestones** arise when compact limestones are naturally ground down by tectonic movements. This is frequently the case in the environment of a large, reversed fault, where the land areas are displaced and grind down the limestone (overthrust nappe).

In northern Italy, for example, about 60 km north of Venice, Cretaceous limestone mountains in the Friaulic pre-Alps were carried onto Tertiary formations in the region of Caneva-Sacile. The white reef limestones of Cretaceous age at the base of the original relief broke up during this process and resulted in a very friable breccia. This rock has been quarried since the beginning of the 20th century in galleries, now in danger of collapse.

In the area around Avezzano in the Abruzzi regions east of Rome, the massive limestones of the Lower Cretaceous period were ground down into breccias (mylonites) up to a depth of several tens of metres along a thrust fault more than ten kilometres long.

Hard limestones

The hard limestones quarried for the production of fillers nearly all originate in the Mesozoic era – from Triassic through to Cretaceous.

- **White, massive limestones**, the "massiccio", are quarried in the mountains of Umbria in the area between Foligno and Nocera, south-east of Perugia. The rock formed between Upper Triassic and Lower Jurassic in a warm, shallow sea, in which many mollusks lived amongst limestone-forming algae. It is strongly consolidated with sparitic cement and its stratification is only slightly visible.

- **Biodetritic sub-reef limestones** of Upper Jurassic and Lower Cretaceous age occur in large beds and are widespread in Europe. But only in a few areas do they achieve sufficient brightness and purity.

  At Belchite in Spain, south-east of Saragossa, for instance, the white limestones of Upper Jurassic are quarried, but they are also found in the south of the country, in the mountains west of Granada. Here the rocks are so compact and solid that they are sawn into blocks, polished and used as "marble".

  In southern Germany two types of white limestones of Upper Jurassic age (Malm) occur: little stratified, massive oolitic limestones and reef limestones in lens

The four grades of metamorphism: schematic pressure-temperature diagram (from Winkler, p. 51).

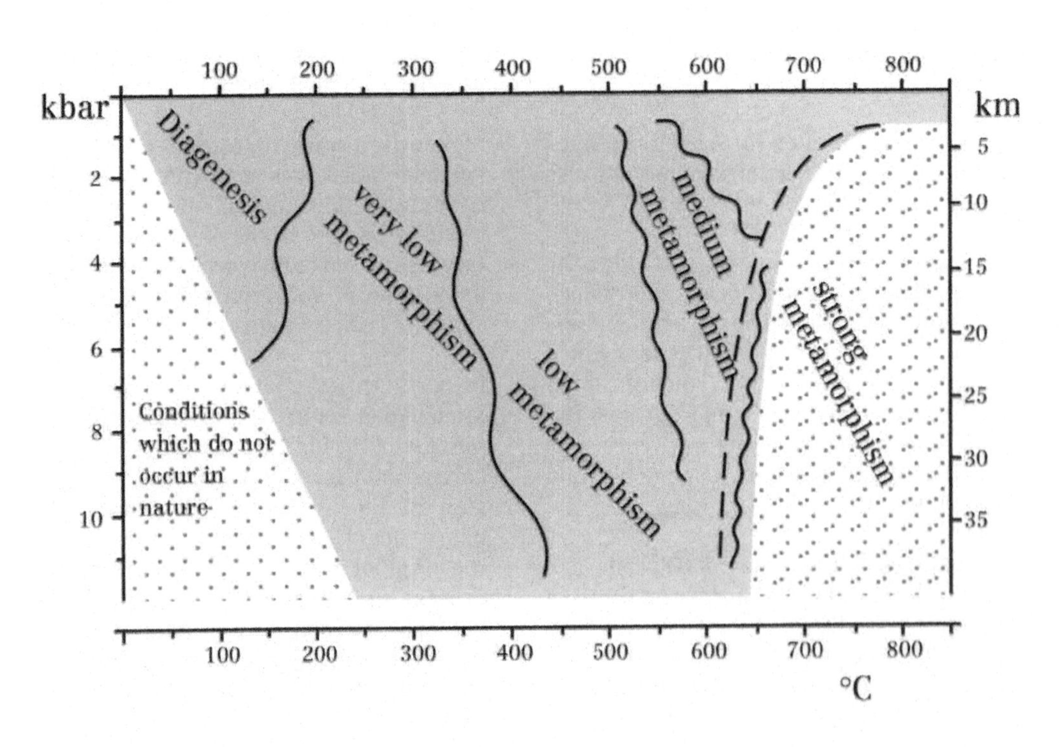

form, which are surrounded by marl limestones. They were deposited in a carbonate platform environment of a warm epicontinental sea and are quarried both in the Swabian Jura and in the Franconian Jura.

Marble deposits

Marbles differ according to the degree of metamorphism and the original composition of the limestone; the multifarious nature of these rocks is considerable. Correspondingly diverse are the possibilities for the subdivision or classification of marbles. In the following, marble deposits are considered according to the degree of re-crystallisation, i.e., according to the intensity of metamorphism. Depending on pressure and temperature, four areas may be differentiated : very low, low, medium and strong metamorphism (see figure).

- In the case of **marbles produced by very low metamorphism,** the re-crystallisation produced calcite crystals of only several hundredths to tenths of millimetres. There are no new minerals, as the temperature of maximum 200 to 250 degrees Celsius was insufficient for a reaction between the calcite and the impurities in the limestone.

  In the eastern Pyrenees the limestones of Upper Jurassic and Lower Cretaceous were transformed into a white, very fractured marble in the course of the Pyrenean orogenesis, 40 million years ago. It is extracted at Tautavel, north-west of Perpignan. Near this village there is also a famous cave, in which human remains, 450,000 to 500,000 years old, have been found.

  About 150 kilometres north of Mexico City, in the Vizarron region, there are limestones which are more than 1000 metres thick. They re-crystallised in Upper Cretaceous, 70 million years ago during the formation of the eastern Sierra Madre. The white layers are extracted at a height of 2000 to 2500 metres above sea level, and some of the quarries originate from the time of the Spanish conquistadors.

  In Korea, limestone layers of Cambrian age underwent a low metamorphism during the granite intrusion in Upper Jurassic, 150 million years ago. These marbles are exploited in the mountains of Kanwon, 200 kilometres east of Seoul.

- **Marbles originating from low and medium metamorphism** are rocks which have been exposed to temperatures from 250 to 500 degrees Celsius and enormous pressure at depths of 5 to 15 kilometres. The crystal structure is therefore more marked : the calcite crystals are visible with the naked eye or with a magnifying glass; often they even reach sizes of several millimetres and have a granular, even saccharoid appearance. The first newly formed minerals of metamorphism are white mica and green chlorite, an aluminium ferric silicate. It is striking that sulphides are concentrated in the form of grains, whereas pyrite rather tends to form small cubes, which quickly oxidise on the surface or inside the marble due to penetrating water.

The best known of these marbles is Carrara marble (see figure, page 48). Since classical antiquity large marble blocks have been excavated in the quarries of the Apuanian Alps east of Carrara and Massa, as a much sought-after material for buildings and sculptures; this marble has been used as a basic material for fillers only since the beginning of the eighties. Only the non-dolomitic, white layers are useable. Large isoclinal folds with parallel flanks repeat the individual layers several times.

In the Sierra de los Filabres east of Granada white and coloured marble layers are inserted in mica schists. These Triassic marbles are extracted in numerous quarries around Macael. In the south of Portugal the region of Estremoz is famous for its pink and white marbles of Cambrian age.

The white marble beds in the US State of Vermont originate from the Ordovician period, 400 to 450 million years ago. These are limestone sediments which were transformed during the orogenesis

Marble quarry in Carrara.

of the Taconic Mountains 380 million years ago. They are quarried from Middlebury in the north to Danby in the south (see figure).

In California the limestones of Monte Cristo (Lower Carboniferous) were deformed and transformed during several periods of the Cretaceous age. Very pure, white marbles resulted, which today are quarried in the San Bernadino Mountains in southern California west of Los Angeles.

Several significant marble deposits in the eastern Alps have undergone two phases of metamorphism. The first of the two phases took place around 300 million years ago during the Hercynian orogenesis, in the course of which the mighty mountain ranges of central Europe arose. The second metamorphism occurred 140 to 30 million years ago during the alpine orogenesis (see illustration). The calcite crystals of these marbles generally reach a size of several millimetres and the marbles can contain numerous minerals, for example, mica and phlogopite, but also sulphides and sometimes even finely distributed graphite.

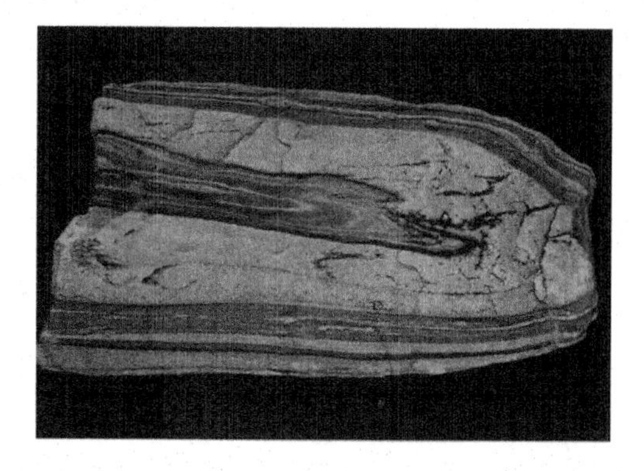

Liassic limestone layers (180 million years), which were deformed in a phase of Mid-Alpine metamorphism 60 million years ago and formed this recumbent isoclinal fold. Brenner region, Austria (original size 30 cm).

Two well-known deposits are situated in southern Tyrol, where a group of marble beds is inserted in gneiss and mica schists of the old Austro-Alpine basement west of the Brenner pass. These are the marbles of Laas in Vinschgau and Mareiterstein near Sterzing.

Somewhat further to the east, in Carinthia and Styria, there are marbles which are also embedded in old, metamorphic series of the Austro-Alpine basement. In Gummern near Villach these marbles are quarried for the production of fillers. Due to the high content of metamorphic minerals certain rock layers must be floated like metal ores so as to eliminate impurities.

Underground mine in Danby. Marble from Danby was used, among other buildings, for the construction of the Supreme Federal Court in Washington (1935) and for the UNO building in New York (1955), but also for the Chiang Kaishek monument in Taiwan (1979) and for the Sama Bank Building in Saudi Arabia (1983).

In the area of Graz in Styria numerous marble deposits were subjected to such extreme folding during the Hercynian and Alpine orogeneses that quarrying is very difficult.

- **Marbles which have undergone strong metamorphism** are coarsely crystalline rocks. Their calcite crystals often reach a size of several centimetres, therefore it is impossible to give these marbles a brilliant surface even if they are intensively polished. Marbles of strong metamorphism are thus not useable in architecture.

Not only the calcite crystals, but all newly formed minerals achieve considerable sizes under strong metamorphism; graphite, for example, is present in millimetre to centimetre sized black flakes. During synthesis of minerals in the laboratory it was possible to determine the extreme pressures and temperatures which are required.

It was shown that the temperature has to rise to over 600 degrees Celsius for a strong metamorphism so as to form the calcium silicate wollastonite, which is present in these marbles.

In Scandinavia most marble deposits originate from the Precambrian era. The more or less pure carbonate rocks which were deposited 2 000 to 400 million years ago have gone through several episodes of metamorphism. During the folding phases they were pulled to great depths of 10 to more than 20 kilometres, where temperatures between 500 and 600 degrees Celsius prevail. These conditions allowed the formation of calcium and magnesium silicates such as amphiboles, pyroxenes, garnets and feldspars with many individual crystals. The proportion of these minerals in the marbles is high, lying between 1 and 10 percent. Filler production therefore requires floatation, the only way to remove silicates, sulphides and graphite.

The youngest of these deposits are located in Norway and belong to the Caledonian mountain chain, which formed 400 million years ago; however, some of these marbles had already undergone metamorphism in an earlier orogenic phase.

In the area of the port of Molde, white marble is combined with eclogites. These silicate rocks, with a high density of 3.2 grams per cubic centimetre, formed under high pressure at depths of at least 30 km. A further important Norwegian deposit is situated in the area of Bodö-Fauske north of the polar circle. Here a metamorphic, white dolomite is quarried.

Swedish marble deposits have often been mineralised by contact with intrusive magma rock. The mineralisations consist either of magnetite (Fe_3O_4), which can easily be eliminated by magnetic separation, or sulphides such as pyrite (FeS_2), galenite (PbS), sphalerite (ZnS) and chalcopyrite ($CuFeS_2$), which have to be separated by flotation.

In Finland two marble deposits should be mentioned: Pargas and Lappeenranta, one 150 kilometres west of Helsinki, and the other 200 kilometres east of Helsinki. In both cases the marbles, which are nearly 2 billion years old, show numerous light and dark intrusion veins. In Lappeenranta silicious limestone layers were transformed into a white marble, which contains approximately 20 percent of wollastonite. This mineral is the actual product of quarrying and is obtained by flotation; calcite results only as a by-product.

Finally, there are numerous marble deposits in Korea and in north-eastern China in the provinces of Liaoning and Jilin, which are intercalated in the granite gneiss rocks of the Pre-Cambrian basement. The individual beds are always considerably folded and crossed by younger granite intrusions. The marble is generally rich in newly formed metamorphic minerals such as silicates, graphite and sulphides: only exceptionally, it can be used without flotation.

Hydrothermal calcite

Deposits of pure calcium carbonate constituted by hydrothermal deposits of calcite are very rare. In general they occur as veins within a rock or as fillings in a karst cavity.

The vein formations always consist of calcite crystals several centimetres to decimetres in size and more or less welded to one another. These crystals were formed by sequential accretions as shown by the zonation due to the presence of iron oxide.

Hydrothermal calcite is deposited under conditions similar to those of sedimentation or petrification; they have very little in common with metamorphism.

In Europe there are only a few deposits worth quarrying for hydrothermal calcite, and, indeed, none has been exploited for a long time. For example, in the Basque Provinces, west of Bilbao, some white calcite veins are found, and in Germany, too, on the south-eastern edge of North Rhine Westphalia, near the town of Brilon, there are vertical veins of an ochre yellow or pink coloured calcite, embedded in grey Devonian limestones, formerly extracted in shafts and galleries. In the south of Ireland, in the County of Clare, veins of white calcite in dark Carboniferous limestones have mineralised into lead sulphide containing varying amounts of silver.

Small quarries extracting hydrothermal calcite are still found today in southern China, where pure white calcite is extracted and processed. The material is sorted by hand, and rock which is coloured yellow or dark brown due to iron oxide is rejected.

On the island of Bawean north of Java a special quality of hydrothermal calcite can be found. Veins of calcite have formed in limestone plateaux, some of which are up to several hundred metres long and several tens of metres wide. The quality of the crystals found there is similar to the famous Iceland spar, that is why they are sometimes used for optical purposes.

Summary

There are countless rocks on earth, but only a few are so closely connected with the development of our planet as limestones. These rocks combine the carbon dioxide of the atmosphere and the calcium of the Earth's mantle rock, which eternal ages ago reached the Earth's surface as a component of red-hot magma. However, to create a connection between a volatile gas and an inert rock, the support of living organisms was required, and therefore the formation of limestones on this planet has always been closely linked with biological activity.

If there had been no plant and animal life on earth, the rocks made of calcium carbonate would not be dispersed over the whole earth as they are today. But if, in the last almost four billion years, 10^{23} grams of carbon dioxide – in words: one hundred million billion tonnes! – had not been bound in the various limestones, then the oxygen content of the atmosphere would still be less than 1 percent, and the development of higher life forms would not have been possible.

So it is not surprising that even today scientists are seriously considering exploiting this process to combat the greenhouse effect. Initial trials in the Pacific around the Galapagos Islands have shown that the algae population of the oceans grows strongly if iron is used as a fertiliser. The "carbon dioxide" consumption of this population increases, and when the algae die off, the carbon disappears into the depths of the ocean, to be deposited as carbonate ooze on the bottom of the sea.

Not only life on earth is continuously changing, but also the surface of our planet has been subjected to persistent transformation. Based on the composition of a carbonate rock a geologist today can reconstruct the original environment in which the rock was formed. He can establish whether it was deposited at the edge of a continent or in the middle of the ocean, at what depth the sedimentation took place and in which geographical latitude.

Due to geodynamics, the inner activity of our planet, the continental masses are continuously in movement. They move, divide and collide, throwing up new mountain ranges. And the carbonate rocks are witnesses to this "eternal dance" of the continents; they show us where once a mountain stood or when a deep ocean silted up to become land.

And today, they give us information on the climate which prevailed at the time of their formation, by their chemical composition, the ratio of aragonite to calcite, the embedded fossils or the striae which made Rüdersdorf limestone a witness of the last Ice Age.

All this makes calcium carbonate rocks one of the most important subjects of geological research. Correct interpretation of their characteristics gives us an insight into the history of our Earth.

II.

THE CULTURAL HISTORY OF LIMESTONE

BY JOHANNES ROHLEDER

Calcium carbonate is a simple chemical compound or, more specifically, a salt consisting of a calcium cation and a carbonate anion. Its molar mass is 100.1 gramme and it has a density in the order of 2.8 gramme per cubic centimeter. When heated to temperatures exceeding 800°C, limestone will dissociate calcium oxide and yield carbon dioxide. In its natural state limestone occurs in three different crystal forms: In the unstable and extremely rare vaterite form, the occasional aragonite form and the predominant calcite form. After quartz, the last-named calcite (calcareous spar) is the most widespread mineral on earth.

Enough said about a detailed scientific definition of calcium carbonate as this does little justice to the importance of the mineral to mankind. Its use as a chemical substance with precisely defined properties is an achievement of recent decades and is described in the chapter on industrial application.

The occurrence of single crystals – also calcite crystals – or an agglomerate of crystals, is very rare in nature. Usually, an innumerable number of individual crystals became compacted into limestone rocks in the course of earth's history. And these limestone deposits, be they in the form of limestone, chalk or marble, all made a significant contribution to past civilizations and to some of the chapters of our cultural history.

Although chemically identical, each of these three minerals have differing properties which have made them so invaluable to mankind right down to the present day. A distinction between marble and limestone can prove to be very difficult, even though geology supplied a clear formation-based scientific classification in the 19th century. As a result of the fluid transition between the simple, sedimentary crystals and the metamorphic crystalline marble, only precise examinations can distinguish between the two. Moreover, present-day definitions are hardly suitable to study the past.

In ancient Rome *marble* was considered to be any ornamental stone that could be easily ground and polished, and this stone even did not have to be limestone! However, to make a distinction between the different stone types, the Romans added the place of origin, colour or pattern: Hence, *marble numidicum* is a yellowish marble which, today, is known as *giallo antico*; *pyrrhopoecilos* is a granite with a reddish hue; finally *leptopsephos* is a white mottled porphyry.

The definition of marble continued to vary throughout history. For instance, in the 7th century AD, the Church teacher St. Isidor of Seville defined marble as "a special stone that is attractive on account of its patterns and colouration". Even today, both in the building trade, and in the vernacular, all solid limestones that can be polished continue to be referred to as marble. Moreover, since both are quarried in the same manner and are used for similar or identical purposes, they are looked upon jointly, though they are not treated indiscriminately.

Chalk, on the other hand, has its own history.

1. The history of chalk

The Anglo-Saxons referred to chalk as "Hwiting-melu" , literally "whitening powder". And this was exactly what chalk was used for over the millennia, be it as a white pigment in paints, primers and plasters, or as an extender for other pigments. Wherever paints were being used there was also chalk.

The Anglo-Saxons quarried chalk all around Dover on England's southern coast. These deposits were only one of many in northern and western Europe – in the Champagne, northern Germany and on the German Isle of Rügen, or in Denmark and Sweden. There are also extensive white chalk deposits in Sicily, Crete and Euboea. This natural pigment can be found in 11th and 12th century wall paintings in India. In medieval Japan chalk was known as "o-go-fun" – one of a total of 21 pigments.

The main preconditions for the widespread use of chalk were the ease with which it could be quarried and processed. The soft rock can be easily quarried with simple tools such as saws and axes. Moreover, depending upon the purity of the deposit, it was usually sufficient to merely crush and grind the chalk chunks to produce a powder in the required quality. Larger impurities, such as flint stones and other minerals, could be easily sorted out.

The mineral name of chalk: "terra creta" – sifted earth – is actually derived from this process. At least this interpretation is more plausible than the link with the island of Crete which St. Isidor of Seville claimed in his "Origines":

"Chalk [creta] was named after the island of Crete where it is better".

Be it as it may, in antiquity they were not concerned with the present-day precise definition that is hardly two hundred years old.

[1] In English prepared chalk, but also ground limestone and marble, continue to be referred to as "whiting".

Be it gypsum or clay, marl or talc, the Romans referred to all minerals that were prepared in the same manner and used for the same purpose as chalk, and the mineral did not even have to be pure white.

The pigments are distinguished by their origin – *creta eretria*: probably a white talc from the south-west coast of Euboea, or by their use, as defined by Pliny The Elder (or Gaius Plinius Secundus as he was known in Latin):

"Another type of chalk is called silver chalk [creta argentarial] because it can give silver its original brightness."

These unsystematized designations make it virtually impossible to determine a mineral by past written sources. Moreover, in the course of time, designations changed. Even today everything is called chalk in painting: "Rock or priming chalk" is powdered dolomite, "hard chalk" is ground limestone, and "Bologna" chalk is gypsum.

Only chemical analysis can reveal the actual pigment that has been used in white paint for a work of art or a simple coating. Since pigment analysis is relatively simple, extensive examinations have covered all important cultural epochs so that precise classification is now possible in most cases. Yet even these results must be treated with a certain measure of scepticism as painters did not actually use everything that can now be chemically detected nowadays.

For instance, calcium carbonate has been detected in nearly all prehistoric cave paintings in the period between 40,000 and 10,000 BC. Yet it was only right at the end of this epoch that chalk and limestone powder were used by the cavemen artists, and even then it was very rare. Three colours were primarily used: red, yellow in different shades of ochre and black from manganese earth or soot. The supposed white in their cave paintings arose over a course of many thousands of years due to the calcareous sediment formed by the evaporation of calci-

Stone Age cave painting in Lascaux, France (from 10,000 to 20,000 years old).

ferous water on the cave walls, similar to dripstone formation. The extremely thin layer of calcareous sediment protected the prehistoric paintings from weathering influences so that they survived undamaged for many thousands of years. However, such cave paintings also became eternally obliterated wherever this "skin" became too thick. As opposed to the first artists who painted on bare rocks, the Egyptians and subsequent civilizations in the Mediterranean region first primed the surface that were to be used for paintings, no matter whether this was a stone or a brick wall, a wooden panel or a sarcophagus (see figure).

Egyptian wall-painting of unknown origin (1552-1306 BC).

At the time priming had the same three purposes as today: Firstly, to produce a smooth and level surface. Secondly, to produce a bright, if possible white reflector to highlight the subsequent colours. Thirdly, as a substance that regulated the absorptive capacity of the painting base.

Primers consisted of a powdered white mineral used as a pigment and filler, and an animal or vegetable glue that was added as a binder. In Egypt gypsum and chalk were the most frequently used minerals for primers, and more rarely white clay or crushed mussel shells. The colour effect of the final picture could even be influenced by selecting the corresponding primer. For instance, chalk gives pictures very airy tones and a delicate matte finish. Moreover, the porous structure is very absorbent.

To include white in their pictures, painters either did not paint over the chalk primer or they subsequently removed the paint to lay bare the white primer underneath. In this manner they were able to add white to their pallet of colours that consisted of red, black and yellow. Especially formulated white paints were not necessary, the only exception being the white wash used by the Egyptians to produce white walls. The calcium hydroxide adhered to the walls without requiring an additional binder, and in the coarse of time this finally set as calcium carbonate.

Micrograph of paint cross-section (50x magnification; top: normal photo, bottom: fluorescent photo). The layered structure of the painting is clearly visible. A soot layer (hardly visible) directly on the canvass, followed by a chalk primer (yellowish colour) and a lead white layer with some red pigment. These are followed by the old varnish with grime deposits on the surface, and finally the painted layers.

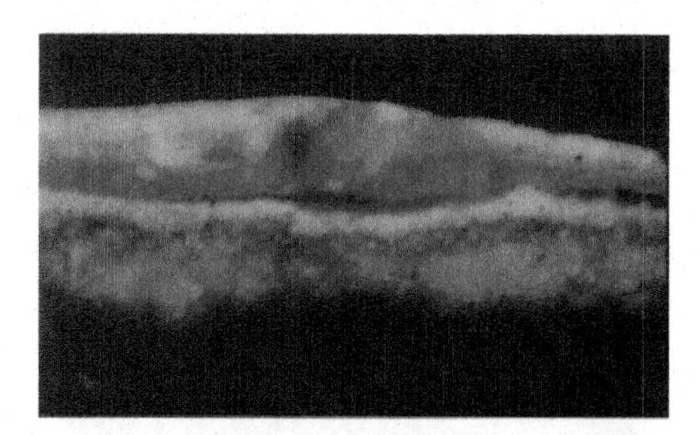

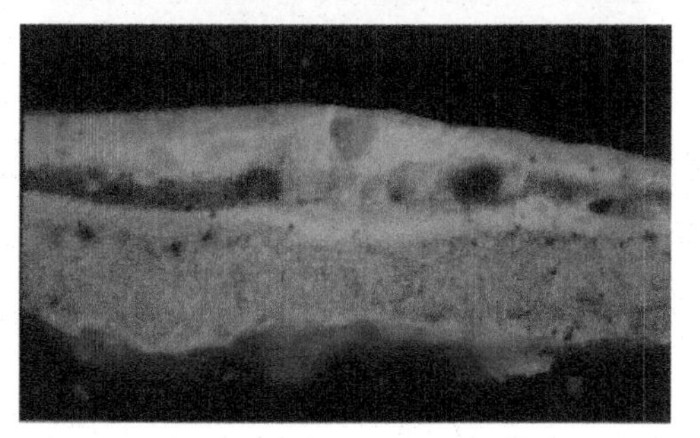

The search for a whiter white

The great Greek painters, among them Apelles and Nikomachos, limited themselves to the use of red, yellow, black and white, at least if one is to believe Pliny. Whether the corresponding sections in his "Natural History", are actually true is somewhat doubtful. For Pliny was not so much concerned with giving an accurate description of Greek paintings but rather with his need to back his critical appraisal of the bright colours used during the Roman period with corresponding historical evidence. Hence:

"Accordingly, everything was better because less was available. The reason for this was that [today] one is more concerned with the value of the material and not with the value of the mind."

PLINY, NATURAL HISTORY, XXXV, 50

One thing is certain – the colours blue and green were introduced into painting towards the end of the 3rd millennium. Already a century before Pliny, Cicero stated that Apelles used more than four colours for his paintings. Yet Pliny did not challenge this nor the fact that the Greeks were searching for a whiter white. In addition to the long known white pigments derived from chalk, gypsum and clay, lead white was first detected in ancient Greek paintings. Lead white is a basic lead carbonate of the following composition:

$$2\ PbCO_3 \cdot Pb(OH)_2$$

Since a compound of this composition is rarely encountered in nature, lead white had to be produced artificially. The recipe for this is given in all handbooks on painting since antiquity: Place metallic lead, best of all strips rolled up into coils, in a stoneware pot in which the bottom is covered with vinegar. Then bed this pot in horse manure. The heat and carbon dioxide generated by the subsequent rotting process transforms the metallic lead into lead white.

Designation	Chemical Composition
Primer chalk	Powdered dolomite rock [$CaCO_3 \cdot CaCO_3 + MgCO_3$]
Bianca San Giovanni	Slip chalk [$CaCO_3$] and sump lime [$Ca(OH)_2$] mixed in a ratio of 7:3
Blanc fixe, baryta white	Precipitated barium sulphate lime [$BaSO_4$]
Cerussa, white lead, Kremnitz white etc.[1, 2]	Basic lead carbonate [$2\ PbCO_3 \cdot Pb(OH)_2$] produced from metallic lead and vinegar
Bolognese chalk	Gypsum [$CaSO_4 \cdot 2\ H_2O$]
Creta anularia, ring chalk[1, 2]	Chalk [$CaCO_3$], mixed with glass powder
Cimolia creta, Cimolic chalk[1]	Chalk or clay-like material
Creta Eretria[1]	Probably a white clay named after a location on the south-west coast of Euboea
Creta Selinusia, Selinusic chalk[1, 2]	Chalk or chalk marl named after a location on Sicily
Egg-shell white	Powder of calcareous shells mixed with vinegar
Lithopone	A mixture of zinc sulphate [ZnS] and barium sulphate
Melinum, White of Melos[1, 2]	Bianca S. Giovanni or white clay
Paraetonium, meerschaumwhite[1, 2]	Lime chalk containing some magnesium phosphate, silicic acid and clay, named after a location in Libya
Creta argentaria, silver chalk[1]	Chalk [$CaCO_3$]
Titanium white	Titanium dioxide [TiO_2]
Zinc white, Chinese white, etc.	Zinc oxide [ZnO]

1 PLINIUS, NATURAL HISTORY, XXXV

2 VITRUV, TEN BOOKS ON ARCHITECTURE, VII

Natural and synthetic white minerals used in paints as a pigment and extender.

This elaborate procedure to produce simple white pigment clearly indicates that the Greeks were no longer satisfied with the natural colours available to them. Most grades of chalk, gypsum and clay, as well as white wash, had little hiding power. Moreover, they often containing polluting matter with the result that organic residues gave colours a dirty grey veil. Traces of iron usually gave white a yellowish tint which contradicted the mythological significance of white as this had to be absolutely pure to indicate chastity. Consequently, the Greeks and all civilizations right into modern times, were constantly endeavouring to replace the imperfect white of natural pigments. And lead white was such a substitute. "Blueing" by adding blue helped to fulfil the quest for a "whiter white". This is because a blue-tinted white appears to be a purer and more radiant white to the human eye.

And what was true of colours also applied to painting bases and plasterwork with the result that the Greeks sometimes replaced chalk or gypsum by marble flour. Although this was not so absorbent, it nevertheless appeared to be whiter and gave the applied colours greater brilliance.

The Romans resolved the problem with a rich choice of natural and artificial pigments. Thus, Pliny reported of eight different white pigments which were being used for different purposes. Vitruvius mentioned five white pigments in his "Ten Books on Architecture" (see figure). However, not all pigments enjoyed the same importance, even though literature may have attributed the same amount of space to each one of them. In the understanding of Pliny he regarded his "Natural History" as a kind of encyclopaedia which was primarily concerned with collecting and not judging the knowledge of his age. Trivialities were therefore given the same treatment as important things.

Consequently, of the eight white pigments existing at the time, only three played a real role in Roman paintings, namely *melinum*, *paraetonium* and *cerussa* lead white. *Melinum* or the "White of Melos" was already being used by the great Greek painters; whether this was chalk or a white talc is just as unclear as the actual chemical composition of *paraetonium* which is named after the location were it was found to the west of Alexandria. On the one hand it contains "minute mussels", thus indicating chalk. Yet Pliny referred to it as being not only the "fattest among the white colours, but also the most durable wall paint on account of its smoothness", thereby suggesting clay.

Correct assignment is only possible with lead white, but this was used only occasionally in Greece and Rome. At least this was what appears to be confirmed by the very comprehensive pigment analyses conducted by the Italian chemist Selim Augusti in Pompeii in 1967. He identified a whole range of chalks containing differing inclusions, yet gypsum and lead white were not detected. From this Augusti concluded that not only *melinum* and *paraetonium*, but also all white pigments used in Rome at the time, were chalk quarried in different regions of the Empire and transported to Rome. If one considers how much effort the Romans invested in marble (see 2.2 "Transport, Organisation, Trade"), then this appears to be very likely even though nothing has been proven.

Be it as it may, developments in painting in the Roman Empire would invariably result in a decline in importance of chalk as a white pigment for paints, so that it would eventually become only one of many fillers used for primers.

The suitability of a paint is determined to a major extent by its hiding power. This is derived directly from the difference between the refractive indexes of pigments and binders: Hence, the greater the difference, the greater the hiding power. Chalk has a refractive index of $n \approx 1.55$, thus positioning it at the lower end of all pigments. As long as water-based binders dominated – an aqueous lime solution has a refractive index of $n \approx 1.35$ – the hiding power of chalk was sufficient. However oil binders have a refractive index of $n \approx 1.48$ so that the hiding power of chalk is no longer sufficient.

Although oil painting was not of any significance in Rome, drying oils, e.g. linseed or nut oil, were used primarily for medical purposes. However, after the Romans had invented the oil-seed screw press (trapetum), linseed oil became more readily available in larger quantities so that the Romans sometimes used it to paint ship planks and utility articles. This is where linseed oil proved to be far superior to lime washes because, as it set, it became firmly secured to the base as a water-proof coat. Since solids are more firmly bonded together than by an aqueous binder, the use of linseed oil as a binder in painting suggested itself.

However a few centuries were to pass before this was generally the case. It is only in the 12th century that linseed oil was more often encountered as a binder and was more frequently referred to by corresponding handbooks. Oil painting only developed into the dominating technique at the height of Dutch Baroque at the outset of the 17th century.

This was when lead white had long asserted itself as an important white pigment on account of its greater hiding power. Moreover, since it was the only white pigment known at the time that could be used in conjunction with oil binders, the dominance of white lead was even greater, and it retained this status right into the 19th century when the choice was extended by a number of new, artificially produced pigments such as zinc white, lithopone and, somewhat later, titanium white.

Chalk had now become a cheap extender for expensive paints, be they white or coloured, to make such paints less expensive. Moreover, as a result of the porous surface, some paints were given a body of adequate hiding power for the very first time. This is due to the fact that if the "materials which give paints their colour are too light – with the exception of some types of ochre – then their consistency will not be increased by adding white. They do not have enough hiding power to cover whatever is to be painted" is what Jean Fèlix Watin wrote in "L'art du peintre, doreur et vernisseur" published in 1744. Moreover, white chalk diluted intense paints so that they could be more easily dosed.

In the shadow of the fresco

The situation was not so clear with priming. Whether or not chalk was used as a filler depended primarily on what kind of surface was to be primed. There were two painting techniques in Rome that required priming: Wood panel and wall or mural painting.

Mural painting was the more important of the two techniques as it could look back on a tradition of several thousand years. The very first drawings produced in caves by Stone Age humans were mural paintings that decorated rooms used for ritual purposes. Thereafter, mural paintings could be frequently encountered in buildings where they often transformed individual architectural elements into an aesthetic entirety. And as architecture progressed so did mural painting with all its colours and primed bases.

Vitruvius dedicated the seventh of his "Ten Books on Architecture" to wall painting. He explained in detail the fresco paintings completed in Rome according to which the paints are applied *al fresco*, i.e. directly on the still moist plaster.

Roman fresco plaster, in itself, was a small work of art which required extensive knowledge and artisan skills as well as considerable experience. At least two layers were applied and in some cases there were as many as five, six or even more. The first one was spray plaster, followed by the first layer – a coarse plaster or *arriccio* consisting of lime and coarse sand as filler. The *arriccio* is followed by the *intonaco* – a fine plaster with a grain that becomes increasingly finer with each layer. The filler used for *intonaco* was often the finest white marble flour that was available in Rome in large quantities as a by-product of marble work. The last painted *intonaco* layer was elaborately smoothed and polished to give the Roman fresco its fabulous mirror-like finish.

Such masterly frescos, however, could only be found in Rome and in the most important provincial cities which had the highly skilled craftsman that could fulfil the most exacting demands, and the patrons who could afford

Section of a fresco in the garden hall of the Villa Livia near Prima Porta, Italy (1st century).

to pay them. Nevertheless, the fresco technique – painting on plaster that was still wet – soon spread to the small less important towns, even though the lime plaster did not satisfy highest expectations. This form of fresco wall painting dominated for the next centuries, at least in the countries south of the Alps. In the north à secco wall painting was preferred where an aqueous lime solution was used as a binder on the dry lime plaster. Whether fresco or dry lime plaster, there was no longer a place in the arts for chalk as both techniques could well do without additional white pigment or paints: the white used here was exclusively bound chalk.

The fact that chalk did not lose its importance as a primer in mural painting is due to the fact that paints did not remain stable on the intensely alkaline base of fresco plaster.

"Paints that love a chalk primer and cannot be applied on a wet base: Purple-red, indigo, sky blue, Melos earth, orpiment, Appian green and lead white."

PLINY, NATURAL HISTORY, XXXV, 49

Consequently, painting on a dry chalk base continued – in antiquity primarily in tempe-

The head of God The Father from the Coronation of the Virgin Mary. Tempera painting on chalk-primed trachyte stone in Cologne cathedral, Germany (1320-1339).

ra with water-soluble binders such as egg, casein and animal glues, and since the end of the Middle Ages also with oil paints.

For two of the paints specified by Pliny, namely purple and indigo, chalk was necessary for a different reason. They are both organic pigments which, contrary to inorganic pigments, are not miscible with water to form a slurry. They must be first fixed, i.e. precipitated on an inorganic pigment (substrate). Only then can they be applied as a pigment/substrate mixture. One of the most used substrates was chalk because it was cheap, readily available and, as a white pigment, it did not influence the pigment and increased its hiding power.

Nowadays substrate pigments have been replaced by lake pigments in which the organic pigments are synthesized in the presence of inorganic substrates and precipitated directly in a ready-to-use form.

Contrary to mural painting, chalk remained one of the most important priming fillers for Roman wood panel painting. However, chalk had to compete with a wide choice of fillers that included marble flour, gypsum, white clay and ground dolomite. The final choice of filler depended on the effect that was to be achieved.

With the collapse of the Roman Empire the comprehensive supply of mineral fillers likewise broke down, thus forcing the painters to rely on local minerals, at least until the industrial revolution in the 19th century. In Italy this was gypsum, whereas white chalk dominated in panel painting in countries with large chalk deposits such as England, France, the Netherlands and Germany. Chalk, known as "Spanish white" also prevailed in "quinoline painting" right into the 18th century. This technique involved the application of a chalk-lime primer on all wooden objects which were then finished with a colourless ethyl alcohol lacquer to protect them from wear (see figure).

The situation was not so straightforward regarding painting on a canvas which emerged towards the end of the 14th and beginning of the 15th century. If the canvas was primed with an aqueous binder such as lime, then painters obviously used chalk as a filler (or gypsum if they worked south of the Alps). However, if linseed oil was used as a binder, then painters resorted to lead white, as was the case with oil paints. Chalk was then only used as an extender for expensive, full-bodied pigments, i.e. pigments with a high hiding power.

Part of a Chinese twelve-section lampshade (around 1700).

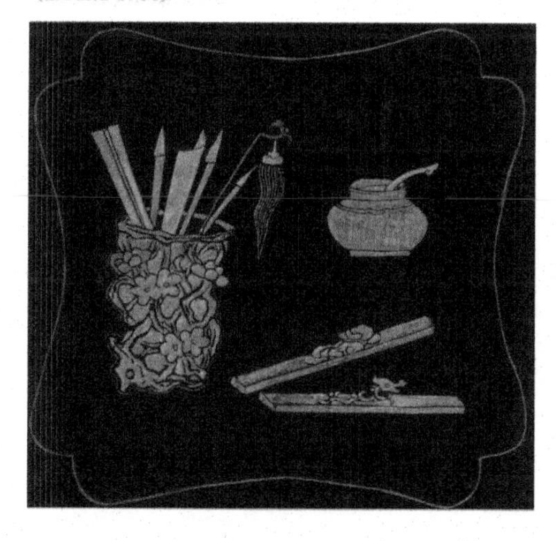

The brief renaissance of pastel chalk

Technical advancements in painting since antiquity always resulted in the replacement of chalk by other better or cheaper pigments and fillers. But there was one exception that gave chalk at least a temporary reprieve. Towards the end of the 16th century the first descriptions can be found of a new painting technique that arose simultaneously in a number of European countries – pastel painting.

Pasta is the Italian term for dough and refers to how painters produced their pastel chalk. They kneaded chalk, pigment powder and an aqueous binder into a uniform dough from which pencils were formed and finally dried. These artificial chalks were then used to outline pictures in simple lines on a rough surface such as paper. This formed the starting point for the actual technique of pastel painting. The painters used their fingers, or bits of cloth or paper, to rub and mix the chalk lines to produce continuous transitions between the individual colours in keeping with painting and thus distinguishing pastel painting from pure chalk drawings.

Actually, this special technique arose nearly a century before when Leonardo da Vinci and his school produced the first pastel drawings, even though this technique – based on natural tools such as red ochre, soot and chalk – was primarily used for realistic studies and sketches of their large mural and oil paintings. This technique then gradually evolved into an independent art form in which complete pictures and no longer mere sketches were produced. The pastel chalks were first based on plant pigments. The range was then extended by synthetic pigments so that the resulting palette soon almost equalled that of oil paints.

The range of different colours matched the diversity of minerals used as substrate. It would be a mistake to assume that pastel chalks were invariably made of natural chalk. This may have been the case in the initial period since they were not mentioned in any sources. However, around 1620, recipes arose according to which the pigment powder was not only produced with chalk and milk but also with pipe clay and water. Chroniclers of the day claimed that the "chalk" produced in this manner was more durable. And at the latest in the Rococo period, at the height of pastel painting, fired gypsum and kaolin were the customary "chalk materials", whereas real chalk became increasingly rarer.

Since then nothing has changed regarding the importance of chalk in the arts. It is still an important extender and is also often used as a filler, but not in any in any significant quantity for paintings. The actual quantities used in former days can only be estimated. Documents on chalk quarrying and trade were only kept towards the end of the Middle Ages.

And this is not surprising in view of the fact that, contrary to the coveted marble, chalk was something quite ordinary and in everyday use so that it was given little thought.

An everyday product

The first document on the sale of chalk is dated 1438. It reports how the Bishop of Châlons required chalk to "white-wash" some of his buildings. Champagne chalk was ideal for this purpose because it was very porous, thus increasing its hiding power, and the few pollutants it contained gave the chalk a pleasant warm tone. Moreover, it could be easily crushed manually. After the coarsest contaminating elements had been sorted out, the chalk powder was finally mixed with lime or glue and water for immediate application with a brush.

Although Pliny described how walls were white-washed with a suspension of Selusian chalk in milk – the casein contained in the milk acts as a binder – few houses were built with stones and bricks in the subsequent

Chalk cliffs on the Island of Rügen.

centuries and huts built of wood and mud were rarely painted. Only then did brick building slowly increase.

It was therefore quite sufficient to extract chalk in small quarries. Simple peasants, often only supported by their wives and children, worked in these quarries. On the German island of Rügen they were referred to as "chalk peasants". Here, as in the Champagne, they only worked in these quarries during the winter months to support their families while all farming work rested.

This form of production dominated well into the 19th century. As late as 1840 more than 300 of such quarries were still being worked in the Marne department, of which 39 were concentrated around Châlons-sur-Marne.

Technical reasons made small quarries ideal. No special skills were required to extract the chalk, and simple tools like hatchets or axes, were perfectly sufficient to break the soft chalk. Only small quantities were extracted, but then there was not much demand for chalk and the markets had to grow in the course of the centuries.

As the transportation of cheap chalk over long distances was not economical, the chalk trade did not go beyond in the Champagne region where it was extracted. When the German poet Johann Wolfgang von Goethe accompanied the Duke of Weimar on his campaign in France in 1792, he remarked that the people, there, were living on chalk:

"A soldier merely had to dig a small hole into the ground to find the cleanest white chalk which he required to polish his equipment. The army actually issued an order instructing the soldiers to provide themselves with as much chalk as possible since it cost nothing."

As nobody buys something that costs nothing, the next documents on the sale of chalk are from the 17th century, and the Bishops of Châlons were once again the buyers.

The initial situation on the island of Rügen was more favourable since the chalk could be shipped across the Baltic Sea to major north German cities so that there was a large market for chalk already at an early date. In Champagne the situation remained unchanged right into the 19th century before the local chalk trade reached out to more distant regions. However, with the advent of cheap transport, Champagne chalk was exported to Germany and even as far as Russia. It was superior in quality to all other chalk grades and it could be sold directly without any previous processing so that, up to a certain point, the cost of transportation remained an economical proposition.

Not only is the Champagne chalk purer than anywhere else, this otherwise soft amorphic mineral is also denser so that it can be cut into blocks for use as a building material.

A remarkable building material

Building with chalk blocks has a long tradition in the Champagne region. It started, how else, in Roman times. The Romans required a large amount of all kinds of building materials for their towns, settlements and forts. And, apart from wood and clay, the Champagne region only had chalk. With the exception of a few traces in the Forum in Reims, no structures from this period have survived. It was only in the 10th century that buildings were once again erected with chalk blocks.

Churches incorporating chalk block walls were then built, but only very few because chalk blocks were very expensive. Chalk used for building had to be extracted by underground mining in order to reach deposits that had never been frozen, not even during the last Ice Age in the Quaternary period. Once chalk has frozen, and this also applies to Champagne chalk, it cracks and becomes brittle, thus rendering it unsuitable as a building material.

The place of entry into underground mines was at those points where the chalk deposits reached the surface. Shafts of up to two meters in diameter were sunk into the soft chalk down to the compact layers where the

St. Pierre Church in Chausée sur Marne (11th-12th century).

chalk was cut into blocks. This continued, layer by layer, with the result that shafts of ten to twenty meters depth were quite common. Large deposits were mined in galleries that were interconnected by passages. Such an extensive network lies underneath Reims. The chalk deposits under this city are riddled by innumerable pits and passages and it is hardly possible to establish the true extent of all these mines. The normal mine, however, was usually a single shaft which, at the base, bulged out into a bottle-shaped bottom (see figure).

The freshly cut chalk was stored underground during the first winter and was only transported to the surface in spring where it remained to dry under the exclusion of all moisture for another year. The next winter was then used for quality testing: If the chalk still contained water then it would freeze, thereby bursting the block. If this was not the case then the chalk was considered to be a suitable building material – a simple yet very effective method which Vitruvius recommended builders to adopt:

"If a building is to be erected, then cut the blocks in the summer and not in winter. The cut blocks should be permanently stored in the open air for two years. Those that are damaged by exposure to the weathering effects in the course of these two years should be used for the foundation work. The other undamaged blocks tested by nature have permanence and can be used above ground."

VITRUVIUS,
TEN BOOKS ON ARCHITECTURE, II, 7

The mining method did not change over the centuries, nor the price. Chalk building blocks remained an exquisite material that only very few could afford. Up to the end of the 15th century chalk blocks were almost exclusively used for churches. No price was too high to honour God for the people of Champagne, yet for their own homes they were content with wood and mud.

With the rise of feudalism, the power, influence and wealth of the aristocracy increased, thereby creating the preconditions for replacing the Church as the most important patron. Apart from palaces, the aristocracy developed a special liking for pigeon houses built with expensive chalk – and they had the means to cultivate such expensive tastes!

The French revolution brought an abrupt end to this epoch. It was no longer the aristocracy and the Church who were the holders of power, but rather the rising bourgeoisie. They were the ones that now used chalk to build their villas and city palaces, and occasionally this exquisite material was even used to build barns and windmills.

However, the use of chalk was never widespread. Even the finest and most compact grades of Champagne chalk were still a soft mineral of poor compressive strength, thus making it unsuitable for supporting elements, even though columns consisting of monolithic blocks weighing several hundred kilos can be found in some churches built in the 13th and 14 centuries. However, this remained the exception, and to this day it still has not been explained how the huge, crack-free blocks were mined. The effort (and price!) must have been enormous. In most buildings only the walls were made of chalk, and even these were thicker than they normally would have been. Other rock material was used for supporting elements and door and window frames. The same applied to foundations because the porous chalk blocks in the moist soil would not have survived the first frost.

Yet it was the porous structure that made chalk such a special and almost modern building material. The pores always contained traces of water which is why chalk walls contributed to a near perfect room climate throughout all seasons of the year. During the summer heat they continuously shed some water from their pores so that the air was pleasantly moist and cool. During the wet and cold winter months the chalk walls absorbed the moisture in the air so that the room remained dry and warm.

However, this was not sufficient to give chalk a secure future as a building material since its disadvantages were too serious and consequential. Even though the amount of chalk blocks used for building in the middle of the 19th century rose once again as the more simple citizens used chalk to improve the appearance of the road-side facades of their houses. Demands then dropped dramatically towards the end of the last century as the market became swamped with cheaper and better building materials. The end of chalk was foreseeable. Mines closed one

Diagram of underground chalk mine (16th-20th century).

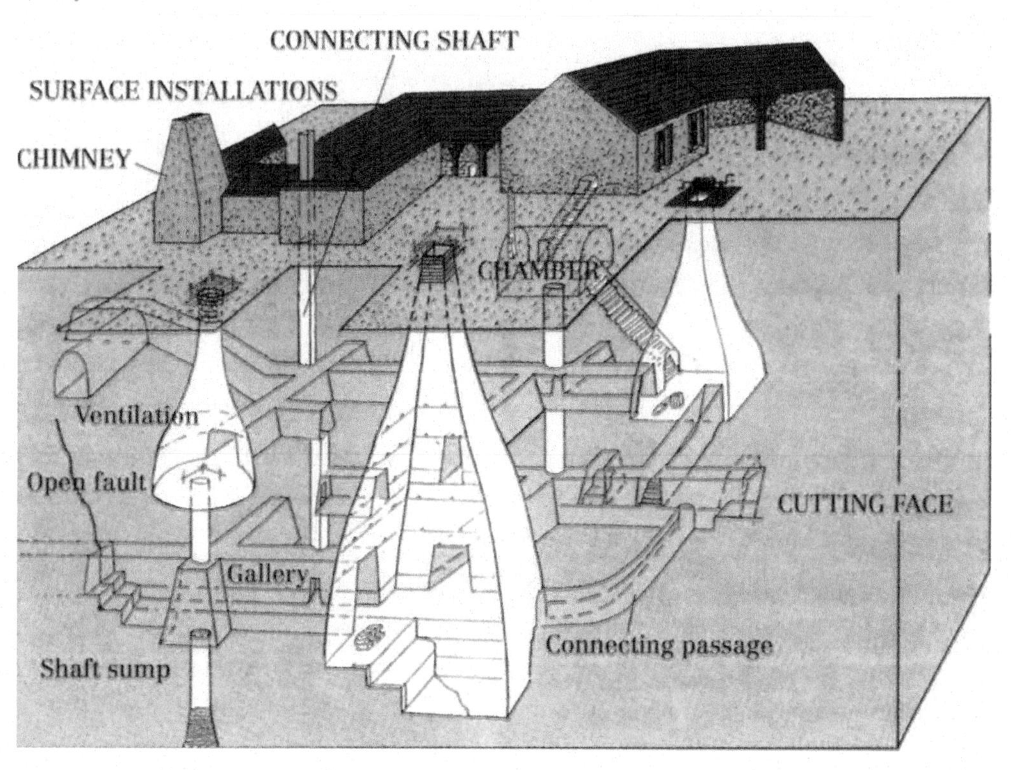

Champagne cellar in Reims.

after the other, the last one in La Chaussée sur Marne in spring 1986.

Nowadays, it is only the champagne cellars in the Reims region that remind us of the great tradition and heyday of chalk: The old mines have now become the storage cellars of the most famous champagne producers such as Taittinger, Piper-Heidsieck or Pommery. There can be nothing better to ferment great champagnes than the climate prevailing inside the chalk mine galleries.

2. Marble and limestone

There are always masses of stones in all shapes and sizes scattered on the ground. If a stone is required for a certain purpose, then it is merely a matter of searching for a matching stone and picking if up. And this is exactly what our ancestors did in prehistoric times. They used stones as tools and ornaments, and they were piled up for protection against wind and weather. Some of them could even be used for drawing. Initially, all these stones were not shaped. They were used in the original shape they were found. It took thousands of years before the first humans started to shape stones into primitive tools. For this purpose they selected stones that came as close as possible to the required shape in order to minimize the necessary shaping work. For a long time this work remained very arduous and was only performed to produce tools, cultic objects and jewellery.

Venus of Willendorf.

One of the oldest objects bearing witness to this age is the "Venus of Willendorf", a small limestone statuette produced by the hundred some 30,000 years ago. Whether these statuettes were an expression of priestess or goddess cult or whether the enormous thighs and breasts symbolized strength, health and fertility, is still unknown. What is known, however, is the time and area in which these Venus statuettes were produced. They date back to the Palaeolithic Age between 28,000 and 12,000 BC and they were distributed throughout Europe from the Atlantic coast to as far as Siberia. They were made of limestone or marble, and some terracotta figures have also been found. The choice of material had not yet become an expression of special esteem. It was rather a pragmatic choice: It had to be easily available and it had to be easy to shape. These criteria continued to prevail right until the early Bronze Age.

In the 5th to 3rd century BC small marble figures and vases were widespread in Greece, on Crete and the Cyclades. The only reason why marble was used was because the material was extensively available to the sculptors in the form of small pieces already weathered into the appropriate shapes. Similar sculptures produced at the same time on Malta were shaped from globigerine limestone simply because half the island consists of this mineral.

The artists were able to contour and decorate the materials using tools made of hard minerals, such as obsidian and corundum, before finally polishing them with sand and pumice. The resulting sculptures testify to the standard of artistic skills, even though they were still severely restrained in terms of size and shape. Since most weathered stones were small, the sculptures were likewise small. The unrestricted use of stones presupposed that they were no longer haphazardly collected but rather that a stone material was specifically selected for a given purpose and then quarried. The Egyptians were the first to adopt such an approach.

2.1 Quarrying stones

Around the 3rd century BC the Egyptians started to quarry rock formations and then to process the quarried material for specific purposes – as building materials, for columns and obelisks, as ornaments, jewellery and vases, or as relief stones. There are two reasons why it was the Egyptians who took the lead. Firstly, they had just learnt to produce, from the long-known copper, stable tools that could be used to shape stones. Secondly, there were extensive rock formations on either side of the Nile. Since limestone dominated in Egypt, and the fact that it can be easily cleaved into blocks on account of its natural layering, limestone became the first mineral to be quarried for a specific purpose.

The first quarrying techniques

Stone quarrying commences with prospecting for a suitable deposit. The first indication could be the presence of the given rock at the surface of the ground. This was when the search ended, at least in Pharaonic Egypt, in view of the immense choice of rocks available for selection.

Once a deposit has been located, it was then necessary to assess the quality of the rock formation. If the colour and texture came up to expectations it was then necessary to establish whether the rock was suitable for the envisaged purpose. The next criterion was the natural cleavage within the rock formation. Normally, numerous fissures pass right through the formation, and often several fissures run parallel alongside each other to form cleavage points (with limestone sediments these are often stratification joints). The position of the principal cleavages determined how easy mining would be and the maximum block size that could be cut. Ideally, there should be a vertical and horizontal cleavage in relation to the planned mining direction. Finally, the thickness of the overlaying strata determined whether mining was an economical proposition. Since most of the mines in Egypt were open-cast, a great deal of arduous work was spent on removing the overburden.

Once a suitable deposit had been selected it was developed into a quarry that was normally worked in benches that followed the natural slope of the hillside terrain, proceeding from the formation floor upwards. If the deposit was very extensive then a trench was dug through the formation so that it could be simultaneously quarried from both sides. Occasionally, the stone blocks were also mined, i.e. the stones were cut out vertically downwards into the depth. However, only a maximum of two layers of blocks could be extracted in this manner because the effort to lift the blocks out of the pit became too great.

Large quarries were worked in a series of terraces arranged at a right angle to each other. This layout was usually adopted for the entire quarry so that the natural course of the rock veins were rarely followed. This enabled more workers to work at the rock face and the yield was higher, but blocks tended to be smaller and the sloping layers were more intensely exposed to weathering effects. To cut large monolithic columns it was necessary to follow the natural course of the layer. This involved more work, but the quality was much better.

Typical quarrying method of Pharaonic Egypt as well as in ancient Greece and Rome.

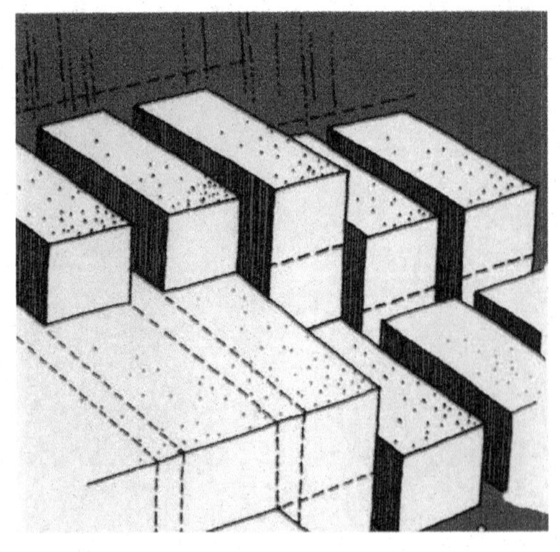

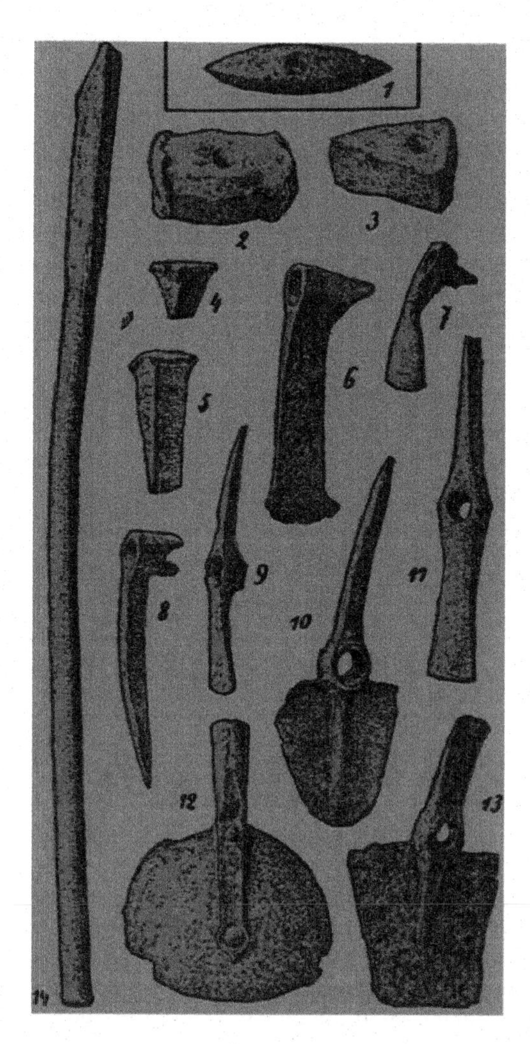

Roman quarrying tools from the 1st and 2nd century AD: iron wedges, axes, sledge hammers, pick-hammers and pick-axes. Probably they differed very little from the tools used in Greek quarries (acc. to Behn, 1926).

Cutting blocks was hard work. First of all the blocks had to be uncovered and cut away from the rock formation on three sides in relation to the hillside. Wherever possible existing cleavages were widened, otherwise a series of channels had to be chipped into the rock. The channel width was usually between 50 to 60 cm, and the tools used for this purpose in Pharaonic Egypt were copper chisels. Even though the cutting edge of these chisels had been toughened by hammering, this was still not sufficient in many cases so that harder minerals, e.g. diorite, had to be used to chip channels out of the soft limestone. A network of channels was cut into the rock deposit when the blocks were quarried next to each other over a wide area. This saved work since each channel simultaneously cut through two block sides.

After a block had been cut away on all three sides, the workers were confronted with the most difficult and dangerous stage. The last side, the block floor, had to be separated from the rock formation. With a bit of luck, or if a favourable location had been selected, then a clay layer or gap along the floor would have made work easier. In most cases, however, a series of wedges had to be driven into the floor with the powerful blows of a hammer, thereby severing the block from the rock formation. This laborious procedure could be reduced if dry wooden wedges were driven into the rock and then soaked with water. The swelling wood would then separate the block from the rock floor. Alternatively, a breakage edge was chiselled under the block which was then broken off by heavy blows on the top surface of the block.

Any one of these methods was not standard practice. There were no guidelines concerning rock cutting. Each time the individual steps had to be newly defined to adapt the work to the given situation – the rock formation, the manner it had to be cut and, last but not least, the required block size. This called for very extensive knowledge and experience. It is therefore very likely that the first form of specialisation arose among quarry workers in Pharaonic Egypt, as was subsequently documented in antiquity, Greece and Rome.

There was the trained stonemason who watched over rock cutting. He passed on his knowledge and skills to the generations within his family and he additionally taught apprentices.

The pickaxe from archaic Greece was very similar to those extensively used in Roman quarries, as is indicated by the scratches on the stones of Roman quarries (acc. to Röder, 1957).

There was also the simple, unskilled worker who was primarily employed to remove the enormous amount of rubble. It is estimated that in Egypt, and throughout antiquity and right into modern times, each cut block produced at least the same amount of rubble. This invariably gave rise to enormous problems. Part of the small fraction of limestone and marble rubble could at least be fired to produce lime, but the majority of rubble had to be removed and disposed off. This resulted in huge overburden and rubble dumps which threatened to stifle the actual quarrying operations. Be it as it may, quarrying stopped when the facing wall of the quarry reached a height of 15-20 meters because then the work needed for rock cutting and block extraction simply became too much.

Driven by the urge to bury their Pharaohs in enormous stone tombs that would survive the ages, the Egyptians developed a new material for mankind. And the rock-cutting methods used to quarry the blocks more than five thousand years ago continued to prevail right into the middle of the 19th century, and in some instances even the 20th century. The blocks were cut free on all four or five sides by channels and then broken off the rock face. The only changes that came about were the materials used to produce the actual quarrying tools. The Egyptians replaced copper with the harder bronze as early as 1500 BC, and the first tools and wedges made of iron appeared around 700 BC.

Continuous quarrying also started in ancient Greece during this period. Although quarrying was practiced in the 2nd century BC during the Minoan culture on Crete and the Myceane culture on the mainland, this quarrying technique disappeared with the collapse of these cultures. The Greek stonemasons, known as *technites*, then introduced a significant innovation which replaced the hitherto customary hammer and chisel by a kind of pick-axe that was used to cut the channels between the blocks. Moreover, this pick-axe was also used to transform the crude blocks into finished stones that had smooth surfaces. The pick-axe soon asserted itself as an all-round tool that prevailed throughout the hayday of medieval stonemasons and continued right on to present-day quarrying where it is still used for soft rocks.

Down into the depths of the earth

Poros, a soft marble-like tufaceous limestone was first quarried in ancient Greece, soon to be followed by normal limestone. Marble emerged towards the end of the 6th century. The oldest Greek marble quarries are located on Paros. This marble was very much in demand throughout Greece on account of its exceptional beauty:

"However, they all only used the white marble from the island of Paros. It was referred to as lychnites because it was cut by lamplight in mines."

PLINY, NATURAL HISTORY, XXXVI, 14

Lychnites is derived from the Greek *lychnos* for lamp. The fact that underground mining was even incorporated in the marble's name clearly indicates the importance of this technique. Normally, the enormous expense and the necessary mining knowledge had made mining impractical in the past. However, as a result of the thick overburden in Paros, the miners were forced to follow the marble vein down into the depths of the earth, particularly since the marble was less fissured there.

In Rome widely differing marble formations were both quarried and mined. The demand for marble was enormous. Innumerable existing deposits throughout their conquered provinces were extensively exploited, but new ones were also developed. Moreover, according to Pliny The Younger (Caius Plinius Caecilius Secondus), prospecting appeared to be more systematic since deposits were difficult to find, even by "specific searching", thereby indicating that some form of prospecting must have been practiced at the time.

Little is known of the prospecting methods practised in antiquity. Scientific disciplines such as geology and mineralogy only came into being in the 18th century. The stonemasons must have had knowledge that had been handed down to them through the generations. This enabled them to determine deposits with considerable accuracy on account of terrain features and rock veins that rose to the surface of the earth. Once a deposit had been discovered, prospecting trenches were dug for more precise exploration of the deposit. Ultimately; however; it appears that chance was the primary factor, but not the way in which marble was discovered in Ephesos, as narrated by the Roman author Vitruvius in his "Ten Books on Architecture":

"I will digress a little and recount how quarries were discovered. [...] But when the inhabitants of Ephesos decided to build a shrine dedicated to Diana with marble from Paros, Proconnesus, Heraclea and Thasos, sheep had been put out to graze here by Pixodaros. Two fighting rams charged but missed each other with the result that one of them hit the rock with his horns with such force that it knocked off a fragment of dazzling white colour."

VITRUVIUS, TEN BOOKS ON ARCHITECTURE, X, 2, 15

Sawing with sand

The Roman contribution to the technique was limited to the more wide-scale use of the stone saw that had already been known for some time. For instance, a porous and very soft white tufaceous limestone in Venetia was directly sawn out of the rock formation. A toothed saw was used, just as for wood, except that this was not suitable for harder rocks like marble. In such cases a different technique had to be applied which resembled more that of abrasive cutting than sawing. Sand that was harder than marble was used for cutting. The sand was rubbed to and fro on the marble block with a dull blade that had no teeth. In this manner the blade worked its way deeper and deeper into the block until the cut was finally completed. Wet sand was used to keep the blade cool, and water was continuously added while sawing. If the grinding sand was sufficiently fine, then it was possible to eliminate the need for subsequent polishing.

Pliny lists in his Natural History five different types of sand used for this purpose. Some sands were even imported from Africa. However, there are only very few finds that testify to the use of such saws in Roman quarries. More common was the use of such saws for marble finishing, particularly to cut it into slabs for wall cladding and floors. Two or more workers had to complete the tedious job of sawing in the quarry, but for marble finishing it was possible to harness the power of water, for instance from the River Mosella (nowadays known as the Mosel), as described by Ausonius in the 4th century AD:

"A torrent of cool water turns the grain grinding stones and draws the screeching saws through the glass-smooth blocks of marble with a continuous noise that could be heard everywhere."

DECIMUS MAGNUS AUSONIUS, MOSELLA, 359-364

The marble was imported and cut to size on the banks of the River Mosella. It was then used for large buildings such as the Emperor's Trier residence as there was no marble in the Mosella area.

A copper engraving (plate XIV) of a quarry by Nicolò Zabaglia from 1743. The methods and tools used in the Middle Ages hardly changed right until the 18th century.

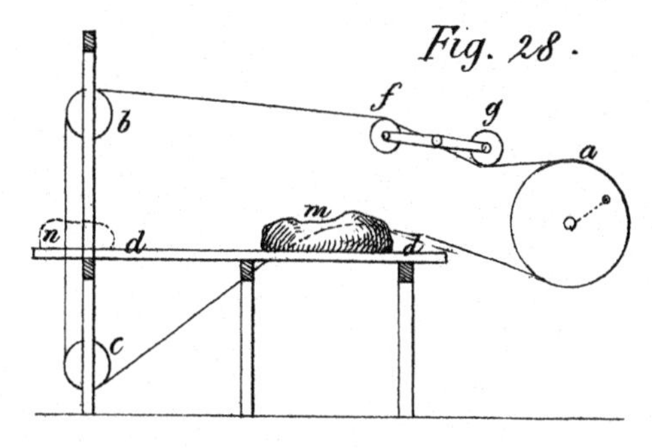

Chevalier's stone saw.

Sawing with water

In technical matters the quarries in the Middle Ages did not go beyond the Roman developments, and in terms of size and extent they were even far short of their Roman counterparts (see figure). Only in the early days of modern times were efforts made to use machines and tools to ease the strenuous physical work in quarries. For instance in 1588 the Italian engineer Agostino Ramelli published his book "Le diverse et artificiose machine del capitano Aogstino Ramelli, dal Ponte della Tresia, ingeniero del christianissimo Re di Francia et di Pollonia" in which he described, among other things, a water-powered saw. The saw built by Ramelli consisted of a frame with several parallel blades, but whether this frame saw was actually ever operated or merely remained an academic concept is still disputed. However, the complicated design must have exceeded the skills of Ramelli's contemporary craftsmen since proof of the existence of such frame saws was first found only in the 19th century.

Just under 50 years later, in 1629, Giovanni Branca described a simple rope saw used by the quarry workers to cut off large blocks manually. The process was very time-consuming and strenuous. It is therefore hardly surprising that many quarry workers and quarry owners placed high hopes in gunpowder.

Although gunpowder had been known in Europe since the 13th century, its first use in quarries was only in the 18th century. The initial euphoria for this new method soon turned into disillusionment. Although blasting did produce a large amount of rocks in a very short time, the yield of large blocks was small and the waste very considerable. All attempts to improve the method failed and a vast amount of rubble covered the surrounding hillsides. Gunpowder ceased to be used to extract marble until the beginning of the last century. However, if the size of the quarried rock material is irrelevant, then blasting has remained the first choice to this day.

Ropes and saws

Only the 19th century brought a decisive breakthrough in combining new driving techniques and materials with the two thousand year old Roman method of stone sawing. As in Rome, the saws were initially used only for marble finishing. A report from France in the year 1829 described a stone saw windmill that produced "marvellous results". On 13th July 1843 the London marble dealer William Hutchinson had a "machine to cut or saw marble and other rocks" patented. However, this machine could only be operated in a workshop.

Some ten years later, in 1854, the Belgian engineer Eugène Chevalier received a patent for a sawing method that was also suitable for use in quarries. The rigid iron saw blade was replaced by a flexible wire rope, and wet quartz sand was used for cutting. The entire assembly was driven by a steam engine. His saw was no longer restricted to a certain location and it could be adapted to local conditions. The sawing speed of just under ten centimeters per hour was not exactly breathtaking, but it had now become possible to saw a block in the required size directly out of the rock face. Chevalier exhibited a refined model of his wire-rope saw at the Paris World Exhibition in 1856 which spread to many marble quarries by the end of the century (see figure). It was first used in Carrara in 1895.

After more than four thousand years the wire-rope saw was the first – and until today the last – radically new method to be devised

Marble saw in Carrara.

to produce marble blocks. At long last the strenuous, time-consuming work to cut the channels had been eliminated, and the amount of rubble dropped to well below 50 per cent. Occasionally, the old quarrying method could still be encountered. However, by the middle of the 20th century this sawing method was practiced in most marble quarries, except that, in the meantime, the steam engine had given way to the combustion engine.

The last innovation to this day dates back to the year 1970. This was the year when German engineers developed a saw that did not require sand or any other abrasive material. A diamond-studded wire rope, driven by a powerful electric motor, cut its way through the rock formation at ten times the speed of conventional saws. The rope had to be cooled with hundreds of litres of water to stop it from being torn apart by the heat. After the usual period of hesitation, these saws spread to the large, productive quarries where they have remained a standard item of equipment to this day. Operations in many small, remote marble quarries still resemble open-air museums of the early days of quarrying rather than profitable production

Mobile saw for underground quarrying, Danby, USA.

centres. Occasionally, one can even encounter quarries where, just as in antiquity, the blocks are still broken off manually – even in Carrara.

The new quarrying techniques greatly reduced the physical strains for the workers, yet their work still remained dangerous. Although the last major accident in Carrara dates back to the year 1913 when fourteen workers were killed by a massive fall of marble, accidents still remain an almost daily occurrence and fear is a constant companion. In the course of 1992, alone, twelve workers were killed by accidents in the Carrara quarries.

By far the biggest source of danger are marble rocks that break off the quarry face and crush the workers underneath. From time to time workers lower themselves down the quarry face on a rope and test the firmness of the rocks with long iron rods and break off those that are dangerously loose. It is solely their experience and not technology that is decisive for the safety of the quarry workers.

There are also other situations where the work in the quarry hardly differs from that

"The Stone Breakers" by Robert Hermann Sterl (1867-1932). The motifs for his paintings of workers were scenes from the Elb sandstone quarries of Saxonian Switzerland.

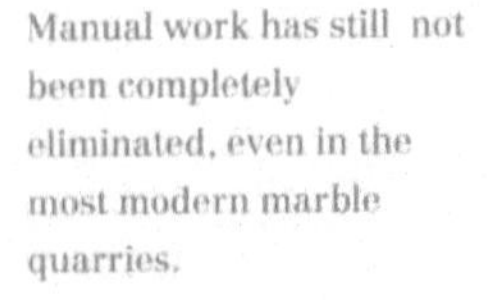

Manual work has still not been completely eliminated, even in the most modern marble quarries.

practised in ancient Rome and Greece. For instance, in the search for a suitable block, the stonemasons are still guided by their instinct. They rely primarily upon the knowledge that has been handed down to them through the generations, and there are still no better methods of detection to locate the finest marble among the 50 grades found in the Carrara mountains.

The ratio between the different forms of marble quarrying has remained the same since antiquity. Conventional bench quarrying still dominates and underground mining is the exception. The only remarkable difference concerns open-cast mining down to deeper floor depths. Although this was known in antiquity, it is only the development of modern lifting gear during the last sixty years that has made it possible to hoist marble blocks to the surface from greater depths.

Nearly five thousand years have passed since the Egyptians first quarried rocks. Yet what may appear to have been incredibly simple methods that were developed in Pharaonic Egypt were, in fact, so ideally adapted to the given task that, for a long time, they were merely refined and advanced, but never abandoned.

The unchanging manner of quarrying contrasts greatly with the extreme fluctuations in the use of the quarried materials. The extent to which sculptors and architects could use rocks was not governed by quarrying techniques but rather by a totally different aspect.

Excavators and crawler-tracked vehicles characterise the picture of the marble quarries of Carrara.

2.2 Transport, organisation and trade

The average weight of marble and limestone is in the order of 2.5 to 3 tons per cubic meter, thus making it relatively light compared with other minerals. However, as a result of the preference of our ancestors for a monolithic manner of construction, the weight of each individual block was still very considerable.

Right up until well into the 19th century, the transportation of heavy loads over long distances was associated with an enormous effort and the employment of considerable manpower. This applied particularly to overland transport which relied on simple rolls, wooden sledges and primitive carts. Consequently, the geological distribution of the minerals exerted a major influence on their selection and use. Locally found rocks were therefore primarily used locally. This principle ended only when the railway became available as a means to transport heavy loads. Until then clay and wood remained the principal materials as they were in plentiful supply almost anywhere. Their use as a building material throughout history was therefore greater than all other natural stones put together.

However, as is the case with most principles, there were exceptions to the rule. These concerned prestigious or sacred buildings such as temples and tombs. In such cases the patrons – for a long time identical with the rulers – imposed their will, thereby overruling all economic considerations. No effort was spared to secure only the very best, and this was usually marble. And the enforcement of their will was all the easier in times when manpower, and even human life, was more or less worthless.

Forced labour for the Pharaohs

Not only are the Egyptian pyramids the oldest intact stone structures but they have also remained, to this day, the largest ones to be built of natural stone. For each pyramid innumerable stones had to be quarried, transported and finished. For instance, Khufu's Pyramid, usually referred to as the "Great Pyramid", consists of 2.5 million limestone blocks, each one weighing an average of 2.5 tons. And this was only one of some 150 pyramids that were erected in Pharaonic Egypt. The provision of the necessary stones in such vast quantities for all these structures required the perfect organisation of quarrying and transport. The pyramids could not be simply hewn out of a corresponding rock formation as was the case with the Great Sphinx. (see figure).

The importance of an effective work organisation becomes apparent when one considers the vast labour force involved. According to Herodotus, up to 100,000 men worked for three months at a time for a period of almost ten years for the construction of the Great Pyramid. However, these figures quoted by Herodotus should be taken "with a pinch of salt". Yet even the very latest calculations still indicate that some 10,000 men worked throughout Khufu's entire 23-year rule in the quarries, for stone transportation and on the building site.

It goes without saying that the quarries in Egypt were the private property of the given Pharaoh. Occasionally Pharaohs, themselves, would check to see how things were proceeding at their quarries and building sites. The immense extravagance of such a venture becomes evident if one realizes that when Rameses IV undertook an "expedition" to a quarry in the 12th century BC he was accompanied by nearly 8400 subjects. However, the day-to-day organisation and control of the work was in the hands of a civil servant from the Pharaonic financial administration. His title was "Head of Quarrying in the Entire Country", and he had to ensure that the enormously heavy monolithic stone blocks that were so popular with the Pharaohs were quarried, transported to the building site and integrated into the pyramids.

In view of the enormous quantity of stones involved, and the limited means of transportation available, it was only natural that appropriate quarries had to be developed as close as possible to the actual site of the pyramids. For instance the limestone blocks

Sphinx and the Khufu pyramid.

for the Giza pyramids were quarried on a plateau in the immediate vicinity and then hauled on wooden sledges to the building site. Each sledge was loaded with only one block, and each sledge had to be hauled by one hundred men. The haulage path was covered with mud to enable the sledges to slide. However, the local Nummulite limestone was not sufficient for all purposes so that an additional, finely finished limestone from Tura had to be shipped across from the east bank of the Nile. The Egyptians made extensive use of the Nile for transportation with the result that many of their quarries were alongside the banks of the river, some as far off as a thousand kilometers from Giza. Stone transport by waterways was superior to any other method, and this did not change until right into the 19th century.

However, not all rock formations are located in the immediate vicinity of a waterway, and certainly not in arid Egypt. Even in the earliest days, a hard rock from the desert between the Nile and the Red Sea was held in high esteem, and enormous effort was invested to transport this rock over distances of two hundred kilometers and more. This was only possible in a country where human labour was not a cost factor. The peasants had to serve their Pharaoh and they were not paid for their hard work. The situation was somewhat different in ancient Greece.

Democracy and reconstruction

It looks as if ancient Greece of the 7th century BC was waiting for the first stimulations from Egypt and Asia Minor to start with the systematic use of natural stones. Until then, at best, only ordinary boulders – *lithoi ligades* – had been used so that a real explosion took place in the development of new quarries. Initially only poros (calci-sinter) and normal limestone were used as these were plentiful in Greece. Towards the end of the 6th century the Greeks took a liking to marble, and at the latest by the beginning of the Classical Epoch at the end of the Persian

Wars, marble had become the ultimate choice for the (re-)construction of their cities.

However, contrary to Egypt, most of the quarries in Greece were privately owned so that entrepreneurial considerations governed the development of mineral deposits. Although the owners employed slaves who did not have to be paid for the quarrying work, they nevertheless supplied the quarried stones to the customers for a payment so that the expenditures for slave labour determined the price. Consequently, considerate treatment of the slaves was in their own interest.

The "location principle" prevailed in Greece to avoid unnecessary transportation over long distances. Quarries were developed in the vicinity of cities that were to be supplied with building material. Normally one quarry supplied the building material for one project so that the customer could select the stones locally. Usually only the very best was good enough. And since the rock formations tended to be readily accessible, it was not uncommon for quarries in Greece to extend over wide areas so that actual "deep" quarrying was quite rare.

However, there were exceptions. For large building projects, such as the Acropolis in Athens, the quarries were owned by the state and they were intensively worked. Thus, over the centuries, more than 400,000 cubic meters were mined in 25 Pentelic marble quarries. However, the transport route was short. The quarries, just as the marble quarries of Hymettos, were in the immediate vicinity of Athens. And in Syracuse on Sicily there were enormous quarries – the latomie – 250 meters long, and at some points the quarry face had a height of up to 35 meters. These quarries even formed part of the city fortifications. It is estimated that several million cubic meters of limestone were extracted here. Most of the work was performed by some 7000 Athenians who had become prisoners of war during the doomed expedition of 413 BC and who all died in these quarries. The "Ear of Dionysos" – a grotto that winds its way in an S-shape through the mountain – is a reminder of their fate. According to the legend the tyrant Dionysos listened to the Athenians at the other end of the grotto – and he could hear every spoken word, even if it was only whispered.

The transportation of columns according to a report by Vitruv. From Walter Ryff's "Vitruvius Teutsch", Nuremberg, 1548, fol. CCXCVIII, woodcut. The illustration of the transport operations is based on Vitruv's report. However, it is doubtful this was actually day-to-day practice.

The greatly prized *lychnites* or *statuario* statue marble from Paros was mined underground and transported over great distances. However, only 30,000 cubic meters of this precious marble was quarried during almost one thousand years – just one per cent of the stone volume used solely for Khufu's Pyramid.

There are other reasons why stone transport in ancient Greece never required such inhuman effort as in Egypt. Firstly, the Greeks soon turned away from monolithic structures involving huge blocks weighing several tons, be it because of the high cost of transport or a change of architectural taste. Secondly, they started to finish the stones roughly in the quarry, leaving just enough to protect the block during transport so that unnecessary weight did not have to be transported.

The means of transport were also primitive in ancient Greece. The marble cut in mountain quarries was brought down into the valley on wooden sledges over haulage paths, or it was loaded onto ox-drawn carts and transported to the point of use. Another technique described by Vitruvius was also widespread. It involved encasing the marble blocks with planks. Wheels were then attached to the planking so that even the largest blocks could be hauled in this manner (see figure.). Apart from statuario marble from Paros, transport by ship was insignificant, both in ancient and classical Greece.

Everything changed with the end of the Greek city states. The Hellenic rulers, starting with Alexander the Great, were greatly attracted by exotic stones, and these were not exactly local. Particularly in the late Hellenic kingdoms coloured marble in differing qualities were transported over long distances. Moreover, the monolithic manner of construction was revived in glorification of the rulers.

These trends were taken up in Rome, initially somewhat hesitantly, but at the latest with the transition from a republic to an imperial empire, the transport of stones took on entirely new and hitherto undreamt-of dimensions.

Perfection and the passion for splendour

Augustus, the first Roman Emperor (31 BC to 14 AD), boasted that he had found "a city in wood and had left one in marble and stone". This is certainly only a half-truth because the first temples in white marble were erected by Q. Caecilius Metellus as early as 146 BC. At the time both the architect and the material had to be imported, and even in Luna, today's Carrara, Greek marble was used in the initial period. Following Augustus' victory at the Battle of Actium (31 BC), marble gained enormous importance as a building material. Rome was now the capital of a world empire, and Augustus the Emperor of an empire – and both had to be expressed commensurately.

Soon large quantities of marble from Luna reached the city, as reported by the Greek geographer and historian Strabo:

"In the vicinity [of the city Luna] there are pits where the white, blue-stained stone can be found which has supplied so many large blocks and columns that it is the most-used material for Rome's architecture and in other cities. This is because these stones can be easily distributed as the pits are located near the harbour from where the ships can cross the sea and sail up the Tiber."

STRABO, GEOGRAPHY, BOOK V, 2.5, CAP. 222

Strabos' remarks also draw attention to an innovative development introduced by Rome: sea transport of marble blocks on ships – the naves lapidariae. This is why the preferred site of Roman quarries was in the vicinity of the sea or rivers. Otherwise, everything else relating to the means of transport and techniques remained unchanged except that, with regard to size and weight, the Romans achieved entirely new dimensions.

For the supporting wall of the Temple of Jupiter in Heliopolis, not far from the present-day Lebanese town of Baalbek, the Roman builders cut and transported the world's three largest stones ever to be cut. Each of these three limestone blocks weighed well over 600 tons, and if stood upright, would have equalled a 5-storey house. How the Romans were able to transport these stones 1.5 km from the quarry to the temple, and then lift them to a height of 6 meters, has remained a mystery. And even this load was still not enough because a fourth block was discovered in the quarry which had been cut but was not transported (see figure). And this is quite understandable considering that some 16,000 workers would have had to pull this block of 22 meters length and 1110 tons weight simultaneously to move it only a few millimeters. There were no machines that could have made the work easier. Two thousand years later NASA developed a transport vehicle that could handle such enormous loads, namely the caterpillar tractor for the Saturn 5 rocket.

Huge monoliths in Rome are rare, but this did not prevent the hauliers from transporting large loads of smaller marble blocks through narrow streets and alleys of the city. The Roman poet Juvenal, a keen observer and critic of the customs of his time, ridiculed the chaotic traffic conditions in Rome:

"If the axle of the vehicle transporting the stone blocks from Liguria [Luna/Carrara] breaks and the entire load falls down on passers-by, then nothing remains of them. Who can then find the limbs and bones? Not only are the people killed, but their bodies are also crushed."

JUVENAL, 3, 252-254

Roman marble transport on open waters were not so much remarkable for technical reasons as the ships had hardly changed from past times. The new element was the political order that made such sea transport possible. As any ship sailing the Mediterranean Sea was exposed to the danger of falling into the hands of pirates, Emperor Augustus enacted the *pax romana* which laid down binding rules which Rome was prepared to control and enforce.

The marble trade soon boomed under Augustus, particularly since rich Roman patricians were responsible for the upsurge of

Hadshar el hibla or the "Stone of the pregnant women". The name of this giant monolith is attributed to a legend according to which all attempts of the Baalbek to remove the stone were of no avail because it was too heavy. A poor pregnant woman claimed that she could tell them how they could remove this stone if they looked after her until she gave birth. This they gladly did and took great care of her. After she had recovered from giving birth she asked the Baalbek people to lift the stone on to her back so that she could carry it away because the stone was too heavy for her to lift and carry away at the same time.

demands. The local marble from Luna no longer satisfied discerning tastes. A penchant for exotic stones had become a matter of status and prestige. Marble from recently annexed Egypt and the Aegean area was pouring into Rome. A marble quarry in the North African Simitthus (present day Chemtou in Tunisia) supplied the greatly cherished yellow *numidicum marble*.

The marble trade was already being speedily pushed to its very limits under Augustus' successor. The upsurge in demands could no longer be fulfilled, even by opening up new quarries. Tiberius (14-37 AD) was forced to enact appropriate measures – the *ratio marmorum* – out of which an entirely new and complex system evolved in the course of several decades that would satisfy the customers in Rome and elsewhere. Six principal elements characterised *ratio marmorum*, some of which are reminiscent of modern economic and production methods. However, all these elements were not introduced at the same time. Some of them evolved only gradually as a result of market demands.

The first step was to place the quarries under **state control**. In the Roman Republic the quarries had been privately owned. In 17 AD Tiberius started his nationalisation campaign as a result of which, by the middle of the century, all the most important quarries throughout the entire empire belonged to the *patrimonium*, i.e. they were imperial property. He had been prompted into such action by the conditions encountered in occupied Egypt where the quarries had been the property of the given ruler ever since Pharaonic times.

His successors, however, did not leave it at that. Large building programmes under the Emperors Domitian (81-96 AD) and Trajan (98-117 AD) soon caused serious marble shortages with the result that Hadrian (117-138 AD) enforced the **reorganisation of quarry operations** with the aim of rationalising production (see figure). The new system was headed by the *procurator marmorum* who, surrounded by a host of officials, administered all state-owned quarries centrally from his seat at the imperial court in Rome. Accordingly, a defined hierarchic structure with a strict division of labour was now in place in all the large quarries with up to five hundred workers. The entire quarry was directed by the *procurator* – a slave or man freed from imperial ownership. The *officinator* – the person responsible for individual quarry sections (officina) – was subordinated to the procurator. In quarries run by the military, the *officinator* held the rank of a Centurio. This as was primarily the case in the unsafe border regions of Gallia and Germania.

Extempores and *lapicidinarii* quarried the stones and removed the overburden. The blocks were cut into slabs by the *serrarii*, while the stonemason's work was completed by the *opifices lapidarii*. Finally, the work of the marmorarii was somewhere midway between that of a stonemason and a sculptor. Furthermore, each quarry employed a faber (blacksmith) to maintain the tools, while the *probator* selected the stones that were to be quarried. This strict division of labour made sense since most work required extensive knowledge and skills which the extempores – the unskilled workers – lacked as they did not require any training for their strenuous and dangerous physical work at the quarry face.

State control, combined with reorganisation, boosted production in the middle of the second century. Moreover, the economic basis of the quarries had changed in the meantime. In the past stones had been primarily supplied for imperial or public works, i.e. the quarries had worked for the state. Now the quarries also had to fulfil the demands of private customers – mostly rich Roman patricians who wanted to show off their wealth in a commensurate manner. This change in business relations invariably influenced quarry operation. In the past the products required by major customers were often made to measure, i.e. customised. The quarries now had to adopt **serial production** methods based on popular trends. Accordingly, columns were now only supplied in standard lengths, the most popular ones being 10, 20, 30, 40 or 50 Roman feet; and sarcophagi were being finished to the point where only the features of the deceased had to be chiselled into the stone coffin locally.

The quantities that were now being produced by far exceeded day-to-day demands. On account of the **new customer/supplier relations** numerous blocks, columns and sarcophagi were being stored in the quarries or, better still, at central market points. Customers no longer selected their marble directly in the quarry. Instead, dealers purchased the products and offered them at the most important market locations. The Marmo-

Organisational set-up of a Roman quarry

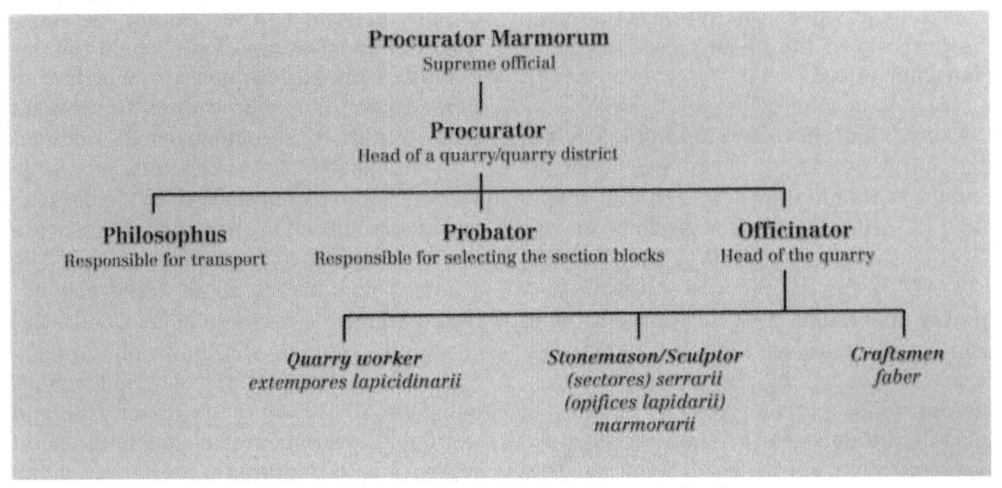

	Deposit	Name (in antiquity)	Description	Use
Greece	Chios (1) *(marmor chium)*	portasanta marble	Mostly brecciated columns	Incrustations and
	Euboea (2)	Cipollino *(carystium)*	White to greenish coarse-grain marble	Columns and incrustations
	Hymettos (3)	*(hymettium)*	Greyish-white, fine crystalline marble	Architecture
	Paros (4)	*(lychnites)*	Somewhat finer than the statuario from Carrara	Sculptures
	Pentelikon (5)	*(marmor pentelicum)*	Fine-grain, slightly yellowish marble	Architecture, sculptures
	Naxos (6)		White, crystalline fine-grain marble	Architecture, sculptures
Italy	Carrara (7)	Carrara marble *(marmor lunense)*	Numerous types - ordinario, statuario, bianco venato and bardiglio	depending on type: for architecture, for sculptures, incrus-tations etc.
	Laas/ Lasa (8) (Südtirol)			
	Tivoli (Rome) (9		Travertin	Architecture
Turkey	Prokonnesos (10) (Marmara Sea)	*(marmor proconnesium or cyzicenum)*	White, light grey marble with medium large crystals	Sculptures, sarcophagi, and Architecture
Tunisia	Chemtou/ Simitthus (11)	giallo antico *(marmor numidicum)*	Compact, fine-grain limestone from ivory to golden-yellow	Incrustations, dishes, less freq. for sculptures
France	St. Beat, Pyrenees (12)		White to whitish grey, fine-grain marble	Sculptures, architecture and sarcophargi
	Bois de Lens, Fons (13)		Urgonic limestone	Architecture
India	Makrana/ Rajastan	(not shown)	White dolomitic marble	Architecture (Taj Mahal)

rata Quarter in Rome was the centre of this new form of marble trade. The Quarter, located on the banks of the River Tiber, had its own quays where the marble was unloaded from the ships, and its own marble depots where the customers selected their blocks and columns.

The responsibility of the quarries no longer ended with the actual sale of the marble. Before the final product could be created out of the supplied marble block, extensive work still had to be performed on the customer's premises in spite of prefabrication. This type of work required **trained craftsmen** or artisans who were familiar with the material and who knew exactly what was possible and what was not. Furthermore, since an ever growing number of deliveries were being effected to areas in which marble and its processing were unknown, the quarries also had to place their artisan workers at the disposal of the customers.

Finally, the changed business relations required new organisational forms. It was no longer sufficient to produce adequate quantities of marble blocks. The transportation and sale of the products had to be organised in such a manner that the deployment of artisan workers could be co-ordinated, even at distances of several hundred kilometers from the quarry. Consequently, the quarries had to set up their **own trade agencies** to give them access to new markets with their regional peculiarities.

These innovations proved to be a tremendous success as vast quantities of marble were soon reaching Rome from all parts of the then known world. And if the procurator marmorum at the imperial court wanted to maintain a close survey, every block had to

Small venus statues carved in yellow marble. These statues were mass produced by slaves in the Simitthus work camp.

be clearly identified by simple symbols and figures or complex identifications involving the quarry name, date, accounting district (ratio), quarry section (officina), precise location (locus) and a given serial number. Blocks that were stored for extended periods had to be correspondingly recorded so that dealers could always monitor the stocks in hand.

After Hadrian had given the decisive impetus by reorganising the quarries, the new system quickly flourished. By the middle of the 2nd century AD an enormous fabrica had been founded in the North African Simitthus which demonstrated in a very impressive manner what the Romans were capable of achieving. This fabrica was a 57,000 square meter complex surrounded by a wall in the immediate vicinity of the imperial quarry. The most important building within the complex was the camp for up to 1200 slaves. There were also dwellings for the imperial officials and soldiers, separate baths, storage rooms and the actual production centre for dishes produced from the locally quarried yellow numidicum marble.

The entire process for items mass-produced from the very finest material (see figure) was subdivided into a series of individual steps, similar to those of a production line. The first step involved the production of marble blanks from which the dishes were then roughly cut, followed by fine chiselling and final grinding and polishing, both inside and outside. Simitthus was certainly exceptional, but it was a well documented quarry. Extensive discoveries at long last furnished comprehensive information on quarrying methods in antiquity, even revealing the finest details to an extent that had never before been uncovered elsewhere.

Although we are now able to give a detailed account of the basic principles of the marble trade at the time, we still depend on estimates when it comes to quantities. Many of the quarries that existed in antiquity are not even known today. This is where different chemical and geological methods have to be combined to narrow down the origin of sculptures and marble columns to specific regions (see box).

The situation is not much different when it comes to marble prices. The sole clue is Diocletian's Price Edict dated 301 AD. State-controlled prices were introduced in a rather futile attempt to stave off rampant inflation. Accordingly, the cost of Pavonazzetta or marble from Docimium, was 200 denarius per cubic foot, whereas marble of equal quality from Skyros was priced at only 40 denarius. Although a modern comparison relating to prices is somewhat difficult, it is nevertheless evident that the selling price was primarily governed by transport costs. Pavonazetto had to be transported over land for distances of several hundred kilometers to reach the sea port, whereas the quarries of Skyros were located directly at the coast.

Knowledge of the working conditions in the quarries of antiquity is particularly scanty. There are very few written reports so that there is much room for interpretation, as is evident from the novel "The Aesthetics of Resistance" ("Ästhetik des Widerstandes") by Peter Weiss:

"In the marble quarries in the mountain slopes to the north of the castle, the sculptors had used long sticks to point out the finest blocks and then they watched the Gallic prisoners at work in the stifling heat ... The defeated warriors driven to work shackled in chains, suspended by ropes at the quarry face, hammering crowbars and wedges into the layers of glistening bluish-white crystalline limestone, and rolling enormous stone blocks on long logs down a meandering path, were notorious for their fierceness and unsavoury customs – and the masters with their entourage would fearfully pass them in the evenings as the slaves, stinking and drunk from cheap booze, camped in a pit."

PETER WEISS,
THE AESTHETICS OF RESISTANCE

Even though the emphasis is on the literary description of an antique theme, it is evident that Peter Weiss had nevertheless studied the scanty available sources very thoroughly.

Indirect confirmation of the exceptionally tough and dangerous working conditions, particularly for the unskilled *extempores*, is obtained when one considers the social

Methods to determine the origins of marble

The interest in the origin of statue marble is a phenomenon of modern times. Until then the people merely enjoyed the unique beauty of marble statues and palaces built with this noble stone and were not concerned with its origin. However, the excavation of Herculaneum and Pompepii in the 18th century initiated the systematic research of antiquity. Seeing was no longer sufficient, it was now a matter of understanding! And when the natural sciences supplied suitable instruments and methods, the research workers were able to satisfy their urge for scientific exactitude. Today the marble used in antiquity and the renaissance is one of the most studied materials. Comprehensive tables with extensive data of all the important marble quarries are now available so that the origin of a given marble can be speedily established.

In 1891 the German geologist Richard Lepsius developed the classic method of determining the origin of marble. It was judged by simple petrographic features, e.g. the colour, structure, mineral constituents and grain size. Appearance had been sufficient for Lepsius, but research workers then used the newly developed thin-film analysis method to study extremely thin marble disks under a microscope.

Since both methods often resulted in errors, trace element analysis was adopted in the 1950s for the identification of marble. Accordingly, the origin of the marble is determined by the amount of sodium, manganese, potassium, strontium and other elements contained in the marble. The established values are then compared with the data of most of the famous marble quarries.

Isotope analysis, developed towards the end of the 1970s, was even more accurate and it has since become the standard method to determine the origin of marble. The method is based on the fact that not all atoms of the same element have the same mass but rather that the weight of the isotopes differs. For instance natural carbon consists of the three isotopes C^{12}, C^{13} and C^{14}, while oxygen in nature is present as the O^{16} and O^{18} isotope. Since not every isotope of an element is equally stable, the ratio changes in the course of time so that the age of a marble deposit can be judged: Each marble quarry has its own characteristic ratio between the carbon and oxygen isotopes. A minute sample from a statue or column is quite sufficient to determine the isotope ratio of the marble under a mass spectrometer. The origin can then be easily and usually accurately established by comparing the result with the data of known quarries. Occasional overlaps between two quarries can be finally eliminated by combining all methods.

Recently it has been hoped that the origin of a given marble can be pin pointed with even greater speed and accuracy by electron spin resonance (ESR) and the thermoluminiscence method.

The determination of the origin of a given marble is always of interest. For instance the examination of marble busts in the Munich Residence in Germany revealed that the Italian renaissance sculptors used marble from Carrara and statuario from Greece, whereas their German contemporaries only worked with statuario from Italy.

origin of these workers. In the Roman Republic and during the first decades of the empire the *extempores* were primarily slaves and occasionally prisoners of war. Although they had no rights, they were still of material value to their owners and were treated accordingly.

The new system initiated by Tiberius required significantly more manpower at a time when the number of slaves was on the decline. Moreover, Rome's wars of conquest were becoming increasingly rarer, thus making it necessary to fall back on a favoured principle in Pharaonic Egypt according to which serious criminals were sentenced to work in quarries. This now became the standard sentence in Rome with the result that the *damnati* – the convicts – became the most important source of new manpower. The sentence, known as 'Ad metalla', came directly after torture and death. Moreover, the quarry operators no longer had to consider the health of their workers. And it was only the plebeians – the *humiliores* – who were sentenced to work in the quarries (or mines). The patricians or *honoratiores* were not sentenced to quarry work for the same crime; they were banished or deported.

The system was so successful that the market for mass-produced articles in Rome became fully saturated by the end of the 2nd century. Stocks were bulging with the result that only the finest quality was in demand in the Roman capital. Everything else was now diverted to the provincial towns which had always been forced to take second place to Rome. This was also the time when the records of central organisations ended, probably because some of the quarries were reprivatised and others were closed. The marble trade in the empire, however, continued to prosper for at least another two hundred years, namely for as long as the *pax romana* guaranteed stable political and social conditions.

At the beginning of the 4th century the demand for marble once again exceeded supplies. However, the entire system collapsed towards the end of the 4th century when the Roman Empire was increasingly shaken by internal conflicts. Remnants of the system were retained in the eastern Roman Empire, but in the western half the heyday of marble was irretrievably over. The last column of white Carrara marble to be supplied to Rome was the Foca Column. It was erected on the Forum Romanum in 608 AD at a time when the town of Luna (Carrara), itself, was merely a conglomerate of wooden huts in the middle of ancient monuments. The Dark Ages had commenced.

The ups and downs

The name Carrara appeared for the first time in history on a gift deed dated 963, but it did not refer to a quarry or marble, but rather to the city which the German Emperor Otto I gave to the Bishops of Luna. The fact that the one-time most important quarry for marble throughout western Europe was never mentioned is typical of the Middle Ages. Quarries, if mentioned at all in documents, were only of local significance, even if they were operated on an extensive scale or underground for large building projects – for instance the limestone massif near Paris permeated by a series of galleries with a total length of 300 kilometres. These are the remains of medieval quarries where limestone was extracted to build the churches and houses of the capital.

The quarries were usually operated as a family business. The quarrying techniques of antiquity continued to be practised, but the trade that, at one time, had spanned the entire Mediterranean Sea area was now replaced by business conducted hardly further than between villages. Rarely did a quarry dispatch its building material beyond regional confines. For instance, in Aachen, Charlemagne erected marble columns from Ravenna (794); the Tower of London incorporated limestone from Caen in Normandy which, following the battle of Hastings (1066), had been imported by Normans in large quantities into England and can be found in many buildings.

However, these were exceptions rather than the norm, and there were two principal reasons for this. Firstly, there was no power base that could maintain order and guarantee a large, well functioning market throughout Europe – even France had only limited access to the numerous marble deposits in southern Europe. Secondly, transport costs were so exorbitant that any commercially minded person would have invariably shrunk back from the thought of using stones that were not regional. An English document from the 12th century confirms that, from a distance of 12 miles onwards, the transport costs for a stone block exceeded the cost of quarrying and cutting that block. Yet, to be still able to build with marble, the people in the Middle Ages developed new sources for this natural stone.

Rome with its one-time million people had shrunk to a city of mere 20,000 inhabitants in the Middle Ages. Many of the buildings were empty with the result that the marble blocks, columns and slabs were removed from the dilapidated buildings and reused

Stone transport carts by **Nicoló** Zabaglia, 1743, **plate** XVI.

The illustrations show the means of transportation over a period of 600 years.

Roman Eifel aqueduct with 30 cm thick fresh-water limestone deposits, Euskirchen-Kreuz-weingarten.

Calcareous sinter column of the Eifel aqueduct in the Chorapsis Collegiate Church in Bad Münstereifel.

for new purposes. These supplies were complemented by the stocks of the Marmorata Quarter which dated back to the golden age of the Roman marble trade.

The situation in the Byzantine Empire, the eastern half of the Roman Empire, was very similar. The Venetians made ample use of this opportunity at the end of the Fourth Crusade (1202-1204) when they held power in Byzantium for 60 years. It is very likely that the Cathedral of St. Mark and other marble buildings in Venice were actually built with marble from other buildings that had previously stood in Byzantium.

After the Ottomans conquered Byzantium in 1453, they invariably continued the Venetian practice. In many instances the columns in their mosques had previously adorned the Christian churches, but with the difference that the former owners had to pay for the material. The Ottomans were even able to perfect their "recycling" system in Egypt where no marble had been quarried since the end of Roman rule in the 7th century AD. Since building material was scarce, the marble elements of torn down or crumbling buildings were stored until they could be used for new buildings.

The reuse of old building stones, particularly in the vicinity of ancient cities, was a normal practice at the time. However, a very special form of limestone extraction could be observed in medieval Cologne. Ever since the 2nd century AD, the Roman Eifel aqueduct had conducted some 20 million litres of drinking water to Cologne each day for the past 200 years. This had resulted in the formation of a 30 centimetre thick limestone crust. Similar to tree rings, a succession of limestone layers had been deposited on top of each other (see figure). The display of colours of this "marble", caused by traces of metal ions, was absolutely fascinating:

"This was cut in the Kriel parish [...] a marble that stood out among Europe's marble for the diversity of its colours"

ÄGIDIUS GELENIUS, DE MAGNITUDINE COLONIAE CLAUDIAE AGRIPPINENSIS, 1645

Between the middle of the 11th century and the middle of the 13th century the aqueduct's limestone encrustation was used primarily in the North Eifel and in Cologne for (decorative) columns, gravestone slabs and sarcophagus lids. As a result of gifts to princes and bishops, the material eventually spread to the Wartburg, Hildesheim and

Brunswick. This calcareous sinter from the Eifel aqueduct can also be found in the Netherlands, Denmark, England and Sweden, even though the total amount of extracted fresh-water limestone could hardly have exceeded 10,000 cubic meters.

Marble was now once again being quarried in Carrara in the late Middle Ages. In 1265, the prominent Italian sculptor Nicola Pisano ordered a large number of columns, slabs and blocks of white statuario for the Cathedral of Siena. This proved to be the starting point for a renewed upsurge following a suspension that had lasted for almost 600 years. It was primarily the rich Italian cities in the north that preferred this stone for their cathedrals, and merchants also used this marble occasionally to demonstrate their wealth. In 1437 three marble columns were traded for 600 lire in Genoa – an amount that was normally paid for an entire house!

There was even a revival of series production for certain marble products. Mortars and floor tiles were extremely popular, and slabs were even being ordered for wall panelling. However, output never regained the quantities of the Roman period. Just under 950 tons of marble were quarried in 1583. The political turmoil that characterised Europe in the 17th century resulted in a drastic decline in the demand for marble. Even the revival in the second half of the 18th century was short-lived. The political conditions prevailing during this epoch prevented a long-lasting increase in production, and the Napoleonic Wars brought marble quarrying to an almost complete standstill.

The advent of technology

At long last peace returned after the territorial map of Europe had been redefined at the Congress of Vienna (1814-15). Production was resumed in Carrara, reaching an output of 9000 tons per year by 1838. And this upward trend continued, as is indicated by the production figures of the subsequent decades: 40,000 tons were quarried in 1857; by 1914 the annual output had risen to 140,000 tons. This was not only the outcome of the more stable political conditions and the neoclassical style that was en vogue in Europe and North America and which had resulted in the rapid rise in the use of white marble. The other major contributing factors were the technical achievements of the industrial revolution which were introduced into quarries with a delay of some fifty years, and the construction of railways for easier, faster and cheaper overland transport of marble blocks. The marble trade soon em-

The Carrara marble train.

At the beginning of the 20th century steam-powered tractors transported the marble from the quarry face to the loading stations for the marble train. However, steam power was soon replaced by more efficient diesel engines.

braced half the world, as was the case in Roman times – except that the world was smaller in those by-gone days.

A trans-shipment quay equipped with cranes was built in the port of Carrara in 1851. The sledges used since antiquity to transport the blocks gave way to inclined hoists and gravity planes while soap-covered wooden sleepers – known as lizzatura – were introduced to slide the blocks down to the quarry floor. Steam power was now being used. A railway line designed exclusively to transport marble was laid in the Carrara mountains between 1876 and 1880. The line covered a distance of 20 kilometers and overcame a difference in height of 450 meters. The line crossed 16 bridges and passed through 15 tunnels, as well as innumerable zigzag shunt-backs, to link the floor stations of the individual quarries with the port (see figure).

The British Empire was the driving force behind the construction of this elaborate and costly railway line. Since the poor roads and paths through the Apuan Alps were unable to cope with the increasing demands, particularly from England, the British East India Company invested in this project for the Carrara region with the justified hope that these investments would be refunded in the foreseeable future. By 1910 this railway link was handling 80% of all marble transports from the quarries in the mountains to the trans-shipment points at the sea.

Developments elsewhere did not proceed at such a fast pace. The demand for this natural stone was never large enough to warrant potential investors to take on the high financial risk of modernisation. One-man operation dominated among the English quarries in the 19th century, and in 1912, only every seventh of the 7100 quarries employed more than ten workers, and only every twentieth more than thirty. Moreover, the quarries were only operated while there was a demand. The quarry was closed as soon as a large building, road or bridge had been completed.

The First World War once against resulted in a drastic decline in demands, although only

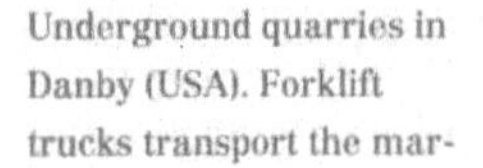

Underground quarries in Danby (USA). Forklift trucks transport the marble blocks through a maze of passages and galleries.

Today lorries transport the marble blocks in Carrara.

for a short period. By 1926 annual output in Carrara had once again risen to 340,000 tons. In the meantime steam-powered tractors were operating in the quarries, soon to be followed by diesel engines (see figure). A modern port for marble was also built in Carrara during this period, thereby concluding the first phase of mechanisation which made marble a mass-produced item.

After the Second World War the half million ton output volume was exceeded for the first time, yet transport still broke down into three distinct steps. Firstly, the transportation of the blocks by inclined hoists and gravity planes down to the quarry floor from where they are forwarded by rail or tractor to the sea and finally shipped to their destination. It was only in the 1960s that production was once again drastically boosted in Carrara through the introduction of cross-country lorries, the transformation of the marble "train link" into asphalt roads, and the extension of even the smallest paths to the remotest quarries. Output in Carrara rose to 1.5 million tons. Lizzatura and gravity planes finally disappeared in the large quarries.

Nowadays it is no longer possible to survive solely from marble, in spite of all the quarrying and transport rationalisation and mechanisation measures in Carrara, and even though the finest marble grades still secured prices in the order of DM 700.- for a ton of finest quality marble. Large building projects using Carrara marble were now far and few between. Only the oil-rich Arab sheiks could afford to order this costly stone in large quantities. Consequently, the wealth of know-how gained in marble processing in Carrara was now also being applied to other natural stones in order to make maximum use of the expensive saws as well as grinding and polishing machines. Today, the Carrara region is the world's largest centre for the processing of all kinds of natural stones. According to current trends the pink-black speckled Sardinian granite – otherwise referred to as Silver Star – is very popular.

"If precious stones were not held in such esteem merely for their price but rather for their noble appearance, then" she continued, "marble would most certainly be a precious stone." I replied that "marble definitely ranks among the precious stones. Even though, as a material, it is not as costly as gems which are only found as small pieces, it is nevertheless so exquisite and beautiful that it is not only treasured in white but also prized in any other colour so that it is transformed into many things, and the ultimate that can be created by human arts is executed in the purity of white marble."

ADALBERT STIFTER, NACHSOMMER, 1858.

2.3 The uses

Marble is undoubtedly the stone of prosperity – a building and decorative material representing power, glory and richness – used for temples, palaces and sculptures. This would be more or less the reply to the question "What is marble?".

These were the same attributes ascribed to marble already in ancient Rome when it was praised for its crystalline beauty, yet simultaneously denounced as an expression of extreme, unrestrained grandeur and decadence. Cool splendour and elegance on the one hand and bad taste and kitsch on the other. And it is between these two extremes that the history of marble and its use has swung to and fro ever since a term was first coined for this stone.

The term marble comes from the Greek *mármaros* which defines a stone or rock block, while the adjective *marmáreos* simultaneously means dazzling. The Greeks used the word *mármaros* for a crystalline lime-

Taj Mahal (1632-1658), built by the Mughal Emperor Shah Jahan in memory of his wife Mumtaz Mahal. Originally, the Emperor intended to have an identical mausoleum built in black marble for himself on the opposite side of the Yamuna River, but this plan was cancelled after this death.

Limestone and marble

Rocks are a diverse material. However, only the rocks of the earth's crust can be used. These rocks are subdivided into 65% magmatic rock, 8% sediments and 27% metamorphic rock. However, the subdivision of the rocks by their formation does not indicate their potential uses. The classification method adopted in the building industry is far more precise. A distinction is made between hard and soft rocks. Consequently, sedimentary limestone is just as much a soft rock as metamorphitic marble. Both of them consist of more than 90% calcium carbonate, and they have the same density of 2.5-3 tons per cubic meter. Moreover, the limestones have homogeneous properties, a typical marbled appearance and are easy to polish. The building industry does not make any other distinctions: Everything is marble! A distinction is only made by the use of the rock.

Statue marble

The demands expected of statue marble (statuario) have not changed since antiquity: Pure white without any veins, a uniform closed structure and a grain that is not too fine. It should be translucent down to a certain depth (nearly 30 centimeters with the very best marble grades) to retain its typical sheen lustre, and it must be completely free of all faults. There are very few quarries that can supply marble in such quality: Pentelikon and Paros in Greece and the Italian Carrara which, since Roman times, have been virtually the exclusive suppliers of statuario.

However, sculptors have to find their particular marble block in Carrara, if possible a freshly quarried "living block" that responds with a clear ring when struck with a hammer. This is the only way the sculptor can be sure that the block does not have any cracks or faults.

When sculptors have found their block they then have to "chip away everything that is superfluous", but "without the possibility of being able to add anything again" - as stressed by Michelangelo.

It is this personal, almost irrational relationship between Man and marble that accounts for the great appeal of working with this stone. Moreover, there is always the fear that right at the end a crack or fault will appear that will ruin everything.

The techniques and tools have hardly changed since antiquity. Even final polishing, waxing and oiling is not new.

Yet the heyday of statuario is over. The choice of sculpturing material is a statement in itself since every material depends on the trends in the arts at a given time: In Egypt they preferred a closed, block-like statue for which granite was used. Since Greek antiquity a naturalistic representation has been favoured, hence marble. Nowadays, sculpturing is abstract with the result that marble has declined to the realms of kitsch.

Architectural marble

Building stones must be easy to process and resistant to mechanical stresses and weathering effects. And this definitely applies to architectural marble, no matter whether limestone or marble is concerned.

Both rocks can be cut and ground into any shape and size, including slabs for facades, tiles for walls and floors and mosaic stones. The compressive strength of marble is comparable to that of concrete and cast iron which is why it is well suited for supporting elements such as columns. Its tensile strength, however, is low so that it is less suited for load bearing elements such as architraves. The Greeks resolved the problem by closely spacing supporting columns to keep architraves short.

Another important reason for the use of marble was its permanence. In bygone days there were no problems in this respect, but the situation has changed in our age with industrial emissions, motorcar exhaust fumes and "acid rain". Minerals containing a high share of calcium carbonate have a low resistance to acids with the result that marble can no longer be used as an external material. Existing marble buildings and sculptures must be impregnated by elaborate and costly processes if they are to withstand aggressive weathering effects. Linseed oil varnish to coat marble was used as early as the 19th century. And if chalk matching the colour of the marble was added, then cracks could be closed. A multitude of impregnating agents are available today, and most of them are silicon-based.

Irrespective of all the functional properties of marble, its selection was always an expression of richness and power. Without this representative character, marble would never have gained such enormous importance. And it was far too expensive to use as a simple building material.

The Temple of the Magician atop a huge pyramid in Uxmal, Mexico (6th-10th century).

stone, usually white, that they used for their sculptures and sacral buildings, thus clearly distinguishing it from normal stone – *lithoi* – for foundations and masonry. With this definition the Greeks elevated marble – and with it any limestone that can be polished – to the status of a very special material, and it has retained this status throughout the ages right to this day.

However, this only applies to the western world where Greek culture prevailed. Marble and limestone never gained such significance in other cultures. It is true that limestone was used for the centre of Maya culture in the north of the Yucatan peninsula, not because the stone was particularly appreciated, but rather because the material was in plentiful supply and could be easily processed.

Initial distrust

The pragmatic approach of the Mayas towards limestone is also reminiscent of the building style in Pharaonic Egypt. Limestone was used as a building material for the first time in ancient Egypt, not because of its fascination but rather because it was readily available in abundant quantities.

Egypt lies on a crystalline stratum, interspersed with granite and diorite, which supports massive limestone layers. And it was this material that the Egyptians used when they started to build their stone buildings during the 2nd dynasty 2800 BC.

Initially, the Egyptian builders distrusted the new material and were not prepared to rely on its load-bearing capacity and permanence. Hence, the Step Pyramid of King Net-

The step pyramid of King Djoser (around 2620 BC).

jerykhet Djoser at Saqqara which was a typical example of the distrust prevalent at the time. Although this pyramid was a solid stone structure – the earliest in its size in the world – the actual stones were treated as if there were clay bricks (see figure). Hardly half a century later these bricks were superseded by huge limestone blocks, each one weighing 2.5 tons, which were used to build the Great Pyramids of Giza. These enormous blocks were cut with such precision that they slotted into place without any mortar!

However, the limestone blocks were only used for the outer casing of the pyramids. The actual inner burial chamber for the kings were built from hard rocks, usually granite which they trusted as an eternal resting place for their dead. The Egyptians also used granite or basalt for their sarcophagi, particularly the red quartzite from Syene (today's Aswan) which was held in particularly high esteem.

The situation was similar with sculptures. The Egyptians insisted on the use of hard rocks, even if they were very difficult to shape. This was a very deliberate choice because the sculptors of the time were already exploiting the properties of a stone to highlight and amplify the expression of a sculpture. This is particularly apparent in the portraits of the Pharaohs, most of which were sculptured out of hard rocks. The portrait was not intended to be a precise likeness of the portrayed person, but rather a stereotype image that would make the individual immortal – an expression that was intensified by the fact that hard stone could only be roughly contoured to create a block-like impression of eternal remoteness in the beholder.

The Egyptians had far too many useful types of stones at their disposal to develop a preference for any particular one, and definitely not the most abundant one – limestone. Famous limestone sculptures, like the painted limestone bust of Nefertiti (see figure), were the exception. The sculptors carved their models in limestone which then functioned as a copy for their actual work or for training. Marble was far too rare for it to play any significant role, other than for small vases and a few statues in the Late Period.

The Cretans, just like the Egyptians, lived on limestone. Their palaces, above all the Palace of Cnossos, were built from limestone blocks and clad with slabs of marble, calcareous sinter or alabaster. The Egyptians opted for the monumental in their buildings, whereas

Limestone bust of Nefertiti (around 1370 BC).

The Palace of Knossos. Crete (around 1500 BC).

the Cretan palaces were of a more playful character. In fact it is easy to see how the ground plan of the Palace of Cnossos could have given rise to the great myth of the Labyrinth and Minotaur of Minos.

The Mycenaeans on the Greek mainland also used limestone for their buildings, but these were far from playful (see figure). The walls encircling the palaces of Mycenae and Tiryns overwhelmed potential attackers by the sheer size of the sandstone blocks used for their construction. Even the great Greek traveller, geographer and writer Pausanias was filled with awe some 1600 years later:

"The wall, which is all that is left of the ruins, was built by the Cyclops from unhewn blocks, each one of such enormous size that even the smallest could not have been moved by a team of mules!"

But neither the Mycenaean monumental manner of construction, nor the playful mode of the Cretans, prevailed. The two civilisations disappeared towards the end of the 2nd century BC, and with them the use of stones for buildings. The first temples built of stone didn't reappear in Greece until four hundred years later.

A country presents itself

Around 800 BC the political, economic and cultural relations between the Greek peoples were deepened, eventually resulting in a common identity. The most remarkable expression of this trend were the Olympic Games held since 776 BC, but a uniform architectural style also arose during this period.

Initially, the Greeks built with wood and bricks, but towards the 7th century BC they turned to stone in their quest for permanence. This was not an easy step considering that stone had not been used on the Greek mainland since the Mycenaean age with the result that the techniques of building with stone had long been forgotten. The Greeks had to arduously relearn all the necessary skills. This was by no means easy since the art of building of the Cretans, Mycenaeans and Egyptians was kept strictly secret, only known by a chosen few. This made a fresh start particularly difficult, but it also gave them the opportunity to create something entirely new.

In the beginning of the archaic epoch, only important buildings, such as temples and theatres which also functioned as cultural centres, were built with stone. The architec-

The Lion Gate of the Acropolis of Mycenae (around 1250 BC).

The Theatre of Epidauros (second half of the 4th century BC) built by Polykleitus The Younger for the Greek poet Pausonias, the most beautiful and best preserved of all Greek theatres.

tural style was harmonious and clear. Each temple was the place of dwelling of a god and so it contained a core area – the cella – that was exclusively reserved for the given god. This is where the statue of the god was erected, and only priests were allowed to enter this room. The actual temple was therefore built around this cella and it consisted of supporting and load-bearing elements: massive columns to support heavy architraves.

The architraves, in particular, were richly embellished with sculptures, reliefs and friezes to transform a Greek temple into an overwhelming structure designed to dazzle the beholder. The interior and its design played only a subordinate role in Greek architecture.

The proportions of the individual building elements and their relation to each other were based on mathematical laws with the result that all dimensions of many temples can be attributed to a 4:9 ratio. These clearly defined architectural principles were in keeping with Greek perceptions. Architecture was not to be governed by obscure, abstract concepts but rather by rules that anyone could follow and understand. This made knowledge freely accessible so that it could spread rapidly throughout the entire Mediterranean region.

The building principles and style were new, whereas the building materials remained the old ones. Just as in Egypt, normal limestone ranked first among the stone used in 7th and 6th century Greece, plus poros which was a tufaceous limestone that could be easily processed. Greece had not yet become the "marble peninsula" of later centuries.

Exceptions were the Cyclade islands of Paros and Naxos which both had rich marble deposits. This is where the first monumental marble statues were created in the 6th century when it was realised that no other material was better suited to model the human body. And sculptors made full use of the opportunities that marble offered to them.

Even the first statues were already endowed with an internal tension that was to become so typical of subsequent periods – bulging muscles gave the statues such a lifelike appearance in clear contrast to their Egyptian predecessors.

This was deliberate because Greek statues intended to emphasize the transitoriness of life. This was also the reason why they were included in tombs. Only the Pharaohs were immortal and god-like. Greek humans only persisted as memories that were kept alive by statues.

The statues from Naxos and Paros were monumental, but rarely monolithic as is the almost ten meter tall Apollo statue hewn out of a single block which the Naxians dedicated to the Delos shrine. The pedestal bears the inscription "I am from one and the same stone – both statue and base." Only the base has survived to this day. The fact that it weighs 25 tons is a clear indication of the enormous size the complete marble block must have been. However, blocks of such size are rare in Greece because the local deposits are fissured. The quarried marble, and therefore the statues, were usually small or assembled, just as the limestone columns.

The statues and columns continued to be assembled also during the classic epoch except that, throughout Greece, they were now all made of marble which was also used to build temples, theatres and other public buildings. The choice of marble was enormous so that there was no uniformity – dazzling white marble with a faint bluish tint from Paros; fine-grained Pentelic marble with a yellowish tint; white, bluish marbled hymettium; and coarse-grained marble from Naxos.

However, there was also a very simple reason why the Greeks only used marble. The Greek cities had to be rebuilt after the Persian wars, and the new material was used everywhere to underline the Greek triumph

The Parthenon Temple,
Athens (447-432 BC).

over the Persians and to indicate the advent of a new era. The Greek cities flourished and their economic and political power had to be presented in a commensurate manner. Hence, the Artemission (450 BC) built by the Ephesians, the Poseidon Temple in the south Italian Paestum (480 BC), the Zeus Temple in Olympia (460 BC), and Athens – the crowning glory of the whole of Greece – where Pericles had magnificent marble buildings erected on the Acropolis. In fact the neighbouring city districts, and eventually the entire city, also became the unsurpassed pinnacle of the Greek classical period.

The Academy, Athens, built in the 4th century BC, were Platon and Aristotle taught.

In Athens of the 5th century BC there were literally hundreds of marble statues, usually as decorative elements of temples and buildings. Free-standing sculptures, as we know today, were the exception in Greek antiquity. The sculptors had only just learnt to detach themselves from the traditional sculpturing techniques. In Egypt, but also in archaic Greece, the statues consisted of a frontal view hewn out of a stone block. The side profile was still finished with a certain amount of care and realism, whereas the

The Temple of Apollo, **Bassai** near Andritsaina (5th century BC).

Sculptor: Artist or craftsman?

A statue produced by Phidias certainly differed from the rough hewn block of a stonemason. This was perfectly apparent to the contemporaries of this great sculptor, yet little distinction was made between the two: Be it a stonemason or a sculptor, for the people, they both worked with the same material and the same tools. Both commanded the same technical skills, and they often completed the same kind of work and depended on orders from moneyed sponsors. Even though many a sculptor in Greece developed a distinct sense of self-awareness and inscribed his signature in his works so as not to be nameless as the Egyptian sculptors, he nevertheless remained a craftsman - banausoi - and he even had to share his protector god, Hephaistos, with the blacksmiths!

The real artists were the poets so that Phidias ranked far down the scale, even though he had been commissioned to take charge of the construction of the Acropolis.

"If a work charms by its exquisiteness, then this still does not mean that its creator merits our profound esteem ... No talented young man, while admiring Zeus in Olympia created by Phidias, or Hera in Agros by Polyclet, will actually want to be Phidias or Polyclet."

PLUTARCH, THE LIFE OF PERICLES

The work was admired whereas its creator was insignificant. And this attitude was even quite justified in antiquity since the individual masterpieces were rarely created by one person. Normally, entire crafts schools worked together on the sculptures of a temple so that the individual remained anonymous. For instance, in the year 408-7 BC more than 50 sculptors and stonemasons were on the payroll for the construction of the Erechtheion Temple on the Acropolis (421-406 BC), and stylistic studies of the Parthenon indicate that a total of about 70 different sculptors worked there.

Such a crafts school was headed by a master who passed on his knowledge and skills to the subsequent generation so that his style would continue. Consequently, the pupils copied the master with such perfection that only a very skilled and knowledgeable eye can distinguish between the copy and the original.

The fact that, nowadays, we focus on the individual sculptor is the result of our perception of the arts. For us an artist is someone who departs from the familiar to create something new and unique - namely a work of art. The masterly technical skills on their own no longer count!

"The sculptor's workshop". This white marble tile originally created by Andrea Pisano for the belltower of Giotto is now preserved in the Museo dell'Opera del Duomo in Florence.

rear view remained rudimentary. Figures were now being sculptured with tremendous care and faithfulness from all sides, enabling the viewer to walk right around a statue to admire it from all sides.

Since buildings and statues formed a common entity, it also meant that they were usually made of the same material. However, this does not mean that the Greeks loved marble to the exclusion of everything else. On the contrary, in Greek antiquity bronze was just as much appreciated for statues as marble. However, bronze was far more expensive which is why only cheap marble copies of bronze originals were often set up in public places. Since the Romans also produced innumerable marble copies of the Greek masterpieces, and the original bronze statues eventually disappeared in melting furnaces over the centuries, our current concept of Greek antiquity is characterised by white marble.

This is an incorrect, one-sided perception. Not only does it pass over bronze, but it also disregards another peculiarity of Greek antiquity: Greek temples and sculptures were coloured! The fact that the marble pieces from antiquity are now white distracts from the fact that they were originally painted in bright or delicate colours to hide the joints of assembled columns, to make the hair and eyes visible or to emphasize the expression of a statue. Remainders of the original paintwork are occasionally still visible today. However, since the Greeks used egg white or glue as binders, the colours were gradually washed off by the rain leaving only the bare marble that we see today.

Marble bust from the Apollo Museum in Olympia, Greece.

Athena Lemnia, the gypsum copy of a coloured Roman statue of 1st-2nd century.

Commenting on a sculpture in the original paintwork, the French writer André Malraux said:

"We do not regard a painted and waxed Greek head as a work that has been revived to new life, but rather as a monstrosity."

The clock could not be turned back with the result that the supposed white marble of Greek antiquity has influenced the arts and culture of the Occident ever since the renaissance. The original motifs of the Greeks only interest us as anecdotes, even though the very reason why the Greeks selected white marble was the fact that it could also "breathe" in an overpainted state without changing the applied colours which remained vivid and rich in contrast. No other stone could compete with marble in this respect.

Be it painted or white, monolithic or assembled, the history of Greece and its culture has been inscribed in marble in the truest sense of the word. And to give "Greek history" permanence, it was actually chiselled by an unknown historian into a slab of Parian marble. This Parian Chronicle, also known as Marble Parium, commences in 1581/80 BC with Cecrops – traditionally the first King of Attica who was represented as a human in the upper part of his body, while the lower part was shaped like a snake – and ends in 264/263 BC, the year in which the chronicle was recorded, and at a time when the Macedonian kings and their successors had taken the place of the Greek city states and when Greek antiquity had long given way to Hellenization.

Intermediate times

The art of the Greek classic period covered the widest area with the conquests of Alexander the Great. Greek architecture and sculptures could be found on the banks of the Indus and within the walls of Alexandria. Yet the apparent triumph of the late classic period already encompassed its decline. The spread of Greek culture to foreign countries was invariably associated with a corresponding return flow into Greece of cultural peculiarities and influences from these countries, resulting in eclecticism where one endeavoured to impress by extreme, unrestrained effects.

This is very much evident in the sculptures and statues of the Hellenist period. Initially, the sculptors remained closely attached to their classic icons, even using clay models of their sculptures for their own work. Yet they did not produce mere copies or imitations because their figures had their own character with special emphasis on human traits. The traditional balance between the divine and human elements in statues shifted in favour of an emphasis on the human side.

The Hermes statue created by Praxiteles around 330 BC is a human with godlike traits, yet his earthly transitoriness is very much apparent. Praxiteles was the first to produce a nude female figure with his Cnidos Aphrodite – a motif that was re-interpreted many times in the course of subsequent centuries. The growing emphasis on the human element was simultaneously

Section of the last frieze of the Pergamon Zeus Altar (ca.180 BC)

Venus of Milo (2nd century BC).

associated with a freedom of expression that was masterly embodied in marble as the most suitable material for such figures. This was also the time when marble was given a more naturalistic, almost deathly appearance by covering it with a wax or oil varnish that was then polished.

Emotionalism soon entered the endeavours to portrait realism. The frieze figures of the great Altar of Pergamom are in restless motion, and the faces are very expressive (see figure), yet at the same time somewhat too sweeping and overacting, and with too much emotionalism in their facial features. Greek Classical Antiquity was now acquiring baroque traits, ultimately culminating in the Laocoon sculpture dating probably around the 1st century BC and the 2nd century AD (see figure, page 116). The group is impressive for the virtuosity of the techniques and masterly finish, but the expression is disappointing as its sole intent is the effect it can achieve without allowing beholders to develop their own impression.

The changes in architecture were just as obvious. The principle of harmony and clarity had to give way to pomp which had become the order of the day. Classical architecture definitely avoided giving temples a spectacular side that could eclipse the effect of its

Colosseum, Rom (72-80).

other sides. The new trends in architecture now focused the beholder's attention on Corinthian columns with richly adorned capitals that lined the facades, while the interior walls were lined with colourful marble slabs and the floors were composed of elaborate mosaics. Large squares, grandiose palaces, column-lined streets and statues of prosperous citizens showed the way into the future. Greek art had come to an end and Rome had become the centre of the world.

The triumph of marble

*„Know'st thou the house, with its turretted walls,
Where the chambers are glancing, and vast are the halls?
Where the figures of marble look on me so mild."*

JOHANN WOLFANG VON GOETHE, WILHELM MEISTER, 1796.

When the great German poet Johann Wolfgang von Goethe visited Italy almost 1500 years after the fall of the Roman Empire, he was so captivated by the marble remains of the ruins he saw that this had a lasting formative effect on his perception of Italy. Yet these ruins can only give us a slight hint of the breathtaking spectacle that must have been created with marble in ancient Rome.

This history of marble in Rome started in the late Republic of the 1st century BC. It was by no means love at first sight. The Romans were initially very wary and sceptical of this new material. The philosophers and writers, in particular, made no secret about their negative attitude towards marble. Cicero complained in his writings that he disliked "the ostentatious villas with their marble floors and coffered ceilings". And Pliny the Elder even demanded that the state should intervene, particularly since it in other respects always put a stop to superfluous luxury:

"There are still censorial laws [...] which [...] prohibit the execution of anything that is insignificant, yet no law has been enacted that prohibits the import of marble and to sail the seas for this purpose."

PLINY, NATURAL HISTORY, XXXVI, 4

The reason for such a vehement reaction was quite predictable. The first marble columns appeared in the private homes of rich Roman patricians who, as governors or high-ranking civil servants, had encountered this fashion in the Hellenist provinces. For convinced republicans such as Cicero and Pliny, who had been educated in the spirit of republicanism, such private luxury was reason for the very greatest concern. They instantly construed in this display of prosperity the intent to grab power in the state. Moreover, stoistic philosophy was on their side as it was constantly preaching that such luxuries were completely irrelevant.

Although they chastised private luxury, the same moralists praised the use of marble in public buildings. Pliny ranked the Basilica Aemilia with its Phrygian marble as one of the wonder buildings of the city. It is hardly surprising that their criticism met with little approval. Most Romans preferred luxury, even though they may have had a guilty conscience. And, in any case, such thoughts disappeared during the Empire period.

Augustus had already started to construct public buildings in marble as a sign of maiestatis imperii, and his successors continued this policy with the result that marble could be encountered everywhere in Rome. A particularly popular variant involved the use of exotic, usually coloured, marble to emphasize Rome's military rule over the then known world orbis terrarum. It was particularly popular to use coloured marble for thresholds so that they could "walk over the defeated enemy" everyday.

However, marble on its own, was unable to meet the demands of the most ambitious building projects so that other stone also had to be used. Travertine from the deposits along the Aniene River near Rome was particularly popular. This calcareous sinter was used to build the Colosseum (see figure). The enormous building programmes under the Emperors Domitian and Trajan created a huge demand for stone. Although every stone was needed, marble still remained the unchallenged crowning glory. Its use, also in private houses, was now explicitly praised. And to be "in" you had to know the origin of the most important types of marble as well as the associated myths.

In his description of the interior finish of a private bath, Statius, the court poet of Emperor Domitian, even listed those stones that were not used:

"Not approved here was Thasos marble and the wavy veined variety from Carystus; onyx grieves to have been left out, while ophite laments that it has been excluded; the sole marble to shine here comes from the yel-

Mosaics and incrustations

"Where were white, green and blue, hangings fastened with cords of fine linen and purple to silver rings and pillars of marble: the beds were of gold and silver, upon a pavement of red and blue, and white and black marble."

This quote from the Book of Esther of the Old Testament is considered to be the first reference to the use of marble for mosaics, and it describes the Shushan Palace of the Persian King Xerxes (519-465 BC) in Susa in present-day Iran. This technique passed through the Hellenistic kingdoms in Asia Minor and Egypt, eventually reaching Rome where stone mosaics reached their first heyday - a process that was favoured by the technique of stone sawing. Accordingly each stone could be quickly and easily cut to any size and shape.

All coloured and cleavable stones are suitable for stone mosaics. However, as in the building industry, cost reasons often made it necessary to use local minerals or to deliberately select marble as a sign of prosperity. Apparently this was much to the disgust of Pliny: "Whoever was the first to have the idea of cutting marble and dividing it up merely to satisfy a lust for splendour certainly had a woeful perception."

Opus tesselatum (cube mosaics) involving the use of cut stone cubes of 1-1.5 cm edge length, and opus vermiculatum (vermiculus = small worm) where the stone cubes have an edge length of only a few millimeters, are the most important forms among the different stone mosaic techniques. The mosaics are laid in a bed of mortar consisting of lime and clay.

Opus sectile (secare = cutting) is the third important mosaic technique and it uses very thin stone fragments, primarily marble. And from this technique it is only a small step to incrustations which originally embraced any kind of wall cladding. However, since the Roman period, the term is usually restricted to wall decorations composed of precious stone panels.

Top: Early Christian floor mosaic from Aquileia, Italy (around 319).

Below: Floor mosaic in a Roman house in Orbe, Switzerland (end of the 2nd century).

low quarries of the Nomads; and solely from the caves of the Phrygian Synnada Attis stained with the glowing drop of its blood..."

STATIUS, SILVAE, Verse 34-40

The importance of marble is also apparent in the different designations for the stone and the stonemasons. The Romans distinguished between the common lapis or lapis quadratus which was hewn by the lapidarii, whereas the precious marmor was processed by the marmorarii. A marmorarius was paid 60 denars per day which was only slightly higher than the customary pay for a

Pantheon, Rome
(188-125).

craftsman. A mosaic worker earned approximately the same, a mural painter slightly more, while a portrait painter was paid two-and-half times the amount. The fact that marble was nevertheless so expensive was not due to the wages but rather the transport costs. Only emperor families and the rich nobility of the city could afford structural work in marble. The patricians in Ostia could not afford such luxury and had to be content with marble facings.

Marble facing involves the plastering of normal brick walls which are then lined with thin slabs of sawn marble. This technique achieves the same effect as using marble blocks, but with much less marble. Mosaic work is closely related to this technique, except that the Romans used sawn mosaic stones (see box) primarily for copying Greek murals.

Mosaics and marble facings in Rome were nearly as old as the use of marble stones. Seneca commented on the penchant of his contemporaries with the words:

"We all admire the walls lined with thin marble slabs, yet we are all well aware of what lies underneath."

SENECA, EPISTULAE MORALES, 115, 9

The lust for luxury did not listen to such criticism with the result that the brick walls in private houses were soon all hidden behind marble facings. The Greek practice of building with natural stone was continued for public buildings except when statics required alternative methods, as was the case with the Pantheon (see figure). The dome, spanning a breathtaking width of 44 meters, was constructed from poured concrete vaults – opus caementitium. And since masonry consisting of simple quarried stones and mortar were not fitting for temples, the walls were faced with marble tiles. The Pantheon did not remain a unique example of make-believe and reality for very long. By the end of the 2nd century virtually no public buildings were built exclusively of marble.

Trajan's Column, Rome (106-113).

Marble facings still required the use of real marble. However, since everyone wanted marble, but only a few could afford it, the advent of imitation marble arose in the 3rd century. Accordingly, the walls were plastered with stucco marble painted to look like real murals. The properties of marble as a building material, which had been so appreciated by the Greeks, were now secondary. The Romans were perfectly happy with imitation.

Imitation also sums up the Roman attitude towards Greek sculpture, at least in the initial period. Ever since the Romans conquered the Greek town of Syracuse on Sicily in 211 BC, a never ending flow of Greek influence entered Rome. And with the fall of Corinth in 146 BC, the Greek city states became a de facto Roman province, thus making it almost impossible to distinguish between Greek and Roman art.

According to Pliny, the Laocoon Group (see figure, page 116) for the Palace of the Roman Emperor Titus (79-81 AD) was a masterpiece of three Rhodian sculptors – so was this Greek or Roman art? There were innumerable marble copies of Greek bronze figures in Rome – once again were they Greek or Roman art? It is exceedingly difficult to draw the line, particularly since those who ordered such figures at the time exerted an equally important influence on the sculptures as the artistic skills of those who created them. Moreover, the change of material from bronze to marble also resulted in significant changes that went far beyond mere copying.

Furthermore, Roman sculptors and those who ordered the sculptures were eclectic in their choice of motifs with the result that late Hellenism was imitated as well as moderate Classicism, and sometimes even both. The only thing that mattered was that the villa or garden was adorned with an appropriate statue – as was the right and proper thing for a rich Roman patrician according to the satirist Juvenal who described the potential consequences of a fire in a typical Roman patrician villa:

"While the fire is still burning, everyone hurries to supply free marble or make a financial contribution to the reconstruction of the home. Others contribute naked statues of white marble ..."

JUVENAL, 2, 210-212

The house owner either purchased the statues from an arts dealer – a profession that had gradually become established – or directly from one of the many sculpture workshops that existed in Rome at the time. And to assert oneself in all this competition and encourage customers to visit the given workshop, the sculptors in Rome started to sign their statues. This told the viewer where more statues were available.

The Colossal Statue, Naples, Museo Archeologico Nazionale (2nd century). The garment and cap are of marble phrygium (pavonazzetto); the head and hands are of black marble.

Although the choice of style may have been dictated by momentary moods and whims, the Romans were always very definite when it was a matter of material: The white marble statues in Rome and throughout the western Empire had to be in statuario from Luna (Carrara) – a choice that remained unchanged for the next two thousand years!

In spite of all the common ground between Greek and Roman sculpture, there were also areas, though small, in which Roman art developed its own originality and where other influences became effective. These differences were nearly always characterised by pronounced realism which had been alien to Greek idealism. This is particularly apparent in the portraits sculptured in marble during the 1st century AD. These portraits continued the traditions of the Etruscan tribal cult according to which wax face masks were cast of deceased heads of families to keep the memories of them alive. These portraits in stone replaced the wax masks. The stone images of Roman patricians were intended as portraits for a private location, while public areas in Rome were reserved for statues of the given emperor.

Each statue was more than a mere object of admiration. The statue acted as a deputy for the emperor with all the political, legal and cultist consequences. It was a significant medium in the emperor's propaganda within an empire which, on account of its sheer size, a ruler was never able to visit all important places during his rule.

The Roman inclination to realism is also apparent in the "narrative relief" which is clearly an autonomous Roman development. Trajan's column, erected in Rome between 106 and 113 to mark the victory over the Dacians, presents an extensive narrative of the Dacian wars (see figure). The individual moments in these wars were captured in marble, just like a report, without laying any claim to being art. Such a narrative relief has nothing in common with the immensely expressive Parthenon frieze of Phidias.

Since the era of Augustus, at the latest, realism induced the Romans to turn to new sculpture materials. The Greeks had used paint to imitate hair, eyes and robes. The Roman sculptors used identically coloured marble for the same purpose. Just as in architecture, sculpture was a means to express Rome's world rule which gave it access to any kind of marble.

Not only were marbles now available in widely differing colours, but since the 2nd century it was also available in sufficiently large blocks. This was an entirely new experience for Roman sculpture which was now able to distance itself somewhat from Greek art.

In Greece, the Roman Republic and the early Empire period, very few statues were monolithic. Usually, only the head, hands and feet were made of statue marble of the very finest quality, while local stone materials were used for the robed body which was then correspondingly overpainted, sometimes with such care that the joints were completely invisible. Even an expert like Pliny praised the Laocoon Group as a masterpiece sculptured out of a single monolithic block. It was only when the sculpture was discovered in a vineyard near Rome in 1506 that, after a very meticulous examination, Michelangelo announced with the greatest of admiration, that the Group had four almost invisible joints (see figure).

The assembly of a sculpture was not a conscious decision like the use of differently coloured marbles, but rather a necessity. Statue marble was rare and large blocks hard to find. This could only be remedied by a well organised quarrying economy. The ability to supply statue marble in any size is reflected in the development of Roman statues with breastplates that were either sculptured from one block or easily produced by a combination of different marbles. In the 1st century the share of sculptures featuring a separate head was almost 90%. Under Hadrian (117-138) only every second statue was assembled, while in the 3rd century almost all figures were sculptured from a single block.

Improved marble supplies went hand in hand with a change in the style of the sculptures. A uniform figure, half clothed and with a closed pose, dominated in the 1st century as this required the least amount of stone material. This contrasts with the figures of the 2nd century which were naked, larger and featured a pedestal. The pose was more open, in keeping with the trends of taste of the time and depending upon the financial status of the customer because large blocks of statuario were still very expensive.

The golden age of Roman sculpture ended with the collapse of the quarry economy. The emphasis was now on the re-use of old sculptures and on processing the extensive

"Lacoon and his twin sons, crushed to death by two great sea serpents sent by Apollo. The statue was hewn out of a single block according to plans of the outstanding Rhodian sculptors Agesander, Polydorus and Athenodorus, all from Crete" (Pliny, Natural History, XXXVI, 37).

The Mausoleum of Theodoric at Ravenna (6th century AD). It was built with limestone from Istria which replaced marble from Carrara at an early date in the cities of upper Adriatic; the heyday of its use was in the late Middle Ages.

existing stocks. And the situation with building materials was also very similar. High-quality marble was becoming increasingly scarce and costly so that patricians had to make do with marble pieces or local stone materials. Local trade was once again the order of the day.

Only a few years after the fall of the Roman Empire, transport costs had become the deciding factor, even for the tomb of Theoderic the Great in Ravenna (see figure). The tomb's dome with a diameter of almost 10 meters was still monolithic, but the 400 ton stone was not Carrara marble. Theoderic, who dealt the deathblow to the Roman Empire with all its marble palaces and statues, had to be content with simple limestone from Istria. Transportation across the Adriatic Sea was cheaper than the 200 kilometers through the Apennines from Carrara.

In the subsequent centuries marble was only of significance in the immediate area around the actual marble deposits. Exceptions, such as the marble for the sarcophagi of the French kings which had to be transported from the Roman quarries in the Pyrenees, could be counted on the fingers of one hand.

Marble sectile of the Cathedral of S. Andrea in Amalfi, southern Italy (12th/13th century).

Campo dei Miracoli, Pisa
(11th-14th century).

Most of the marble quarries fell into disuse. Genuine interest in marble and limestone now only prevailed in lime works, however not so much for the quarried stone but rather for old slabs, columns and statues which disappeared by the hundred into the furnaces. The heyday of marble was over, and a new golden age was not in sight.

Rediscovery

"We climbed on the first stone, snow-white marble, so shiny on all sides that I could see my reflection."

DANTE ALIGHIERI, DIVINA COMMEDIA; PURGATORIUM 9, 94-96

Dante may have got his inspiration for his description of the steps to purgatory from walks in his home town of Florence where, since the 11th century, this rising Tuscan metropolis was using marble on a major scale – Carrara was only 125 kilometers away. For instance, the San Miniato al Monte Church was faced with 4-5 centimeter long panels of white Carrara marble, and the old Roman technique was also revived for other buildings (see figure).

The Florentine protorenaissance of the 11th and 12th century as an architectural style remained limited to Florence. However, neighbouring towns like Pisa also rediscovered Carrara marble, particularly since the quarries were only 55 kilometers away. The Campo dei Miracoli, consisting of the cathedral, baptistery and campanile, were all built in white and coloured marble. And it is this marble that gave the ensemble its uniform appearance, even though it took three hundred years to complete (see figure).

Pisa and protorenaissance were only a faint foreboding of what was to come three hundred years later, particularly since their effect hardly went beyond Tuscany. The rest of Europe still remained in the late Middle Ages.

The importance of stone as a building material was gradually increasing as not only churches and cathedrals, but also palaces and even simple houses were being built of stone, primarily to offset the danger of fires in cities and to make up for the growing scarcity of wood. The choice of stone was limited on account of the high transport costs. One had to be content with local stone materials, as was the case in the Franconian

Jura where even the roofs of the houses were made of limestone. There was plenty of marble in northern Italy, but it was only affordable locally. Although no effort was otherwise spared for the construction of the Gothic Cologne cathedral, only trachyte from the local quarries could be afforded as building material.

Ca' d'Oro, Venice (1421-1440).

The building material always reflected the financial situation of those who were having the given buildings built. If they were wealthy then better quality stone from a more distant quarry was used, while those who had little money had to be content with local stone materials. In northern Germany a separate style evolved which became known as red-brick Gothic architecture because there were no suitable natural stone deposits in the local sandy soil. Customers were not even prepared to pay for quality stone for figures so that sculptures were made of cast stone consisting of a combination of limestone, gypsum and mortar.

While limestone and sandstone still dominated north of the Alps, builders in the north Italian cities were once again using marble for their cathedrals, palaces and houses (see figure). Strengthened by economic success, the ruling bourgeoisie and city nobility were no longer prepared to accept the unchallenged leadership role of the Church. They were no longer prepared to wait for salvation in the kingdom to come, but rather enjoy the fruits of their success during their earthly existence.

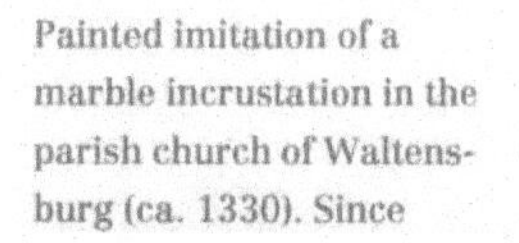

Painted imitation of a marble incrustation in the parish church of Waltensburg (ca. 1330). Since marble was far too expensive it was imitated by impressive painting.

The spiritual basis for the new sense of awareness was provided by humanism which, in clear disassociation from medieval scholasticism, was concerned with the philosophers and writers of antiquity and placed Man at the centre of its deliberations. Renaissance was the artistic response to this development. By turning to the painting, architecture and sculpture of antiquity, filling these with new life, the Renaissance fulfilled the abstract ideals of humanism – even though when Erasmus of Rotterdam saw the Pavia Cathedral he complained that the visitors were more impressed by the marble inlay work than by its religious motifs.

This movement was headed by Florence – the "New Athens" – where not only the Medici were patrons who generously promoted young artists and sculptors. In 1480 there were 54 sculptor's and stonemason's workshops but, at best, only every fourth was directed by a well known personality. Most of those who worked in these workshops were simple craftsmen who produced columns, capitals and mortars for day-to-day use. Not everyone was a Michelangelo!

Michelangelo Buonarroti was a universal genius. His fresco "The Last Judgement" in the Sistine Chapel is just as world famous as his architectural masterpiece the Biblioteca Medicea Laurenziana (Laurentian Library) and the Mortuary Chapel of the Medici family. But sculpture was his greatest love:

"If there is something good in my skills, then this is due to the fact that [...] with the milk of my wet nurse I was also given a hammer and chisel with which I create my figures."

GIORGIO VASARI, VITE DE'PIÙ ECCELENTI ARCHITETTI, PITTORI ED SCULTORI ITALIANI

Although Michelangelo was joking when he said this, it nevertheless contains a grain of truth. His wet nurse was the daughter of a stonemason, and she was married to a stonemason. Michelangelo grew up among stonemasons and came into contact with all their skills at a very early age, and he was to profit from this later on in life when he created new forms of expression in stone.

It may have been the tough, simple life during his childhood that gave him the name "loner" in the courts of the powerful, or something else, but Michelangelo was only happy when he was faced with challenges that he had to overcome. And ever since the morning of 16th August in 1501, his greatest challenges have been marble blocks.

This was the day when he was commissioned to sculpt "David". He had completed a number of marble reliefs and sculptures in the past, but this commission was something totally new – a marble block of enormous proportions, possibly a little too slender for the enormous length, had been waiting for its master in the Florentine cathedral construction workshop for the past 40 years. Several sculptors had failed before Michelangelo accepted the commission that everyone else had rejected. He had instantly seen in the block his David who was waiting to be liberated from his "marble captivity".

He immediately started to work without a model – "a concept is alien to the great artist" – and after 28 months "David" had been liberated: 6.5 tons remained of the original 34 ton block. "David" was immediately hailed by Florentines as "Il gigante" – the giant 5.35 meter marble statue was a "true son of the earth" which is the translation of the Greek *gigas* (see figure).

David in marble (1501/04). Michelangelo is "perhaps the only one of whom it can be said that he reached antiquity, but only in strong muscular figures of the heroic age." (Johann Joachim Winkelmann).

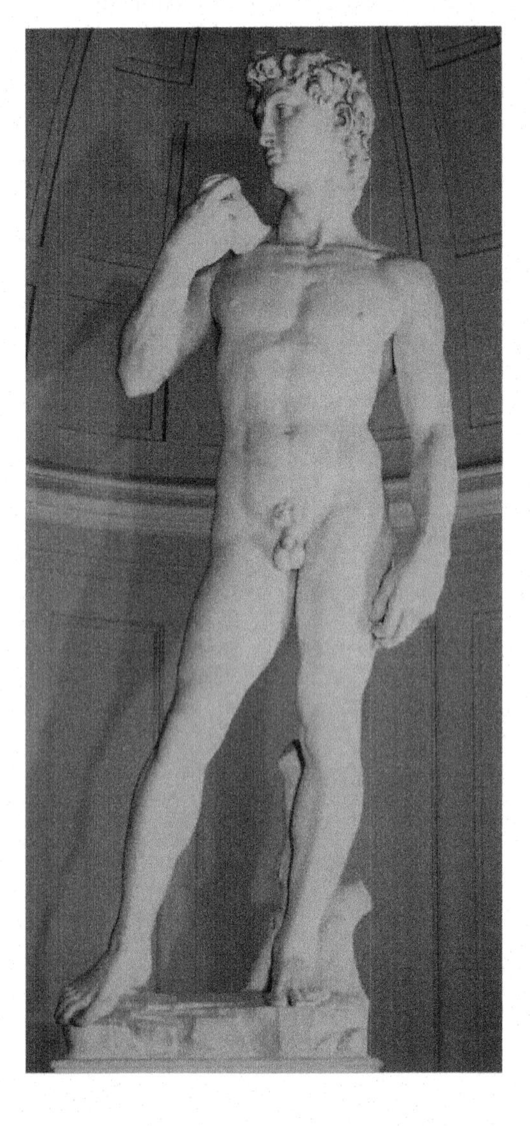

Michelangelo in a marble quarry, a painting by Antonio Puccinelli (1822-1897).

Someone else had selected the marble block for Michelangelo's David, but this was to change. Thereafter Michelangelo nearly always went to Carrara himself to select the marble blocks for his subsequent projects. For instance, in 1505, he was in Carrara eight times to purchase the marble for a sepulchre which Pope Julius II had commissioned him to do. He had received 1000 ducats, equivalent to almost DM 100,000, to buy marble. But Michelangelo was not happy with just any type of marble, no ordinario, no bardiglio, no arabescato or fantastico. He insisted on his statuario from the remotest and least accessible quarries in Monte Altissimo.

The fact that he had spent his childhood days among the quarry workers was now of enormous benefit. He was able to guide, sometimes even pushed them in their work without consideration to his own life and limb. On two occasions Michelangelo miraculously missed being crushed to death by falling marble blocks. He eventually found the statuario which he used, ten years later, to create "Moses" to adorn Pope Julius' tomb. The *cavatori* – the quarry workers – could not decide whether they hated or loved Michelangelo, but they were definitely very deeply impressed by him, as is indicated by what an unnamed cavatore said:

"Michelangelo is the god of Carrara – with the hands of a quarry worker, the eyes of an insane and a yearning for the infinite."

Even though Michelangelo dedicated the last ten years of his life almost totally to the Sistine Chapel, the statues he created during the first decades have influenced the work of successive generations of sculptors throughout Europe, even the whole of the western world. And, thereafter, statuario has always been regarded as the ultimate sculpture material.

Villa Rotonda, Vicenza (1566-1569).

Stucco marble

The history of stucco marble is nearly as old as the history of marble itself. It is closely associated with the time when pomp and grandeur were the order of the day. Before stucco marble reached its absolute pinnacle in the baroque period, it was already widespread in late Hellenism and late Roman antiquity, and subsequently rediscovered in the renaissance.

Italy and southern Germany were the main centres of stucco marble during the baroque period. Stucco gave rise to the plasterer's craft. Journeyman plasterers spread stucco marble to all European countries, and by the beginning of the 19th century the craft was also being practised in New York. However, this heyday was soon over with stucco marble being only occasionally encountered, for instance in Otto Wagner's Art Nouveau architecture in Vienna at the turn of the century.

Stucco marble is plaster of Paris, yet this comparatively cheap marble imitation often exceeds the diversity of colours and mottled effects of real marble. Gypsum, sizing water and colouring pigments were needed to produce stucco marble. This mixture was applied in a 1.5 cm thick layer which was then ground and smoothed up to eight times and finally oiled, polished and waxed. Success depended on the patience and skill of the plasterer so that this was where experience counted.

In the early days the art of producing stucco marble was a closely guarded secret with the result that the craft has almost died out. As there are only very few plasters left who are masters of this craft, seminars for stucco marble are being held nowadays in an attempt to maintain and revive stucco marble. Occasionally it can be encountered in new buildings. For instance, the "marble" columns in the Berlin Postal Museum in Germany are in fact made of gypsum.

What Michelangelo was to sculpture, Andrea Palladio was to architecture. In fact he is regarded as the greatest architect in 16th-century northern Italy. In his youth Palladio was apprenticed to a sculptor and stonemason. He enrolled in the corresponding guild and for the next 25 years he worked as a mason in workshops specialising in monuments and decorative sculptures. He became an architect at the age of 40 after he had travelled extensively throughout Italy where he had measured innumerable buildings of antiquity as a source of inspiration for his own work. His style of architecture was the ideal oneness of Renaissance and antiquity. Palladio did not propagate an anaemic, academic style of building. He took over the forms of antiquity and adapted them to his own times. His principal buildings were churches as well as palaces and villas for the urban rich and rural gentry – secular buildings that enjoyed an almost equal status to ecclesiastic and public buildings during the Renaissance (see figure).

Palladio not only rediscovered the art of building in antiquity. He even shaped the appearance of our present-day cities. His facades, ornamented with columns and temple gables, have left their mark everywhere, determining the Classicism of the 18th century just as much as the Historism of the 19th century, even influencing our perception of buildings in antiquity.

However, he was unable to re-establish the original importance that marble enjoyed in antiquity. Marble was merely an occasional decorative element for floors and wall facings. Either the patrons could not afford marble or there was a general lack of suitable craftsmen for this sensitive material. Be it as it may, Palladio's buildings had imitation marble facades, i.e. stucco marble:- brick walls covered with plaster that simulated marble.

More illusion than reality

Stucco marble is characteristic of the Baroque. Starved by the deprivations of the Thirty Years War, European aristocracy, or what was left of it, wanted to express its newly gained power and strength in a commensurate manner – be it the potentate of a small German principality or an absolutist ruler like Louis XIV who, with Versailles, created the ultimate of baroque art. Everywhere the baroque rulers were commissioning the most extensive building activity since the Roman Empire.

The demand for natural stone and other building materials increased enormously, and new knowledge of the properties of different materials resulted in an informed choice. Gone were the days when everything was used that could be found. The material was now being used in keeping with its properties: compression-resistant stones as supporting elements and tension-resistant materials such as wood for load-bearing elements. Marble, however, was rarely included. Suitable local stone was preferred, for instance in Salzburg the Untersberg and Adneter limestones as these could be easily polished to simulate expensive marble.

In spite of all their craving for grandeur and splendour, these absolutist rulers still followed economic concepts. Real marble was

Piazza Navona in Rome with the Fountain of the Four Rivers and the Triton Fountain are an impressive example of Baroque architecture. The piazza was created in a large sculptors' workshop directed by Gian Lorenzo Bernini.

expensive, and even its favourable properties as a building material could not lessen this fact. Hence, stucco marble was used for decorative elements inside and outside buildings. Not only was it significantly cheaper and easier to manage, but it also gave craftsman greater creative freedom. They could determine the colour and marbled effect and adapt the shape of the stucco ornaments to their concepts. Even mistakes made during work could be easily remedied. These were invaluable advan tages in the attempt to fulfil the most eccentric demands of their patrons. Consequently, anyone who now enters a baroque building can be certain that the supposed beautiful marble column at the back of the hall is actually gypsum. Make belief or illusion always triumphed over reality in Baroque.

This applies also to a certain extent to sculptures. Although still sculpted from real marble, the figures gradually lost their original, underlying character. Both columns and sculptures were reduced to mere building ornaments. Not only did they once again form an integral part of the architectural concept, they were also only conceived for viewing from one or two sides for the sole purpose of creating an impression as a pleasant ornamental element.

This showy, outwardly projected art was a reflection of the greatly inflated self-importance of the rulers. It is therefore hardly surprising that the economically successful bourgeoisie of the end of the 18th century increasingly regarded Baroque as unrestrained over-indulgence. By inventing Classicism, the enlightened bourgeoisie was able to disassociate itself from the aristocracy.

The spirit of antiquity

The excavations in Herculaneum (1738) and neighbouring Pompeii (1748) caused a sensation throughout Europe. The interest in antiquity was reawakened overnight, except that this time it was not restricted to the Latin heritage, as had been the case during the Renaissance, but rather to the rediscovery of all "Mediterranean styles". However, this did not go so far as to inject new life in marble as the material of antiquity. Cost reasons persuaded a more spiritual rediscovery, while stucco marble remained the order of the day.

Educational trips to Greece and Italy rapidly became a must for the wealthy bourgeoisie. The "History of the Art of Antiquity", published by Johann Joachim Winckelmann in 1764, soon became the standard literature of upper-class society. The "noble simplicity and serene greatness" which Winckelmann discovered in Greek art provided the aspired basis for a demarcation from the aristocratic lust for splendour and rapidly became the guideline for bourgeois architecture and sculpture.

The Marble Palace in Potsdam (started in 1787). Originally the brick building was to be entirely clad with Silesian marble, but the material proved to be too difficult to work with. Consequently, the facade relief was worked in sandstone and then overpainted to imitate marble. The belvedere was even a painted wooden structure.

The Glyptothek in Munich (1816-34) built by the architect Leo von Klenze, an ardent champion of classical architecture: "There has been, and there continues to be, only one architecture, and there will only be one architecture in the future, namely that which achieved a state of perfection in the Greek age of history and education."

The urge for demarcation was greater than the endeavour to actually follow the models of antiquity. Whereas the Greek temples and theatres were richly ornamented with sculptures, reliefs and friezes, all decorative elements with columns and gables were now dispensed with, thereby creating their own image of antiquity.

Since the statues and buildings excavated in Herculaneum and Pompeii consisted of white marble, Classicism advocated an antiquity that radiated in the gleam of white marble. Nor was this notion changed by Gottfried Sempers' paper "Preliminary Remarks Concerning Painted Architecture and Figures in Antiquity". Although the stucco marble facades were once again coloured, the concept of white marble sculptures has remained with us to this day.

Our concept of the style of sculpture in antiquity was also formed two hundred years ago. In spite of all their ornateness, baroque sculptors still developed their own style, whereas classicist sculpture merely copied ancient art. Although superbly accomplished, the immaculate beauty of the sculptures was often cold and empty.

The dilemma of the sculptors now become very much apparent. In the course of two thousand years of marble sculpture everything had already been done before. Sculptors had to break completely with tradition in order to create something really new. A new viewpoint, a new concept, that went far beyond mere copying of nature had to be found. But this also meant that sculptors had to detach themselves from their patrons. The sculptors were still wage-earners and not free artists.

Painters offered their uncommissioned pictures on the free market, as opposed to sculptors who were not only commissioned but also had to complete the sculptures according to precisely defined instructions. The patrons usually followed general trends of taste, i.e. Winckelmann and his Classicism, or they wanted to leave a memorial to themselves – as Napoleon Bonaparte who founded an academy for sculpture in Carrara in 1800.

The time had arrived for a break with tradition. After five thousand years, sculptors were now subdivided into craftsmen and artists. The craftsmen continued to be commissioned and were ranked among the stonemasons or were referred to as copyists. Their material continued to be marble and their style was naturalistic. Artists continued to call themselves sculptors, but they broke away from everything else. They no longer worked to order, nor did they copy nature but rather developed their own styles. They no longer worked exclusively with marble and selected widely differing materials. They often even rejected marble to ensure that they would not be mistaken for the copyists.

The vestibule of the Semper Opera, Dresden, Germany (started in 1870). The styles of historicism "have been borrowed and stolen, they do not belong to us" wrote Gottfied Semper. Yet this did not stop him from using a great diversity of wall and ceiling cladding: Marble incrustations alternate with Stucco lustro as well as oil and tempera paintings.

Modern without marble

The 19th century and thereafter is characterised by a diversity of styles in sculpture, painting and architecture. Baroque was the last uniform style. Classicism was no longer definitive for all areas. At the close of Classicism almost everything was permissible, as strikingly formulated in a circumscription of Historism: "The house is finished! Which style is to be added?".

The rich choice of styles was now matched by an equally wide range of building materials. In the middle of the 19th century the railways revolutionised transport and drastically diminished costs. Natural stone was now affordable, and a host of new building materials such as steel, concrete and glass, entered the market. The boom in the building industry and the number of new buildings achieved proportions that dwarfed the building activity of the baroque rulers.

These could have been the best preconditions for a marble renaissance considering that its properties as a building material were superior to those of all other stones. This renaissance happened, and yet it didn't!

It is true that the amount of quarried marble increased with each passing year, but the typical "local petrographic setting" of towns and rural communities of by-gone days was gradually lost as the new houses were no longer built with the same local stone. Yet marble was far from being once again the dominating element of facades and interior design, let alone for the building of entire houses. Even in upper class circles marble was only occasionally used for portals and stairways.

Commenting on this point, the contemporary German writer and journalist Victor Auburtin remarked:

"The feeling when walking on such Italian marble stairways is always very soul uplifting. You feel bigger and better when walking on marble. Weary of walking the paved streets, the soul soars to new heights, enabling you to breathe the cool palace air."

Stucco marble continued to dominate the facades, walls and ceilings. And in the event that an unusual material was to be used, then clients and architects usually opted for steel or concrete. Initially these were hidden by stone facings but, gradually, they came out of hiding and become the dominating element of facades and interiors. Steel and concrete, plus glass, were the materials with which the future was being built. Marble was a material of the past – a material reserved for memories.

Would-be emperors

It was by no means a coincidence that Benito Mussolini loved marble. The "Duce", the title assumed by the Italian dictator, dreamt of the glory of by-gone days and regarded himself a worthy successor to the Roman emperors. And since their glory had been eternalized in the white marble from Carrara, it was only natural that Mussolini insisted on having his "greatness" hewn in marble. Almost 1500 years after the fall of the

Roman Empire, Mussolini had a 17 meter, almost 300 ton marble block cut out of the "La Carbonera" quarry in Carrara. This monolithic block was intended for a memorial in Rome (see figure).

A wooden cage protected the monoliths on their long journey down into the valley.

The gigantic proportions of this monolith exceeded the capacity of the new diesel tractors. Consequently, the transport workers had to resort to the old methods of antiquity. Innumerable workers drove 30 pairs of oxen day and night through the mountains and used 70,000 litres of liquid soap to lubricate the wooden planks over which the block was slipped down to the sea – a painfully slow procedure that took half a year. The stone was loaded onto several barges that had been secured together and shipped to Rome where it was erected, together with 60 statues of athletes, on the new Forum. It was the world's largest monolith ever to have been sculptured from Carrara marble.

Mussolini's love of marble was also the outcome of rational considerations. As a result of the growing isolation and economic boycott of fascist Italy, marble production collapsed after its zenith in 1926 with the result that all work in the quarries came to a complete standstill. In such a situation the state invariably became the saviour of the Italian marble industry and the principal Carrara marble buyer for the next 15 years.

EUR, Palazzo della Civiltá e Lavaro, Rome.

Mussolini and his vassals inundated the entire country with their ostentatious buildings: be it the Esposizione Universale di Roma and Forum Italicum in Rome or the central law courts and railway station in Milan, marble facades had become the synonym of fascist monumental architecture.

In spite of any retrogressive perception of fascist architecture it was still characterised by certain modern traits in the initial years. This can be attributed to a certain extent to the fact that Tommaso Marinetti, one of the founders and spokesman of futurism, became a minister under Mussolini. This trend only ended with the growing influence of German nationalism in the 1930s. The ultimate in bad taste that was characteristic of totalitarian states such as Nazi Germany, but also the Soviet Union, had now also triumphed in Italy.

The buildings designed by Albert Speer, Hitler's chief architect, were primarily influenced by antiquity as interpreted by Neoclassicism in 19th century Germany.

This architecture was already quite void in spite of its elaborate forms, yet under Speer it was elevated to new, unprecedented levels of emptiness. The grandiose monumental structures impressed merely by their enormity, and their message was equally banal: This is the beginning of the thousand-year Reich with all its power and strength which will continue right into eternity. Even the impression that these "super" buildings would make as ruins had been taken into account – just in case the Reich did not last for a thousand years.

Albert Speer preferred natural stone for his buildings. Although the German Reich lacked noteworthy marble deposits, it could draw upon a diversity of limestones, all of which enjoyed a long tradition as a building material and had been the craze for facades in Berlin since the beginning of the century. Speer was quite happy to use local limestone for this purposes, particularly since it underlined the national character of the new movement, whereas steel and reinforced concrete were required for the war effort to build weapons and bunkers.

The greed of the Nazis for building material resulted in the virtual exhaustion of many limestone deposits, particularly since neither the time or the manpower were available to develop new deposits during the war years. The brief heyday of limestone in Germany, as well as marble in Italy, came to an abrupt end with the total annihilation of the Nazis and their allies.

The victory of the allies was also a victory for cultural diversity over the supposed national character of the arts preached by the Nazis and the fascists. The newly gained freedom enabled different concepts and styles to prosper side by side. And the same applied

Statues, wrapped in sheeting to protect them from dirt, awaiting their dispatch.

The birth of a house - a sculpture of Wilhelm Scherübl at the Krastaler Sculptors' Symposium, Gross Glockner 1999.

This sculpture in local marble in the Austrian Krastal forms the focal point of the annual Sculptors' Symposium.

to the great diversity of materials with which architects and sculptors endeavoured to express themselves. Marble had now become just one of many materials, even though it was tainted with the aftertaste of ostentatious luxury.

The end of a myth

The artists' relationship with marble was, and has remained, very mixed. "Nothing can follow Michelangelo! – Marble is both noble and presumptuous. – As a result of its history marble is permanently associated with sacral and glorifying works." This is how, in 1969, Italian sculptors justified their rejection of the material that was predestined for sculptures and statues, and this attitude has prevailed to this day. Only now is marble regaining a foothold in an area in which it had dominated for so many centuries.

Copies of old originals and bric-à-brac are still the primary reason for resorting to statuario – primarily from Carrara which accounts for almost 90 per cent of all the statue marble used worldwide and where the best copyists live and work in the immediate vicinity of these deposits. For instance, the Instituto d'Arte S. Stagi at Pietrasanta still teaches all the skills and techniques that stonemasons and sculptors require to work with marble.

Even a simple copy of Michelangelo's marble statue of David requires the very highest skills to ensure that a sculptor can maintain full order books: 1 "David" for a cemetery monument in Los Angeles (to replace the one destroyed by an earthquake; 1 "David" for a shopping centre in Australia; 1 "David, together with 1 "Pieta" and 1 "Pauline" of Antonio Canova, for the private museum of a millionaire in Taiwan.

As modern artists are always overshadowed by the old masters, it is therefore hardly surprising that they only occasionally use Carrara marble. For many years Henry Moore lived in the vicinity of Forte dei Marmi; Niki de Saint-Phalle placed a "White Nana" of Carrara marble alongside her "Black Nana" painted polyester sculpture; and kitsch artists like Jeff Koons also 'discovered' marble.

Only few artists work with marble since Statuario is also a cost factor. Simple marble from Carrara can cost as much as DM 700.- per ton, and there are virtually no limits for Statuario. The individual blocks are traded like the greatest vintage wines and, just as the location of a vineyard, so does the location of a quarry determine the price.

Long forgotten works of the Renaissance and Baroque suddenly re-emerged, for instance when the great sponsor Lorenzo de Medici commissioned Michelangelo to produce a tomb. Today the sponsor is called Silvio Berlusconi, the sculptor is Pietro Cascella, and the tomb has become a small mausoleum in Arcore, a village near Milan.

La Grande Arche, Paris.

Statuario and the fine arts are one aspect, but to earn money with marble is a totally different matter. In spite of the astronomical prices for individual blocks, statue marble is more a recollection of long past ages rather than a profitable business. This is because the discovery and quarrying of suitable blocks is far too complex, and the quantities sold far too small. This contrasts with the vast quantities of "normal" marble that are quarried and sold, and the amount is continuously increasing with each passing year. The Carrara marble industry is healthy and there are two principal reasons for this situation.

Ever more living space is required to keep pace with the continued growth of the world's population. The building industry is booming, and the demands expected of the quality and size of homes are increasing with each generation, at least in the USA and western Europe. Consequently, the demand for all types of building materials, including marble, is increasing. And the days when marble was restricted to the western cultural hemisphere have long gone. To the same extent that the western market economy conquered the world, so did the cultural peculiarities of the West reach into the remotest corners of the world where marble became the building material of the ruling classes.

It is therefore hardly surprising that since the middle of the 1970s countries like Saudi Arabia, Kuwait and Oman rank at the top of the list of marble consumers. This is where medieval political structures with their autocratic feudal lords still prevail. Oil has made these countries immensely rich, and marble is the building material of their choice. For instance 150,000 square meters of marble panels were used for the Dshidda airport in Saudi-Arabia. Even the USA and western Europe cannot keep up with this, in spite of the fact that the concrete years are over and the importance of natural stones is once again being revived. What were once the Basilica of St. Peter and the Hagia Sophia are now the palaces, banks and casinos in steel and concrete clad with slabs of white Carrara marble.

"Marble ennobles" is the first of the "three basic principles of sentimental architecture" as formulated by the architecture critic Benedikt Loderer. And this ironic esteem obviously applies to the white marble from Carrara as is evidenced by such structures as "La Grande Arche" in Paris which French President François Mitterand had built in 1986 in the fifth year of his rule. Or the Shri Swaminarayan Mandir Hindu temple, built in the north-west of London and completed in 1995, which the BBC called "England's Taj Mahal" on account of the 2000 tons of Carrara marble that were used for its construction.

Rich bourgeois clients assigned the same objectives to simple marble, be it coloured, veined or mottled. Just as their 19th century ancestors, modern citizens wanted to have their economic success expressed or "ennobled" in a commensurate manner to indicate the rank they hold in society, and marble helped them to achieve this aim. To make sure that these affluent clients can enjoy their marble for a long time they are nowadays supplied with a guarantee. All large marble producers have their own laboratories which supply precise values for the resistance of the given marble to weathering influences and exhaust fumes, and to develop appropriate protective measures.

The number of social climbers is considerable, yet even greater is their hunger for the exceptional, as embodied by a stone that continues to sustain the myth of richness and power. However, something that everyone wants to own can no longer be exquisite, and vice versa. Every newly quarried piece of marble breaks down the marble myth in favour of mass production. And since annual quarrying records are no longer sufficient, the Italians broadened the definition of marble just as the Romans did in by gone days. Nowadays, any stone that is reasonably polishable is considered to be "marble": dolomite, alabaster, onyx, travertine – the list is endless. And clients are not worried as long as it is a natural stone that is beautiful, ennobling and not too cheap. And even the actual use is not really important. The people living in Carrara have carried this attitude to extremes. Everything in the villages around the quarries is made of marble, be it fireplaces, window ledges, stairs, floors and even the feeding dishes for cats and dogs. After all everyone is surrounded by marble.

Even this somewhat remarkable, sometimes nonsensical and usually ostentatious use of marble as an everyday commodity is not new in the long history of this stone. It has all happened before since the scope for the use of marble is much wider than just a material for sculptors and architects.

Marble as a material for everyday use

The Egyptians had very little marble, and what they did have was used to produce small vases and cult objects. More than one thousand years later the Cretans conducted

Fragments of the world's oldest limestone dish from a farming settlement in the pre-ceramic neolithic period (6000 BC) found in Wadi Ghuweir, Jordan.

their cool mountain water over marble slabs to give them the very first "refrigerators". For the next thousand years there are no records of the day-to-day use of marble. Nor have any objects of everyday life from ancient Greece been found, even though they used large quantities of marble. Perhaps they preferred ceramics and earthenware in their homes. It was the Romans who used marble as an everyday commodity and therefore introduced mass production.

Bowls, mortars and worktop slabs were the most important everyday objects for which marble was ideally suited. On the one hand it was sufficiently soft to be easily shaped into any required form, yet on the other hand it was hard enough to permanently withstand normal mechanical stresses. Moreover, simple polishing gave it a smooth, closed surface that could be easily kept clean.

This was one of the reasons why marble was so extensively used in Roman baths. Another reason was the penchant of the Romans for luxury:

"You feel poor and miserable when the walls do not radiate with precious round disc ornaments, if Alexandrian marble does not alternate with Numidian mosaic panels and everything is not surrounded by rich painting-like adornments, if the vaults are not hidden behind crystal, if the baths in which we cleanse our bodies after intense sweating cures are not enclosed in white marble from Thasos – at one time seen only occasionally in one or the other temple – and if the water does not flow out of silver taps."

SENECA, EPISTULAE MORALES, 86.6

Onyx marble is neither onyx (a crystalline quartz) nor marble, yet this fresh-water calc sinter was widely used in ancient Rome. The mineral could be cut into discs that were so thin that they appeared to be almost transparent. The rich patricians used the likewise transparent gypseous spar mineral as a substitute for window panes – a form of use that continued right into the Middle Ages when glass finally replaced all other materials. Ever since then transparent onyx marble has only been used in lamps and for art and crafts objects.

Canon balls made of white marble were cheaper than those made of iron. Moreover they were better than other stone balls since their homogeneous structure prevented them from splintering when they were fired. Hand-made canon balls have been serially produced in Carrara since the 15th century and they were sold individually.

The use of marble for practical purposes also started during this period. Thus, in the 16th century, mortars and floor tiles where the most important products that were made from Carrara marble. However, output cannot have been extensive since marble production during that century fluctuated between 500 and 1000 tons per year – and this figure includes the canon balls that were being made in Carrara since the 15th century (see figure).

In medieval Germany the balls were smaller, in fact less than 1/2 inch in diameter, and they were called "Marbeln" in Old High German to distinguish them from the cheap clay marbles with which children played. The special marble mills evolved into an entire industry on the southern slopes of the Thuringian forest. But just like window panes, glass proved to be a more appropriate material for 'marbles' so that the output of marble balls came to an end by the 19th century.

Even so general marble quarrying continued to reach new record levels. With the growth of prosperity and the increasing de-

Solnhofer limestone and lithography

Litera scripta menta" - That which is written down is permanent. However, permanence depends greatly on the material so that it is hardly surprising that, right from the outset, people wrote on marble. With the spread of cheap paper that was far more practical for this purpose, marble lost its importance as a writing base. Nowadays, only tombstones and inscription panels are a reminder of a great tradition of bygone days. Marble, or in this case polished limestone, can not only bear inscriptions but its function can also be reversed, i.e. it can be used to "write"! This was demonstrated by Alois Senefelder in 1796 when he developed lithography - to 'write' or print from stone - thereby revolutionising printing technology.

A fossil longbill fish (Aspidorhynchus acutirostris), found in Solnhofen; length 76 cm.

Senefelder's invention involved the use of a lipophilic (grease soluble) ink to engrave the text that is to be printed on a limestone plate. The engraved text held the lipophile ink, whereas the untreated moist parts of the stone repelled the printing ink. Senefelder was thus able to produce printing and non-printing areas on a printing plate. Until then this could only be accomplished with the very elaborate gravure and letterpress printing processes. Moreover, a lithographic stone could be ground down so that it could be used several times, thereby reducing printing costs still further.

Senefelder found the material for his printing plates in Solnhofener limestone which could be easily quarried for plates of varying thickness, similar to slate. The Solnhofer deposits had long been known. The Romans used this stone, and in the Middle Ages plates from Solnhofen were used for the floor of Hagia Sophia. The favourable location in the Altmühl valley provided a direct waterway link to the Danube so that transport costs were low.

Only plates of bluish-grey colour and at least 5 centimeter thickness were suitable for lithography because they were particularly hard and had a very fine, uniform grain. To find a suitable lithography stone the workers struck the stone to establish its tone. Just as in sculpture work, if it rang with a light clear tone then this indicated a faultless stone which fetched the highest prices. In 1920 one had to pay $ 540 in the USA for a plate of 125 x 175 cm size. Such large formats were used to print maps and posters. However, such stone plates were available in different sizes, the smallest ones being 14 x 16 cm (see figure).

The trade in lithography limestone was profitable. Almost 9,400 tons of this material were sold alone in the record year of 1907. However, the development of offset printing quickly displaced this antiquated technique. However, the quarry owners still had another attractive and quite unusual source of income. Solnhofener slaty limestone proved to be a worldwide unique location to find all kinds of fossils. They are in big demand among museums and collectors on account of their outstanding condition. The eight Archeopteryx skeletons found here hitherto certainly resulted in a 'windfall' profit for the owners.

A lithographic stone with a caricature of the Swedish cartoonist Janne Graffmann (1871-unknown).

mands expected of comfort and hygiene, the bathroom in the houses of the rich became the focal point of attention. Wash-stands with a marble top became a must, and those who could afford it used marble for the bathroom walls and floor. The hospitals contributed significantly to these increased demands since without the wide scale use of polished, easy-to-clean marble it was hardly possible to fulfil the new more stringent hygiene regulations.

Marble – and this includes limestone that can be polished – experienced a boom when drastically reduced transport costs and the systematic development of new deposits made marble affordable. This was the age when the craftsmen were specialised in working with marble. Marble turners and marble grinders were two of the new trades that arose at the time. Moreover, new stone sawing mills were being equipped with especially developed marble grinding and polishing machines. In addition to bathroom furnishings, these mills produced tabletops, worktops, wall panels, mortars and even switch panels. The product range also included christening fonts and tombstones.

The heyday of marble as an everyday utility material was in the years following the First World War. The Second World War resulted in a dramatic collapse since plastics and stainless steel were then being used in hospitals, kitchens and bathrooms, while linoleum and tiles replaced marble on the floor and walls. Marble was unable to compete with such materials. The marble trades disappeared and the independent Association of German Marble Works was integrated into the Natural Stone Association.

Recent years have seen a small revival with the result that everything from the washbasin to complete bathrooms is now once again available in marble. However, this upsurge is not the result of its functional properties but rather the desire for luxury. Consequently, the interior architect and designer is now concerned with marble and no longer the craftsman.

The filled grinding passage of a marble mill. "The little glass or clay balls with which children play are called marbles because, as the name suggests, they were originally made of marble" (from the German Dictionary of Jacob and Wilhelm Grimm).

The future of marble

Is marble to be a stone of luxury for the chosen few or to be transformed into a mass produced commodity without any aesthetic appeal? The future of marble is uncertain, even though there are no alternatives looming on the horizon. Consequently, if this stone is to regain its prestige value which does justice to its natural beauty, then different courses must be adopted. The functional properties of marble must be once again combined with its decorative and ornamental elements. It must be used sparingly in harmony with its environment. It must be given space so that it can "shine"! Perhaps marble may have to disappear for a while so that, later on, we can rediscover it – just like travertine.

Travertine was a greatly coveted building material in ancient Rome. This material was used for the Colosseum and the original St. Peter's Church built under Constantine. In the subsequent centuries it became just one

Getty Center, Los Angeles.

of many stone materials, but it has now made a great comeback.

On 13th December 1997 the Getty Center opened its doors. It was built by the top American architect Richard Meier between the slopes of Los Angeles and the Pacific coast. He selected Italian travertine building material. Almost 100,000 cubic meters of this rough, honey-coloured stone had to be shipped half way around the world to build this new centre for the world's largest private art foundation.

The expenditure was well worth it as the building is now surrounded by an aura of timeless elegance, and comparisons are being made with the Acropolis and Hadrian's Villa. Some are now looking upon California as the new country "where the lemon trees bloom" as Italy was once referred to by Goethe.

It is difficult to predict whether, after a period of abstinence, marble will 'flower' once again. In any case an extended interval would help to preserve the supply of high-quality marble. Estimates suggest that Carrara marble supplies could last for anything between 400 and 10,000 years. Moreover, not all types of marble from Carrara can satisfy the most fastidious demands. Further more the growth of the world's population makes it extremely difficult, if not impossible, to make any forecasts. However, one thing is certain: Pliny's hopes were very much mistaken:

"Among the many other wonders of Italy, Papirius Fabianus – a man who was very experienced in natural history – reported that the marble in quarries is increasing; even those who quarry the stone claim that each quarry will eventually fill up again. Should this be true then there remains the hope that there will never be a lack of this material for luxury."

PLINY, NATURAL HISTORY, XXXXVI, 125

III.

Calcium carbonate – a modern resource

by Johannes Rohleder and Eberhard Huwald

The demands expected of modern products are continuously increasing. This applies to high-tech equipment just as much as to everyday products. For instance, paper is expected to be ever whiter and glossier, and plastics ever thinner, yet still very strong – this list could be continued indefinitely. However, every improvement requires new machinery and processes, as well as raw materials that are optimally adapted to the new challenges. Calcium carbonate is such a modern raw material and it has become indispensable as a filler and coating pigment.

This mineral gained immense importance in the years after World War II, primarily as a result of the rapid progress made in processing techniques. Without new crushers, sifters and mills the super-fine calcium carbonate powder would be unthinkable. However, the beginnings of calcium carbonate as a filler go back a long way.

1. The beginnings: Calcium carbonate in glazing putty and rubber

The Industrial Revolution had many features, one of them being the steady growth of consumption of all kinds of raw materials. Consequently the demand for chalk increased continuously during the first half of the 19th century. In Champagne there was a big demand for chalk building blocks, and the sale of chalk as a cheap colorant also reached considerable proportions. Dyehouses and printers were demanding chalk on an ever-growing scale, and the chemical industry was using this mineral to neutralise acids or to produce carbon dioxide.

Demands were now on the increase. However, to meet these continuously rising demands chalk quarrying had to be adapted to the new economic situation. It was no longer

Draining the sedimentation basins on the Island of Rügen. Chalk drying shelves can be seen in the background.

sufficient for local peasants to extract chalk at times when they did not have to work on the farm. Large chalk quarries had to be developed by entrepreneur operators who were exclusively concerned with quarrying chalk.

1.1 A chalk industry is born

Friedrich von Hagenow was the first to venture into this line of business when, in 1832, he acquired the sole rights to the Stubnitz chalk quarries on the German island of Rügen in the Baltic Sea. Documentary evidence of the first business activities in Champagne and Söhlde near Hanover dates back to the years between 1820 and 1840. And these new quarry operators not only extracted the chalk on a large scale, they also started to process the chalk on the site or at least in the immediate vicinity of the quarries.

This was a gradual process. For instance, in the beginning of the 19th century there were still numerous processing plants in Berlin. However, by 1843, they had all disappeared. As the chalk suppliers adopted industrial production methods, the processing plants were relocated to the large quarries. Consequently, Friedrich von Hagenow operated his own plants on Rügen, and at Stettin and Greifswald, where his Rügen chalk was processed.

This was a profitable business since the price for processed chalk was significantly higher than for crude chalk. One hundredweight of dry chalk cost 16 groschen or ½ thaler in 1790, as opposed to crushed chalk which cost 1 thaler and 8 groschen, and chalk slip or "whiting" 2 ½ thaler for one hundredweight. The cost of processing was not particularly high.

Processing

Plants for the production of fine chalk powder are relatively simple. Chalk is a natural mineral, often containing unwanted clay or marl layers, as well as flint and silicates. All these impurities must be first removed by a simple washing process which exploits the fact that the polluting minerals have a higher density than chalk.

The crude chalk is first crushed by large hammers. Some plants also operated pounding mills for this purpose. Workers then sorted out the coarse impurities manually, after which the crushed chalk was sifted and finally fed into a mill with plenty of water. The stone mill ground the crushed chalk down to a particle size that was so fine that it remained suspended in the water. This homogeneous suspension of chalk and water was conducted through zig-zag shaped wooden troughs. The heavy polluting particles sank down while the suspended chalk flowed away into large settling basins. The chalk can be classified into fineness grades by a series of settling basins arranged one behind the other. The largest chalk particles settle in the first basin, while each subsequent basin holds chalk of increasing fineness.

These basins were often simple pits in the ground or wooden barrels which held the chalk suspension until all chalk had settled at the bottom so that the now almost clear water could be returned to the washing process. The workers then formed the thick chalk slurry into loaf-like blocks and placed these on dry chalk blocks to remove part of the water contained in the slurry. The predried chalk blocks were then stacked on drying shelves where the remaining water evaporated out of the blocks in the course of days or weeks, depending upon the weather and season. The chalk slurry was then either sold en bloc or broken up in an edge mill and packed into sacks or barrels.

Apart from some minor and major deviations, this process was more or less standard in all chalk quarrying areas. English chalk factories, for instance, reduced the chalk in edge mills rather than with grinding stones. Moreover, in English, the processed chalk was then referred to as "whiting". In Rügen, on the other hand, chalk processing was based on a thin slurry process in which the mill was replaced by a large mechanical stirrer. This consisted of a large vat with a mechanically powered cross that carried heavy

Slurry edge mill (1902).

iron slurry hooks which turned inside the vat. The vat was filled by more than 80% with water, thereby pulverising the chalk (see figure).

As simple and effective as this slurry process may have been, it had one major disadvantage. Since neither the crushing facilities nor the drying rooms were heated, slurry processing could only proceed during the frost-free months. During the winter months it was only possible to quarry and store chalk. Otherwise production was at a standstill.

Slurry machine (1902).

In Champagne a different processing method was practiced. Since the local chalk deposits are almost free of impurities, it was quite sufficient to weather the quarried blocks to dry them out. The blocks were then brushed with large metal brushes to produce a fine chalk powder for immediate use. This process is known as "dusting" and it does not require any water. Consequently, production can continue throughout the year provided that sufficient quantities of chalk were dried during the summer months.

Windmill equipped with a dry edge mill to grind chalk, Söhlde (1914).

Irrespective of the practiced processing method, the level of mechanisation in the factories was very low. Only the grinding and pounding mills were driven by horses – occasionally by water or wind power (see figure).

And it was not until the year 1873 that the first steam engine was used in chalk processing.

For a long time chalk continued to be quarried manually, be it by workers suspended down the face of the quarries as on the island of Rügen or standing on the quarried benches in Champagne and Söhlde.

Chalk quarry in Söhlde (1908).

The funnel slot chute process was the customary method of chalk quarrying on Rügen.

Initially, the situation was not so favourable in Champagne. Although most chalk quarries were located alongside the River Marne, thus ensuring cheap and fast transportation to Paris (see figure), the route to the Atlantic was long so that there was little access to foreign markets. This only changed with the opening of the Rhine-Marne Canal in 1850. This gave the French quarry operators direct access to the German market. The number of factories rose rapidly, simultaneously multiplying the associated problems.

Favourable transport conditions were an essential precondition for the speedy development of the chalk industry. The chalk regions in England and on Rügen enjoyed the enormous benefit of having direct access to the sea. The chalk quarries on Rügen and the large processing plants in Greifswald and Stettin were all located in the immediate vicinity of the Baltic Sea. In England the chalk quarries of Gravesend in Kent and the processing plants were all located around the Thames estuary for direct access to the North Sea.

Transshipping chalk from canal to Rhine barge.

The search for markets

Even though the market for chalk and chalk products increased, the number of suppliers did not grow correspondingly at the same speed. The lower transport costs enabled individual quarries to supply more distant markets. Competition became fiercer and prices dropped. For instance the price of one hundredweight of medium quality whiting in Berlin was 2 ½ thaler in 1790. By 1830 the price averaged 1 ½ thaler, and ten years later it had dropped to 20 to 25 Silver groschen, i.e. it was little more than half its original price. This greatly diminished the profit margin so that only those quarries that achieved high turnover rates year after year could survive.

This was not a problem on Rügen. After the chemist Hermann Bleibtreu set up Germany's first portland cement works in Stettin in 1855, rising sales were guaranteed for the forthcoming decades. More than 80 per cent of the chalk quarried on Rügen was used by the cement industry around Stettin. Since the chalk was also processed locally, it was also possible to save the cost of wash-milling.

The other chalk regions did not have such a clearly defined long-term market. They remained divided up into innumerable small segments right into the second half of the 19th century. Chalk could be encountered virtually everywhere: as a pigment or extender in water, and occasionally also in oil paints; bonded with glue to produce drawing and blackboard chalks; in innumerable cleaning agents on its own or in combination with soap; in roofing felt and art paper; in numerous products of the chemical and printing industries; and occasionally chalk powder was even used as a cheap means to forge expensive white powdered substances such as flour or super-phosphate fertilizers.

However, none of these markets offered the prospect of successful long-term growth because sales fluctuated within wide limits and the competition resulting from newly founded companies was fierce and ruinous. New long-term growth markets were necessary for the further development of the chalk industry. And these arose with the industrial manufacture of glazing putty and the rapidly expanding rubber industry.

1.2 Rubber and glazing putty

Top quality glazing putty consists of a mixture of whiting and linseed oil. And this simple formulation has remained unchanged, even after large scale industrial production ceased and only restorers occasionally require small quantities of putty for their work. When this formulation first arose remains unknown but it must definitely be a very long time ago.

The customary method

Both chalk and linseed oil were being used by medieval painters so that, at some time, they must have noticed that when they mixed their paints, chalk and linseed oil formed a plastic substance that became hard within days or weeks. However, this substance was rarely used in painting because of its low hiding power. And the fact that its plasticity was ideal to seal windows against rain, dust and air was irrelevant. Glass windows were not widespread and glazing putty unknown. In his paper "Schedula diversarium arium" published in 1170, Theophilus Presbyter never referred to putty even though he described in great detail everything associated with glass painting.

Whether or not lead framing in medieval and early modern times was sealed is still disputed. Some art historians claim that glaziers only placed glass panes in lead frames which were then sealed by pressing the lead fold together and by adding a few drops of molten lead. Others maintain that lead framing also had to be sealed and that clay or resin, soaked bread or paper strips, were pressed into the gap between the pane and frame. However, since all church windows – other windows were irrelevant in this context – from medieval and early modern times have all been renewed on frequent occasions in the course of the centuries, and the restorers used the customary sealing materials of their given period, this question can no longer be clarified.

The first written mention of a sealing substance is recorded in the accounting books of Salisbury Cathedral in the south of England. Accordingly, the cost in 1531 was as follows: "For settyng 48 ft. of old glas in new lead, the price of a foot with sement 2 d., and without sement 1 1/2 d." The composition of "sement" is not described, but in view of the significant costs associated with sealing, it was not used whenever it could be dispensed with.

These accounts remained the only source in literature until the 18th century when manuals on glass painting, technological encyclopaedias and polytechnical papers, all suddenly referred to window putty. The first reference can be found in the second volume of "Detailed Instructions on Bourgeois Building Art" published by Johann Friedrich Penther in 1745:

"To prevent pelting rain, air and the cold from penetrating the space between the glass pane and the wood in which the glass is framed, the English have invented a putty that is brushed into the gap so as to exclude rain, air and the cold."

Eighteen years later the "Leipziger Intelligenz-Blatt" described a recipe for the production of window putty:

"Parisian window putty is produced in the following manner: 7 pounds of linseed oil, together with 4 ounces of ground umber, are intensely boiled; and while this is still hot add 2 ounces of yellow wax and reheat everything once again. And then knead in 5 1/2 pounds of ground white chalk and 11 pounds of lead white."

As a result of the high share of lead white and other drying agents, these early types of putty dried very rapidly with the result that the panes could be subjected to considerable tensions and even be shattered. Gradually the share of chalk increased, the putty become more pliable and cheaper. And when, in the 19th century, more and more bourgeois houses had glass windows fitted, and wood replaced lead framing, the demand for cheap glazing putty rose sharply. The age of chalk had finally arrived.

The "Technological Encyclopaedia" of Johann Josef Prechtl from the year 1836 gives a detailed description of how glazing putty is produced from linseed oil and chalk and how it is properly stored:

"Sealing (now the customary method) is usually completed with glazing putty produced from old linseed-oil varnish (linseed oil boiled with red lead or lead oxide) that is kneaded with finely crushed chalk in a mortar until, after a short while, it reaches a certain measure of hardness. Kneading in the mortar must be continued until the substances are intensely mixed and has reached a viscosity that can still be easily kneaded and applied with the fingers without crumb-Oling."

The finished putty should be moulded into "lumps and stored in a cool place tightly wrapped in wet canvas or a wet ox bladder". Before it is used again it should be heated and kneaded; some linseed oil varnish can be added to make it pliable again.

Sealing with putty had now become an everyday job and glazing putty made of chalk and linseed oil was the most frequently used type.

But just like the painters who mixed their chalk colours themselves, so did the glaziers produce their own putty in conformity with their requirements. Industrially produced putty that could be purchased and stocked until it was needed had not yet appeared on the market.

However, this was only a matter of time as the market for glazing putty grew. Moreover, the work and expenses associated with industrially produced putty were low as the process did not require any special skills nor complicated, multi-stage production plants. All that was needed was the ability to mix the right quantities of chalk and linseed oil into a homogeneous mass of the right consistency – and this requires only simple machines.

View of a putty factory around 1930 (Switzerland).

Glazing putty as a mass-produced product

The initial equipment used for putty manufacture followed the manual method so closely that it could hardly be considered to be a machine. In fact it was an over-dimensioned mortar consisting of an iron-clad mixing trough and a large, elastically suspended iron pestle so that each time the pestle was pushed down it sprang back to its initial position. Even though this eliminated one operation, the amount of manual work still remained very high. And since only 20-50 kg of putty could be produced per batch, three to four pestles were soon arranged one behind the other, similar to a pounding mill, and the entire assembly was powered by steam or water. The putty was finally passed through a roller mill to produce a uniform, well kneaded putty. The disadvantage of this method was the very considerable operating noise and that one worker still had to stand alongside each pestle in order to replenish spattered material.

The first fully automatic machines appeared on the market towards the end of the 19th century. One of them was the kneading machine made by Werner & Pfleiderer in Cannstatt which was similar in design as the dough-making machine used by bakers. The machine consisted of a large mixing trough in which a rotating mixing blade kneaded the putty. The chalk and linseed oil were jointly fed into the machine, and the lid remained closed during operation. Kneading continued for some 30 minutes after which the mass was passed through a roller mill to produce a smooth putty.

The last operation was eliminated when an edge-runner mill (see figure) was used for mixing and kneading. As is the case with all edge mills, two granite edge runners project into the dish-shaped grinding track. Each rotation not only mixed the putty but also rolls it into a smooth consistency. Moreover, a major advantage of this machine is that large chunks of chalk can be fed in to grind them down to fine powder before the required quantity of linseed oil is added.

The finished putty is finally transferred to a moulding machine where it is pressed into ½, 1, 2 and 5 kg blocks ready for dispatch. Large wooden crates were normally used to dispatch 5, 10, 12 ½, 25 and 50 kg. To exclude the air from the putty during transport, the blocks were wrapped in wet parchment paper, and the crates were additionally sealed with a coat of water-glass paint. For extended storage the putty had to be kept in a cool and dry place.

Until now the experience of the workers was the decisive factor in the manufacture of quality putty. However, optimal operation of new machines presupposed a systematic approach to the problem. Glazing putty is a sealing compound that dries by oxidation. Its cementing effect is the result of a reaction between the linseed oil and the oxygen in the air by which a densely cross-linked resin-like polymer is formed. The purpose of the chalk in this substance is to transform the liquid oil into a readily pliable substance that retains its moulded shape. Moreover, it functions as a filler to give the polymer body and, finally, to reduce the share of expensive linseed oil in the putty.

Consequently, it is the aim of all putty manufacturers to incorporate as much chalk as possible in the putty. However, if the putty contains too much chalk it will crumble and can no longer be used. If the putty does not contain enough chalk then the outside will

Edge mill with rotating bowl and granite rollers manufactured by J.M. Lehmann.

N: 1.

Fabrikationbuch. 1906.

Kittproben 16. 3. 06

N: 46.

25 gr. Leinöl à 45. } = 16% Oel
20 " Kittöl Wertheimer }
235 " Blanc de Troyes à 3. 84% Kr.
280 " Kitt
7 " Kreide 100 k = 8.36
287 gr. Kitt

Scheint gut aber fataler Geruch

Bemerkungen: 17.4.06 bleibt weich ist gut. Geruch verschwunden. Farbe weiss

N: 47.

15 gr. Leinöl à 45 = 6.75
15 " N: 58 = Ölfirnis à 46 = 6.90
20 " Vas. Oel à 24 = 4.80 } = Oel 13.6%
10 " " Kuhn M B à 30 = 3.–
2 " Seife à 35 = .07
30 " Baryt à 30 = .90 } = B. 6.4%
368 " Bl. à 3 = 11.04 } = Kr. 80%
460
33.46 : 460 = 7.28

100 k = 7.28

Bemerkungen: 16.4.06 Außen fest und innen ... Farbe gelblich

Analysis of a typical glazing putty formulation. The examined sample contained 84% chalk (blanc de Troyes) and 16% oil.

harden very quickly, but the inside will remain soft for many months with the result that even the slightest pressure will deform its shape.

Good glazing putty consisted of 84-88 per cent chalk and 12-16 per cent linseed oil (see figure). 0.1 per cent phenol was additionally added to the putty to prevent mould formation. Small quantities of siccative, e.g. manganese borate, lead oxide, red lead or zinc white, were added to the putty to produce rapidly drying grades of putty. To produce black putty some of the chalk was substituted for manganese dioxide (brownstone); and red lead was added to produce red putty.

Roller mill to crush linseeds (1894).

The reason why the recipe did not specify exact quantities is due to the fact that chalk and linseed oil are natural substances. Consequently, their composition and purity can change with each delivery. Nevertheless, this mixing ratio did permit continuous operation of the machines. The workers no longer had to repeatedly add chalk or linseed oil during the mixing process. And each time chalk or linseed oil was added, the machine had to be stopped. Now it was only necessary to check the consistency of the product once towards the end of manufacture.

As the amount of equipment needed for putty manufacture was minimal, the economic success of a company depended primarily on how it was supplied with raw materials. Linseed oil from expressed linseeds was supplied by oil mills (see figure). The seeds were first ground in edge or roller mills and then transferred to connected presses. The linseed oil expressed in this manner was then either used directly, or boiled together with siccatives to obtain boiled linseed oil for the production of fast-drying putty. Pure linseed oil putty was easier to use and it became just as hard after a corresponding drying period.

Good quality chalk has to be an amorphous, soft and mellow fine slip and absolutely dry with a high oil absorption value so that it can bind as much oil as possible. That is why the chemically identical calcareous spar could not be used as a substitute for chalk because, even when ground to a very fine powder, it still remained crystalline so that it could only absorb a relatively small amount of oil with the result that the putty became brittle. Champagne supplied the finest chalk for putty, but Rügener and Söhlder chalk was also suitable provided that it was properly processed.

To maintain a continuous supply of high-quality raw materials, the manufacturers of glazing putty acquired chalk quarries together with the local processing plants, and they also operated their own oil mills. In this manner they were always able to supply customers with putty of unvarying quality. By the beginning of the 20th century industrially manufactured glazing putty had asserted itself on the market.

Masticating and vulcanising

This period was also associated with a rapid increase in the consumption of natural rubber which was processed with chalk that was used as a filler. Caoutchouc, also known as latex, is the dried sap from a number of tropical plants, the most important one being the Hevea brasiliensis tree, also known as Hevea. During his travels through America, Christopher Columbus became familiar with an elastic resin that the local Indians produced from Ca-o-chu, the "crying tree" (see figure). Columbus brought it back to Europe, but this sticky substance did not keep and was soon forgotten.

Interest in this substance was again renewed at the end of the 18th and beginning of the 19th century when Thomas Hancock discovered the mastication of caoutchouc in 1820, thereby creating a market for rubber articles. Mastication involved mixing, kneading and rolling the crude caoutchouc in a rolling or mixing machine to make the substance soft and elastic. At the same time the masticated caoutchouc could take up fillers and other auxiliaries. This property was exploited in order to extend an expensive raw material with cheap minerals. And the most commonly used extenders were chalk and kaolin.

However, caoutchouc or rubber still did not achieve a real breakthrough. Caoutchouc first had to be heated before it could be processed and moulded into corresponding shapes by rollers or other machines (see figure). However, crude caoutchouc and masticated caoutchouc lose their elasticity if they are heated to temperatures in excess of 150-160°C. All that is left is a brittle, totally useless mass. Only the invention of the vulcanisation of caoutchouc with sulphur by Charles Goodyear in 1839 made problem-free processing of caoutchouc at higher temperatures possible.

The properties of rubber, namely extensibility, elasticity and tenacity, made it a very interesting commercial material for widely

Instruction chart "Pepper and Rubber" from the series "Foreign Crop Plants" (end of the 19th century).

Four-roll rubber calender (1929).

differing products in equally differing forms: rubber shoes, rubberised weather-proof clothes, all types of hoses and, as of 1888, bicycle tubes manufactured by Dunlop, all gained a significant share of the market. Consumption gradually increased, and by 1900 more than 48,000 tons of natural caoutchouc were being processed worldwide. Proceeding from the fact that the share of chalk in most of these products was in the order of 15 per cent by weight, and that the caoutchouc share was around 50 per cent, then this meant that the rubber industry was consuming annually up to 15,000 tons of chalk as filler.

By 1911 the production of natural rubber, worldwide, had risen to 75,000 tons, after which output progressed by leaps and bounds – 1914: 120,000 tons of natural rubber; 1916: 200,000 tons; and 1920: 295,000 tons. The few thousand tons of synthetic rubber produced on an industrial scale since 1912, primarily in Germany, fade into insignificance compared with the output of natural rubber.

Production of rubber shoes in the 1920s. The rubber formulation used for the shoes contained up to 40 per cent whiting.

Responsible for this tremendous upsurge was the tyre industry for motor vehicles which had been booming since 1910, accounting for more than 80 per cent of the total output of rubber. However, this did not result in new quarrying records for the chalk industry as it played only a minor role as a filler in the production of tyres. Its share in the mixture was less than 10 per cent since carbon black became the principal filler in 1912. This was the year when a US American tyre manufacturer looked for ways and means to distinguish his tyres from those of competitors with the result that his company was the first to market black car tyres. Consequently carbon black was used as a filler. In retrospect this advertising idea proved to be an ingenious move: Black tyres were not only different, they were also better. The carbon black coloured and extended the rubber and it endowed tyres with very important properties, including a much higher resistance to abrasion.

The radical change

The fact that fillers could not only be an extender but also a reinforcing agent had been discovered earlier. Thus, zinc oxide was the standard reinforcing filler for rubber prior to the advent of carbon black. Zinc oxide shortened the vulcanisation time of the rubber and also slightly increased its abrasion resistance.

Chalk, on the other hand, had no reinforcing properties. It was merely used as a cheap, inactive filler for undemanding rubber products. This concept was even reinforced when, in 1910, it was discovered that even small traces of manganese, copper or iron oxide acted as "poisons" and destroyed the rubber. Since chalk, as a natural product, was often contaminated with iron oxide, considerable expense and effort were necessary to remove all traces of this contaminant with the result that chalk's share of the market declined still further.

This development in the rubber market was symptomatic for other areas in which chalk, used as a filler or extender, held a significant share of the market. It was used wherever possible simply because it was so cheap but, otherwise, it was merely regarded as a filler without any outstanding properties. There were many such competing filler materials – be it kaolin, talc, silicic acid or heavy spar. In fact most minerals were used as fillers if their refractive index was too small to be used as pigments. These filler materials were also supplemented by a number of artificially produced minerals such as lithopone or zinc oxide, as well as sawdust, shredded paper and cork meal.

Moreover the individual chalk regions were competing with each other, as was apparent in Germany. Although there were significant deposits in Rügen, Söhlde and Lägerdorf in Schleswig-Holstein, the more pure grades of chalk were being imported from Sweden, Denmark and France. Consequently most German chalk quarries were struggling to survive.

Even the apparently safe glazing putty trade began to totter during and after the first World War when raw material supplies became increasingly problematic. Trade journals, including the "Seifensieder-Zeitung" (soap-boilers' journal), were now publishing the first substitute recipes for putty. Although the primary purpose of this development was to replace the expensive linseed oil with fish and palm oil, the chalk monopoly also began to tumble.

Temporary bottle necks and financial considerations made many putty manufacturers look around for other fillers such as diatomite and alumina which were easily available, or heavy spar to reduce production costs. The latter was extensively used because it drastically increased the weight of putty which was sold by weight. But it also had a lower oil absorption value so that less expensive linseed oil was needed. The fact that heavy spar putty was not pliable but brittle and crumbled easily was accepted.

A difficult economic situation was not new to the chalk industry. The need to make the first adjustments in keeping with the market had already arisen at the beginning of the 20th century. At the time it was primarily the small family businesses that had to close down, either because the size and quality of their deposits were no longer sufficient for profitable operation or the transport conditions were inadequate and costly. Only those that were able to harmonise sales and investments were able to survive, and this has not changed in the meantime. But to the attentive observer of the fillers industry it was clear that a radical change was about to take place throughout this industry.

New dimensions

The first scientific papers concerned with the influence of the structure and chemical composition of fillers on their functional properties were published at the beginning of the 1920s. Until then fillers had merely been regarded as cheap extenders. The discovery and use of their reinforcing properties, as was the case with carbon black and zinc oxide, was purely co-incidental. Since there was absolutely no understanding of the properties and their functions at the time, these studies were specifically targeted at establishing the optimal filler for a given application.

For instance, William B. Wiegand published a paper in the "India Rubber Journal" in 1920 in which he detailed the influence of different fillers such as carbon black, heavy spar and chalk, on the strength of rubber. The most important result of this publication was the discovery of the marked interdependency that existed between particle size and reinforcing effect. Accordingly, the finer the particle size, the greater the reinforcing effect. It was then up to the filler manufacturers to implement these scientific findings in practice.

How important such findings could be for the successful expansion of an industry was demonstrated by an example in the chemical industry. After prominent chemists, among them Fritz Haber, Walter Nernst and Wilhelm Ostwald, had defined at the beginning of the 20th century the decisive interdependencies in reaction kinetics and catalysis, the chemical industry rapidly applied these findings in the large-scale synthesis of important chemicals. Possibly the most famous example of this development was ammonia synthesis according to the Haber-Bosch process. It was first tested in a BASF pilot plant in Ludwigsburg, Germany, in 1913. By the beginning of the 1920s it was being applied on a worldwide scale.

Obviously, exclusive mechanical processing of fillers cannot be compared with the often complex synthesis performed in the chemical industry. However, all manufacturers, including the chalk industry, had to act accordingly in order to assert themselves in an increasingly competitive market.

If chalk was to remain an important filler it was essential that its position in the old markets had to be strengthened and if possible, extended. This was still at a time when chalk was an important raw material for the manufacture of paints, art paper, wall paper and glazing putty. And it was still possible to win markets which, until then, had not existed or had refrained from the use of chalk. However, this aim could only be achieved if chalk filler was further developed to satisfy the specific requirements of customers.

This opportunity arose as a result of the slight lead enjoyed by the chalk industry in the demand for ever greater degrees of fineness. Chalk is, by its very nature, microcrystalline. In fact, with the exception of kaolin, it is significantly finer than any other mineral. However, the use of new grinding techniques threatened to rapidly change this situation. Consequently, there was no alternative for the chalk producers – either invest or close down. Which decision was the right one would only become apparent as a result of developments during the subsequent years.

The criteria of success

To minimise the risk a company had to be critically examined to establish whether the preconditions for success existed. The most important criteria of such an analysis were defined in 1928 by André Moussy in his lecture "La craie et l'industrie du blanc dans le départment de la Marne" which he held at the Academy of Sciences in Châlons-sur-

Excavators and lorries were operated for the first time in the chalk quarries of Champagne during the 1920s and 1930s.

Marne. Moussy listed four primary criteria for success that not only applied to Champagne but also to chalk quarrying in general, and they have remained valid to this day.

First: Suitable chalk deposits are necessary. Suitable in this context means that the chalk can be easily extracted, preferably by opencast methods, and it should contain few contaminating minerals in order to keep the cost of quarrying and processing as low as possible. Furthermore, the deposit must be sufficiently large to ensure that the processing equipment can be continuously operated to full capacity.

Second: A good transport infrastructure must exist. At the time good implied the presence of a waterway in the immediate vicinity of the quarry and processing plant, or railway sidings had to link the works with the railway network. Transport by road was too expensive for a cheap product like chalk where the transport costs account for the main share of the selling price.

Third: Comprehensive technical further development was imperative. This applied to virtually all operations within the works because they had remained virtually unchanged since the beginnings of the chalk industry some 70 years ago. Although there were some works that even operated cableways to transport the freshly extracted chalk out of the quarry to the processing plant, transport within the works was still not automated. There were no shaking or belt conveyors for horizontal transport, nor bucket elevators for vertical transport, as were already quite common in the cement industry at the time, let alone the pneumatic conveyors that were being operated in USA cement mills to convey the finely ground powder through pipes by compressed air. The fully automatic packing systems for cement also originated in the USA. They consisted of a filling machine with a valve

Crushers, mills and sifters

The development of processing techniques and the corresponding machinery was primarily a matter of the availability of the given materials. Be it grinding or crushing, all reduction processes subject crusher and grinding mill materials to extreme stresses.

It is therefore hardly surprising that the edge and grinding mills of antiquity were still being used, virtually unchanged, in the 19th century because not only did they produce a fine flour, but their grinding stones operated virtually wear-free. Moreover, these mills could easily be used for multi-stage pro-cesses in combination with pounding mills and settling basins. However, their crucial drawback was a very low throughput. Even steam-driven mills could just about manage to turn out 1 tonne of fine flour in one hour.

Higher throughput rates required other machines. Moreover their design was no longer a problem since adequate quantities of cast iron became available in the 19th century. As early as 1858 Eli Whitney Blake was able to patent a jaw crusher (see figure). This was a very simple, yet highly effective machine consisting of two hard-metal jaws positioned in a V-shaped arrangement in relation to each other. One jaw remained stationary, while the other was moved up and down by a lever, thereby crushing the material.

The robust Blake crusher rapidly became the accepted standard for coarse reduction while, in the 1870s, the ball mill replaced the edge mill for fine reduction. Ball mills feature a rotating grinding drum filled with small steel balls, together with the material that is to be reduced. The tumbling action of the steel balls crushes and pulverises the material as the drum rotates.

There are two possibilities to achieve the required grinding fineness: The grinding drum can be lengthened to extend the grinding course, as is the case with tube mills (see figure) or a sifter can be incorporated in the mill, as is the case with the first screen ball mill built in 1876. The grinding drum of this mill was totally enclosed by a wire screen so that the finely ground particles could drop through the screen while the coarse particles remained in the grinding process. However, screens only allowed the classification of particle sizes of 0.2 mm. A higher fineness required a wind sifter, as was first developed by Robert Moodie in 1888.

Wind sifters use a current of air to separate the particles by size. The particles are lifted to different heights in relation to their size and weight, so that they can be easily separated accordingly. The velocity of the blowing air can be varied within fine limits to adjust micrometer fineness rates with the utmost precision.

The wind sifter also offer other significant advantages: It achieves a higher capacity and takes up less space, its design is very simple and wear is minimal. The fact that the ground particles are continuously removed also increases the mill's throughput capacity.

Since 1900 wind sifters, in combination with ball mills, have been asserting themselves particularly in the cement industry where very fine particle sizes are expected in conjunction with high throughput rates. And these requirements are also increasingly being expected in the chalk industry.

Multi-chamber tube mill (1930). This mill is 18 m long with a drum diameter of 2 m, and it was used in the cement industry.

sack that could pack between 500 and 600 50-kg sacks of cement per hour. This contrasts with the European chalk industry where, at the time, barrels and sacks were still being filled with chalk by hand.

Manual methods in chalk processing had to be reduced. The ultimate aim had to be a fully automatic process in which chalk lumps were fed into the system and fine pure powder was discharged by the grinding installation. This required the operation of modern machines such as jaw crushers, ball and tube mills and air classifiers.

The steel ball mill with air classifier that had become standard in the cement industry represented an enormous step forward in chalk processing because they enabled fine grinding and classification by particle size. The combination of ball mills and air classifiers, however, presupposed dry processing. A ball mill with a normal screen had been available for wet processing since 1900. It was able to grind thick slurries containing up to 60 per cent chalk. This eliminated the time-consuming procedure of classifying by means of settling tanks.

Fine grinding was not the sole problem, particularly since natural chalk already has a mean particle diameter of 5 micrometers on account of its microcrystalline structure. Far more important was the need to make the chalk industry independent of the weather, particularly since the dry process – namely "dusting" – offered significant advantages. By building heatable drying rooms in which the freshly quarried chalk lumps could be 'weathered', operation throughout the year would become possible. However, the operation of rotary drying kilns, as used in the cement industry, did not prove to be successful because the chalk was occasionally blackened by soot or firing transformed it into lime (see figure).

Chalk drying in a rotary kiln, Champagne (around 1930).

The decisive advantage of wet grinding was the greater purity of the product due to the elaborate slurry processes. Very few deposits were pure enough to warrant dry processing. The big problem remained the industry's dependance on the weather. Even the use of filtering presses to drain some of the water in the whiting was of little use. Ultimately an efficient wet process required heatable mills and drying rooms, but they consumed a great deal of energy so that this was only profitable with very high throughput rates.

For long-term further development it also became necessary to set up laboratories which could examine the structural properties of the chalk. It was essential to understand these properties, and possibly even measure them in quantifiable terms, as this was the principal precondition for developing new applications for filler chalk.

Fourth: The previously mentioned improvements required adequate financial leeway. A large amount of money was required to purchase suitable deposits, establish an efficient transport organisation and to introduce the latest technical standards in all production processes in the quarry and the processing works. Moreover, considerable staying powers were necessary for chalk to become established on the market. The competition was fierce and the profit margin very small because the price continued to be the deciding sales factor for any filler.

The USA as the forerunner

The European suppliers were able to learn from the USA how a new filler could assert itself on the market in spite of all adversities and even though it was a product of the chalk industry that had suffered such drastic losses as a result of the new fillers.

Up until the beginning of the First World War the USA was the world's largest importer of European chalk due to the fact that it had few chalk deposits of its own that could be used as a source for the production of fillers. The only few deposits of commercial significance in the whole of the USA were in Kansas. The World War cut off the USA from the European markets. Chalk supplies faltered so that it became necessary to look for alternatives.

Since the USA had large deposits of marble and limestone – both of them chemically identical to chalk – it was only natural that ways and means were sought to use these minerals as fillers. However, both limestone and marble are significantly harder than soft chalk. This made it necessary to introduce processing techniques which, until then, were restricted to ore processing. These techniques and machines merely had to be adapted to the new requirements.

Initially the quality of the substitute limestone and marble fillers was clearly lower than that of chalk. However, the problem was soon overcome by selecting better, but particularly purer deposits and by introducing new grinding techniques to produce the same levels of fineness as with natural chalk. It was therefore quite consistent that the new ground limestone and marble were likewise referred to as *whiting*, just as chalk. They then became established as fillers on the market, together with chalk, and they retained the share of the market they had gained, even after trade relations with Europe were revived after the end of the World War.

It was primarily the high transport costs that prevented chalk from regaining the position it had enjoyed on the American filler market before the World War. European chalk was now only used in the paints industry and to produce glazing putty. The new limestone and marble fillers which had ousted chalk were now being used as coating pigments for art paper and as a filler for rubber. Eventually, a certain amount of limestone powder was added to glazing putty in place of chalk. And when artificially precipitated calcium carbonate (PCC) appeared on the market, chalk lost even more ground merely because it was not competitive.

The development in the USA definitely underscored Moussy's postulates in a very impressive manner. Starting almost at zero, marble and limestone powder became the most important filler on the American market due to adequate high-quality deposits at favourable locations with regard to transport. Moreover, improved processing techniques made it possible to offer this product in any required fineness at a competitive price. Even an artificially produced filler, such as PCC, was able to assert itself because its structural properties could be precisely controlled during the production process. In this manner buyers were guaranteed unvarying quality standards at all times.

1.3 From chalk to calcium carbonate

As painful as the loss of the American market was for the European industry, but particularly for the English chalk industry, this pain could still be beneficial if the manufacturers drew the right conclusions from this development. It was imperative that they acted promptly to ensure that chalk would also continue to play an important role on the European market in the future. The preconditions for this already existed in Europe.

England, France and Germany, the three most important industrial countries, had adequate chalk deposits, and other significant deposits were located in neighbouring Belgium and Denmark. The soft chalk could be processed by relatively simple techniques at a much cheaper rate. Limestone grinding, therefore, was only practiced in those countries that were too distant from the chalk regions – for instance in Sweden ground limestone was marketed as "Motele chalk". The expensive PCC was only used in Europe where considerable importance was attributed to super-pure raw materials, for instance for cosmetics and toothpaste.

The way was then free for chalk, particularly since the first scientific studies of the properties of chalk had been conducted, for instance the treatise of Erich Kindscher published in 1937: "Boundary surfaces and their importance for industry". Kindscher had produced putty using chalk from different origins, but always the same linseed oil, yet he encountered serious differences that could not be explained by the fineness of the material. He concluded from this that the chalk particles and the condition of the particle surface were responsible for their characteristic effect as a filler; the size was only of secondary importance.

World War II put an end to the efforts of the chalk producers, particularly since the economic relations in Europe experienced a radical restriction as a result of the strict self-sufficiency policy enforced during the German Third Reich. The German market that had been of such importance to the French chalk industry suddenly ceased to exist overnight. By the time the war was over the European chalk industry had been set back to the state that prevailed in the late 1920s and early 1930s.

A new beginning

After the end of the World War, the new start for Rügen was characterised by changed political and economic conditions. The island was now part of the Soviet occupation zone which meant that it was cut off from the western European markets, but on the other hand it faced virtually no competition from eastern Europe which had very few chalk deposits. Consequently, these two circumstances meant that there was little economic pressure on the Rügen chalk industry which would have made it necessary to make technical improvements to ensure survival. In 1946 manual quarrying was resumed and the total annual output was in the order of 3,000 tons. The first excavators appeared in the Rügen chalk quarries in 1949, and another 13 years passed before machines were used for processing, packaging and dispatch.

Apart from Lägerdorf, Söhlde was the sole source of chalk in West Germany in the immediate post-war period, particularly since no Champagne chalk was being imported into Germany. As on Rügen, most of the quarrying was manual during the initial post-war years. However, explosives were used in the Söhlde quarries since the chalk

Chalk quarry in Söhlde. After the war a great deal of work was completed manually.

was harder and could not be readily excavated or cut manually. The workers sorted the chalk lumps by size and purity and transferred the resulting individual grades separately to the processing works.

The large blocks were processed into crude chalk. First they were air-dried before the chalk was crushed, ground and sifted to obtain fine powder. The main product, however, continued to be whiting. The whiting was packed in jute sacks and sold for DM 4.00 for 100 kg. The price of chalk dust was only DM 3.30.

The principal buyers of Söhlder chalk were large companies of the West German rubber industry, e.g. the Continental rubber works in Hanover, the Phoenix rubber works in Hamburg and the textile and rubber works of Vorwerk in Wuppertal. IG Farben was another large-scale buyer in 1948.

Chalk quarrying in Söhlde was still organised along the lines of the previous century, as opposed to the consumers which were all modern high-tech industries. Chalk was still primarily quarried by farmers as a side-line job. Only a few works were dedicated exclusively to chalk production, and these did not produce more than 4,000 to 5,000 tons per year. This was due to the fact that quarrying in Söhlde depended on the season of the year. Since, during the immediate post-war years, fuel was not available to heat the mills and drying rooms, everything came to a complete standstill during the winter months.

The chalk industry in Great Britain was far more advanced at the time. The British Whiting Association, which represented more than 90 per cent of all whiting producers, founded a Research Council as early as 1947. The Council recognised that the chalk industry could not survive fierce competition without applied research. A Research Institute was founded in Welwyn, some 30 kilometers from London, with the support of the British Government. More than 20 scientists were exclusively concerned with developing applications for chalk in the different industries. The production of glazing putty on a scientific basis was first developed here in Welwyn.

With the backing of extensive research, whiting was able to assert itself on the British filler market. Wherever a cheap neutral filler was required companies turned to British chalk, be it in the wallpaper and linoleum industry, rubber factories or for the production of paints and sealants. The British whiting producers were even able to export significant quantities of chalk to the USA.

Three rocks – One product

Eventually, chalk regained a secure place among the fillers in the other European countries. However, in spite of annually increasing quarried quantities, the actual share of chalk within the total quantity of consumed fillers still declined. This may sound paradoxical but this was merely the logical consequence of rapid economic growth which started with the boom throughout the 1950s. The demand for fillers rose drastically within a very short time.

The rapidly expanding plastics industry was primarily the driving force behind the development of fillers production. As early as the 1920s Hermann Staudinger had established the scientific principles in his "Theory or Macro Molecules" which formed the basis for the systematic production of plastics. The first processes for large-scale manufacture of polyvinyl chloride (PVC), polystyrene and polyethylene (PE) in the 1930s were the logical consequence. World War II had put a sudden end to this development. By then, however, the triumphal advance of plastics was unstoppable as they rapidly entered the realms of everyday life, be it as nylon stockings, artificial leather handbags made of PVC or non-iron shirts made of polyester. Life without plastics had become unthinkable, especially in the automobile and building industries where vast quantities of plastics were being used to produce cables and electrical equipment.

Plastics production in the German Federal Republic in 1951 was no more than 81,000 tons. By 1960 this had risen to 610,000 tons and in 1965 output had clearly exceeded the millions barrier with a total volume of 1.35 million tons. The amount of fillers varied between 10 and 70 per cent, depending upon the type of plastic. It was clear that the demand for fillers would soon outstrip available supplies, particularly since other industries were also demanding ever-increasing quantities of fillers. Consequently, new fillers had to be found.

The criteria for this search were soon formulated. The chemical behaviour was of little concern. A filler had to be chemically inert. The physical properties such as the surface, particle size and the distribution of the particle sizes within the filler proved to be of primary importance. Yet these could still be varied relatively freely for almost all rocks and minerals by new grinding techniques that had already proved their worth in ore dressing.

Consequently, the following points were really decisive: If possible the rock had to be white or at least not have a disruptive colour; the contaminating matter in the deposits had to be minimal to avoid complicated and costly separating processes; suitable deposits had to be relatively abundant; and transport costs had to be minimal.

These three criteria applied primarily to two rocks which had long established themselves on the US market, namely marble and limestone. Both of them consist of the white calcite mineral, they can be found in many countries, and they had both captured the filler market. As calcium carbonate was the chemical basis of calcite, the new filler become known as calcium carbonate irrespective of whether the initial material was limestone or marble. Apart from minor contaminations, the chemical composition of both minerals is identical with the result that, after processing, the petrographic differences become almost indistinguishable.

Chalk, the third natural limestone, is also a calcium carbonate from the chemical standpoint, and it was now also being sold under this name. Initially, chalk was well able to compete with marble and limestone, but then its share in the total annual output of calcium carbonate declined continuously.

However, chalk still continued to prevail, even though glazing putty had long disappeared from the market. This was due to the fact that some of chalk's specific properties were still required. Consequently, pure high-quality chalk remains an important raw material for the paints and varnishes industry. It is still used as a filler in cable manufacture. Even modern grades of paper require chalk. Obviously, natural chalk continues to be used in low value-added areas, for instance in the production of fertilizers or for flue-gas cleaning.

Even the old drying processes were retained right into the 1960s. The whiting was first predried on stones (below), and then the moist chalk cakes were placed on a drying frame (above).

2. Calcium carbonate – pigment and filler

The use of minerals is an essential part of daily life. For thousands of years an extremely wide range of rocks and comminuted minerals has been used in various areas and today the level of production and consumption of "nonmetallic raw materials" is even used as an indicator of a country's development status (see figure). Industrial minerals have become a symbol of wealth and none of these mineral materials has as much significance and such a wide range of applications as calcium carbonate.

Chalk, limestone and marble have a long history (vid. Chapter II. "Cultural History"), and calcium carbonates are still to be found in the most diverse applications: From sculpture through the construction industry to chemistry, from fillers in paper, paints and plastics to pharmaceuticals (see figure)

The enormous economic importance of calcium carbonate minerals becomes clear when we look at the volumes of chalk, limestone and marble produced year by year. In 1994 alone around 4.6 billion tonnes of calcareous rocks were extracted worldwide (see figure). The significance of calcareous rocks can be seen in a comparison with total nonmetallic mineral consumption. In Germany, for instance, calcareous rocks accounted for a good 7% of industrial minerals consumed in 1994, or in absolute figures: The total of 920 million tonnes of "nonmetallic minerals" included 65.4 million tonnes of calcium carbonate. If dolomite and other carbonate rocks are added, as well as the limestone not consumed by companies in the lime industry, the proportion was as high as 16 percent or 144 million tonnes.

It is typical for the highly industrialized economies of today that the nonmetallic raw materials clearly outstrip the metals in importance (Source: Federal Institute for Geosciences, Braunschweig).

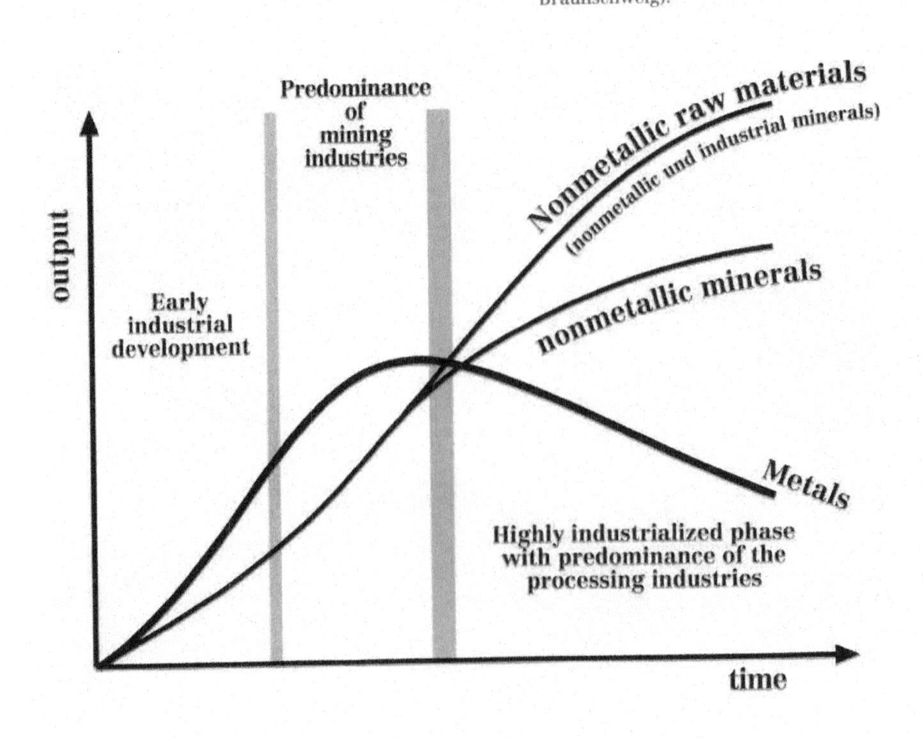

Market segment	Germany		Global	
	[Mio t/year]	[%]	[Mio t/year]	[%]
Building industry (structural and civil engineering)	23.2	35.4	(k.A.)	
Cement	33.3	51.0	1 420	31.5
Agriculture	1.4	2.1	(k.A.)	
Iron and steel	3.1	4.75	(k.A.)	
Environmental protection	1.3	2.0	(k.A.)	
Fillers/coating pigments	2.0	3.0	20	0.4
Others	1.1	1.75	(k.A.)	
Total	65.4	100	4 500	100

Consumption of limestone in Germany and worldwide. The figures given are estimates since exact data are not available, particularly for the worldwide consumption (Source: Oates, 1998; Federation of the German Lime Industry).

By far the biggest amount of limestone goes to the construction industry: as building stone, cement or as crushed stone for road construction. By comparison, the proportion of fillers is low, but in absolute terms still considerable. In 1997, for example, 9.8 million tonnes of calcium carbonate was consumed by the paper industry, around 7 million tonnes by the plastics industry and more than 4.8 million tonnes by the paint and coatings industry.

And in contrast to construction raw materials fillers are high-tech products whose production and processing involve elaborate processes and the deployment of expensive machinery. Apart from the world-famous "statuario" of Carrara and other relatively exclusive grades of marble, the highest prices are attained for calcium carbonate if it is used as a filler (see figure).

Compared with the cost of other raw materials, however, the price of a filler is low. Calcium carbonate fillers in paper, for example, can replace part of the required cellulose which is around ten times more expensive. The more mineral raw materials a paper

Average 1997 prices in Great Britain for limestone products. Although the individual prices can fluctuate considerably, in their general trend they can be transferred to other countries (Source: Oates, 1998).

	Building material	Production of caustic lime and glass/flue gas desulphurization	Filler and slurry pigment	PCC
Preis [£/t]	2 - 5	5 - 10	25 - 250	250 - 1 000

Uses of calcium carbonate. The fields of application charted are in summarised form. The field "chemical industry", for example, comprises a host of different applications.

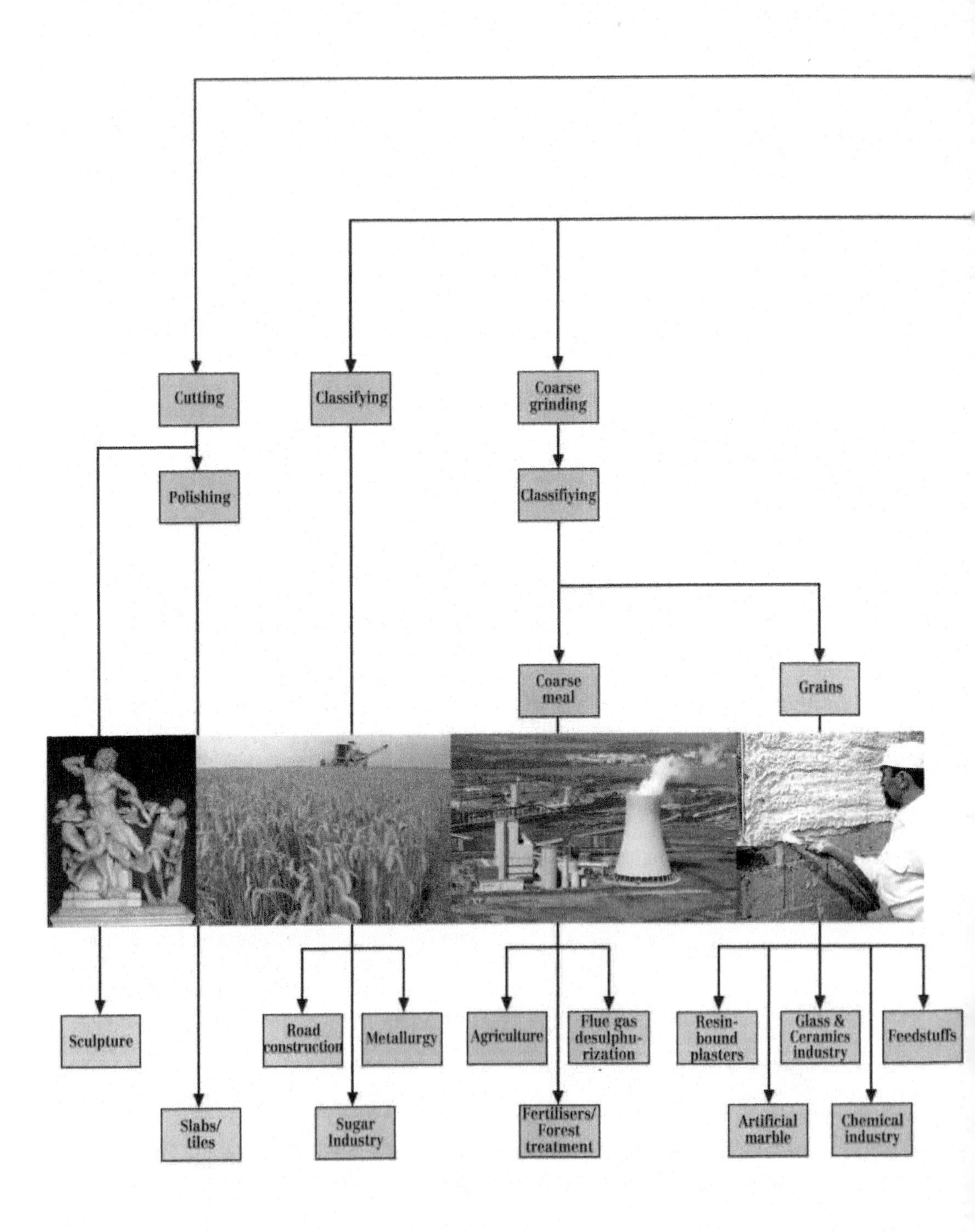

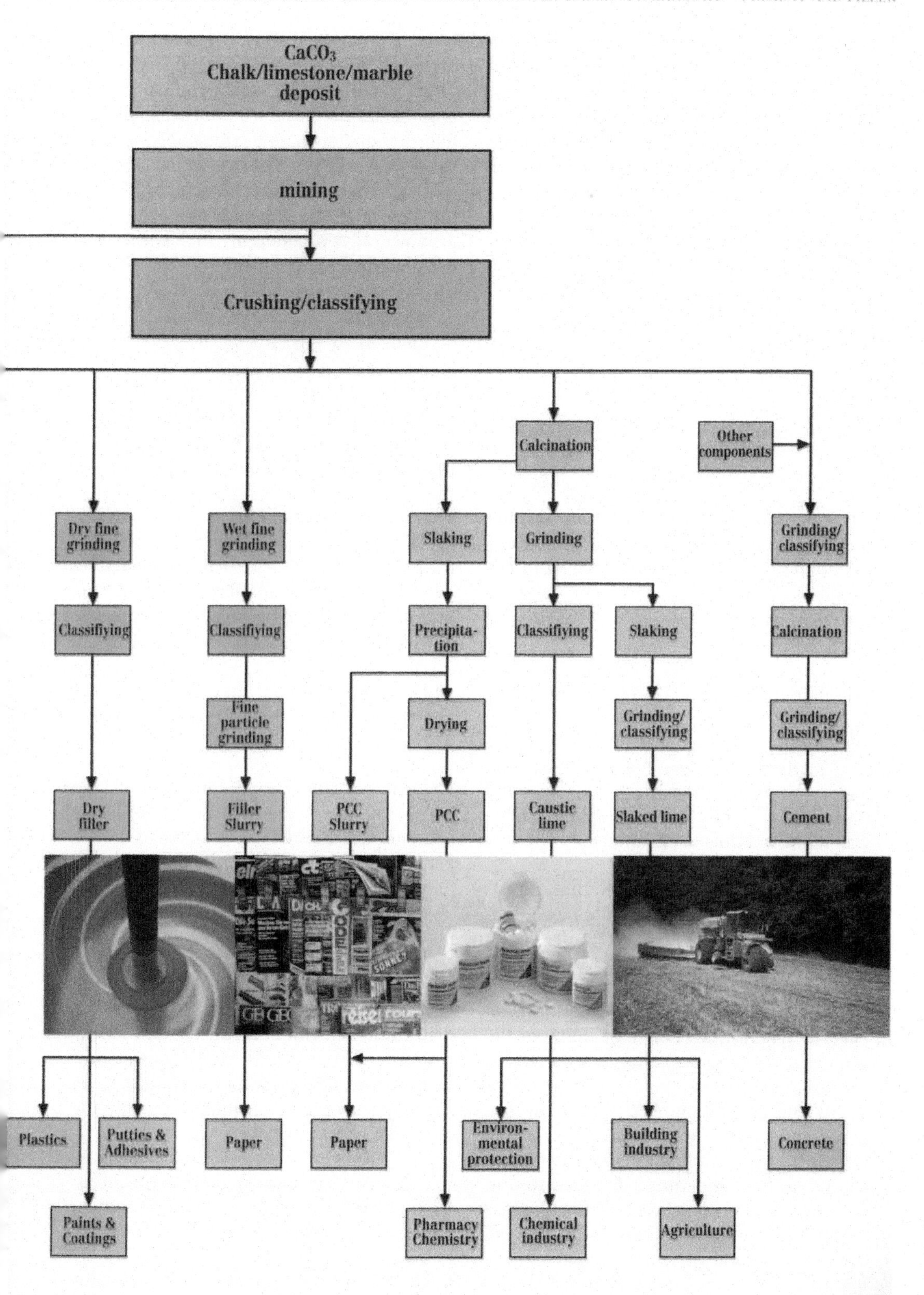
$CaCO_3$
Chalk/limestone/marble deposit
mining
Crushing/classifying
Calcination
Other components
Dry fine grinding
Wet fine grinding
Slaking
Grinding
Grinding/ classifying
Classifiying
Classifiying
Precipita- tion
Classifiying
Slaking
Calcination
Fine particle grinding
Drying
Grinding/ classifying
Grinding/ classifying
Dry filler
Filler Slurry
PCC Slurry
PCC
Caustic lime
Slaked lime
Cement
Plastics
Putties & Adhesives
Paper
Paper
Environ- mental protection
Building industry
Concrete
Paints & Coatings
Pharmacy Chemistry
Chemical industry
Agriculture

contains, the cheaper it is to produce. This does not mean, however, that highly filled papers are always cheaper. On the contrary, the very high-quality papers in particular often contain large quantities of mineral raw materials. And in the paint and coatings or in the plastics industry fine mineral powders contribute significantly to the value-added of end products. Moreover, for technical reasons many products today are hardly feasible without fillers.

2.1 Properties and effects of a filler

"A filler is a substance consisting of particles which is virtually insoluble in the application medium and which is used to enlarge the volume, to achieve or improve technical properties and/or to influence optical characteristics."

The definition to DIN 55943 not only highlights the key role of modern fillers, i.e. improving "technical properties", but by adhering to the term "fillers" also reflects the history of these substances. Because the first and for a time the only task of a filler was to fill and thus to cheapen.

But it was gradually recognised that the addition of a filler could contribute more than just "filling". Just as in concrete the cheap "filler" gravel not only helps to save on expensive cement but also significantly influences its strength properties, fillers can be used to specifically change or improve the product characteristics of the filled materials.

Today the criteria applied to a filler are many and varied. For instance, fillers have to exhibit high availability, and their price should be low. They must disperse well in the most diverse application media, possess high purity and be non-toxic. Depending on the branch of industry, further criteria are applied. Fillers in papers should have as high a refractive index as possible and high whiteness. In the paint industry weathering resistance and rust inhibition are required. But not all properties are equally important. Some, such as adequate availability or low price, are fulfilled by so many minerals that they are hardly a factor in the selection of a filler; others, such as health safety, by contrast, have an exclusive impact. Thus, asbestos materials have disappeared from the filler market completely and even finely ground quartz products are in the process of disappearing because they can cause silicosis.

If we consider only the properties that can be modified by processing, it is mainly the fineness and particle size distribution (to some extent also the whiteness) of a filler which alongside the natural properties such a pH value, density and particle shape which determine its application.

The coating material for carpets, for example, requires only a relatively coarse filler with a broad particle size distribution curve, and whiteness does not play a decisive role. In order to increase the strength of a plastic for window profiles and to guarantee an evenly white colour, however, a fine and white filler is required. In paper even finer and whiter fillers are used; apart from high whiteness, opacity is also essential here.

Regardless of the area of application, the degree of filling should always be high, in order to save as much as possible of the expensive base material such as plastic or cellulose. Moreover, a high degree of filling has a positive influence on the product properties such as material strength and processability, while also improving the optical characteristics. A high degree of filling can, however, only be achieved if the particle size distribution is optimally tailored to the special case. This imposes high requirements on the processing technology, and a precisely tuned sequence of grinding and classifying of the raw material is essential.

At the same time the customers in the industries concerned must be assured that a filler with constant properties is always available in adequate quantities so that they do not have to constantly adapt their production processes to changing filler qualities. What sounds rather trite, represents a major challenge for the filler industry which can only

	Refractive index n_D	Density (g/cm²)	Hardness (Mohs)	Crystal class/ habit
Barium sulphate	1.64	4.5	3-3.5	rhombohedral/ cubic
Calcium carbonat				
- Aragonite	*1.63*	*2.95*	*3.5-4*	*rhombohedral-dipyramidal*
- Calcite	*1.6*	*2.6-2.8*	*3*	*rhombohedral/ cubic or needle-shaped*
Dolomite	1.60-1.62	2.85-2.95	3.5-4	rhombohedral/ cubic
Kaolin	1.57	2.60-2.63	2	triclinic/ flaky platelets
Quartz	1.549	2.65	7	trigonal/ lamellar
Talc	1.57	2.7-2.8	1	monoprismatic/ splintery
Titan dioxide				tetragonal
- Anatase	*2.55*	*3.87*	*5.5-6*	
- Brookite	*–*	*4.17*	*5.5-6*	
- Rutile	*2.75*	*4.26*	*6-6.5*	

Physical properties of typical fillers and pigments.

be met by a high level of technology. Because as natural products a key feature of mineral raw materials is that they exhibit high variability, no deposit is like another.

If all criteria are summarised, there are still numerous minerals which are especially suited for use as fillers. These primarily include kaolin, talc and calcium carbonate, but magnesium carbonate, barium and calcium sulphate also have interesting properties. A comparison of the volumes processed and sold worldwide year by year, however, shows that the calcium carbonates are clearly in the lead (see figure). There are good reasons for this.

2.2 Chalk, limestone, marble, PCC – common features and differences

The mineral calcium carbonate is available on earth in apparently random quantity. Of all mineral raw materials only quartz, or silicon dioxide, is found more frequently in the outer crust of the earth. And all natural

Paper fillers and coating pigments

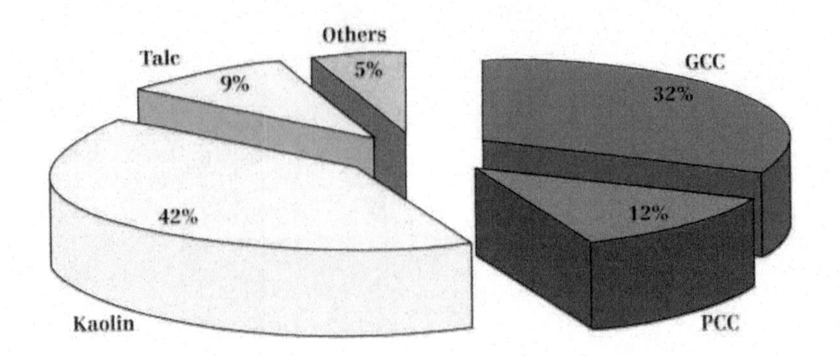

Plastic fillers

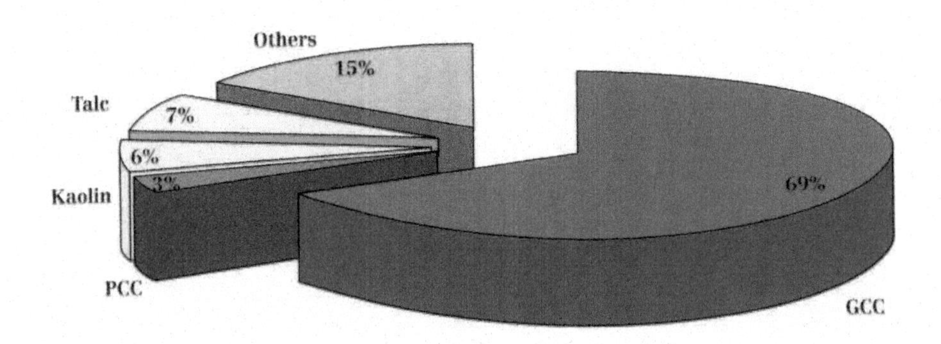

Paint and coating fillers

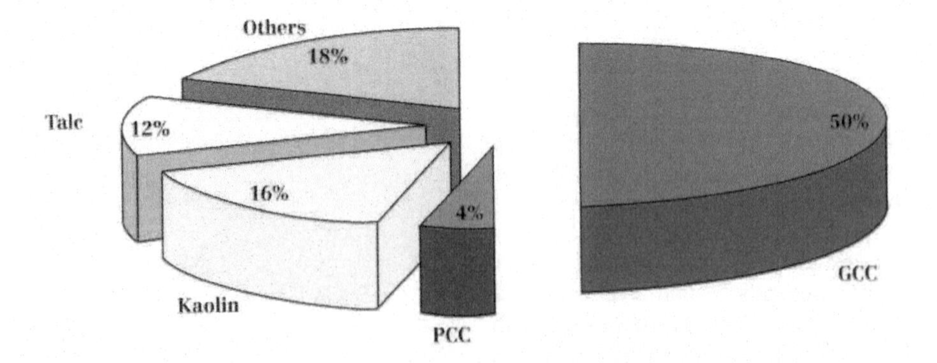

Market shares of major fillers and coating pigments.

calcium carbonate varieties are suitable for use as filler: chalk, limestone and marble. In addition, there is the artificially produced precipitated calcium carbonate (PCC). Although all varieties are chemically identical, each one of them possesses specific features which differentiate them from the others.

2.2.1 Chalk

In natural form chalk occurs as a finely dispersed resource. It was for this reason that chalk came to be used in early times as the first calcium carbonate filler, especially in paints and putties. But the variety of the range of applications for chalk fillers was and remains incomparably larger. The following characteristics in particular favour this mineral:

- No risk to health (a key requirement for use in household products such as toothpaste, cosmetics and cleaning agents)
- Easy processability
- Whiteness
- Platelet-shaped particle structure
- Good coatability
- Extremely good wear properties which have a positive impact on the service life of the processing and production machinery

Frequent impurities in chalk are silicon dioxide in the form of flint, organic residues, and the sulphites (mostly gypsum) and sulphides present as pyrite. But apart from a few disruptive sulphites, the impurities can be simply separated by "slurrying" (see section 1.1 "A chalk industry evolves") a process still used today. Depending on the deposit and the eventual area of application, either directly after washing or later in the dried condition the chalk is divided up according to grain sizes, by which the filler's characteristics are specifically matched to the application.

Owing to its biogenesis and low metamorphosis chalk can never achieve the highest possible whiteness of a calcium carbonate. In the formation of chalk, organic compounds become deposited so to speak as a reinforcement and adhesive for the calcium carbonate in and between the coccolith platelets and in this way increase the strength of the shells. At the same time these inclusions reduce the whiteness of the chalk.

2.2.2 Limestone and marble

When firstly the paint and coatings industry and then also the paper industry expressed the wish for increasingly whiter products the filler producers had to look around for new starting materials made of calcium carbonate, because chalk could not meet this requirement. The suitability of limestone and even marble for use as fillers was therefore investigated as they both display a much higher degree of whiteness than chalk.

However, both limestone and marble have to be processed in most cases to attain the required high degrees of whiteness. As a result limestone achieves degrees of whiteness which in the most favourable case are TAPPI R457=92 or Ry=94 (see Appendix "Definitions and Measurement Methods"). If even whiter products are to be produced, marble has to be used as the raw material. Owing to the strong metamorphosis of this mineral the impurities in the raw rock are separated and present in such particle sizes that that they can be removed by suitable processes. Products with degrees of whiteness of TAPPI R457=95 or Ry=96 are thus possible.

But limestone and marble differ from chalk not just in the degree of whiteness. Particle size distribution, abrasivity and opacity of the respective fillers are also different. A chalk naturally exhibits a very broad particle size distribution with a large specific surface. With a crystalline marble or limestone it is very difficult to achieve such a particle size distribution, because even with intensive grinding and classifying the particle size distribution of a comminuted mineral can only be influenced within certain limits and in the final analysis depends on the crystalline structure of the starting material. The calcium carbonate fillers from the crystalline rocks marble and limestone thus have a tighter particle size distribution than the fillers produced from chalk, and their specific surface is also smaller.

The abrasivity of a filler depends decisively on the particle shape. The coccoliths in chalk are built up from individual calcium carbonate platelets. Under the scanning electron microscope it can be recognised that these platelets have rounded edges (see figure). By contrast, a filler ground from marble consists of sharp-edged individual particles. These sharp edges and the steeper particle size distribution lead to more pronounced abrasivity of the fillers made from limestone and marble although the Mohs' hardness for all three varieties is the same. If, however, the particles can be ground to below a defined particle size, even marble loses its abrasivity.

Chalk (a), GCC (b) and PCC (c) under the scanning electron.

As a result of its biogenic structure – whether through the slightest impurities or optical irregularities – chalk has a higher opacity than crystalline marble and limestone. This is because the better formed the crystals, i.e. the purer the calcium carbonate, the higher the degree of whiteness but the lower the opacity. This tendency can be reduced by special grinding processes; if, however, products of the same particle size distribution but of different crystallinity are compared with each other their opacity differs.

2.2.3 PCC

A further variety of calcium carbonate is precipitated calcium carbonate or PCC (see figure). The value of this artificially produced calcium carbonate for example in the paper industry stems mainly from the fact that freshly precipitated PCC has a higher volume than natural calcium carbonate (GCC = ground calcium carbonate), but this volume disappears as soon as the PCC is exposed to a higher pressure during calendering of the finished paper. On the other hand, the conditions of petrogenesis or metamorphosis of the marble cannot be completely represented in the laboratory.

If the PCC fillers are included there are today four calcium carbonate varieties on the market which enable calcium carbonate as a filler to be optimally adapted to the particular conditions applying and thus enable it to be used in a wide range of industries.

2.2.4 Areas of application

In principle, fillers made from the four calcium carbonate varieties can be used for the same products. But certain product properties can be achieved more easily or better with the one or the other starting material – each calcium carbonate has its own specific area of application (see figure). It would, however, be uneconomic and nonsensical to crush a marble at great expense in order to replace chalk.

Over the course of time the proportions of calcium carbonate varieties in total produc-

Mean particle diameter [µm]												
	1.0	2.0	2.5	5	10	15	30	50	70	90	160	≥500
Paints and coatings												
Dispersion paints	■	■	■	■	■	■	■	■	■			
Decorator's paints	■○	■○	○	○								
Powder coatings	■	■	■	■	■							
Industrial paints	■○	■○	○									
Road marking	■○			■		■		■	■		■	
Printing inks	■											
Primers	■	■○	○	■○								
Stoppers and fillers			■	○				■	■	■		
Silicone resin paints			■	■	■							
Plasters												■
Plastics and adhesives												
PU, PE, PVC		■○	■○	■○								
Cable sheathings	■○											
Undersealing				■○							■○	
Latex coatings					■○	■○	■○				■○	
Calender film		■○	■○									
Floorings						■○	■○				■○	
Adhesives			■	■	■	■		■		■		
Sealing compounds			■	■	■	■	■	■				
Others												
Writing chalk			■									
Toothpaste			■	■								
Cleansers			■		■	■	■	■	■	■	■	
Pharmaceuticals		■	■									

■ - untreated calcium carbonate grades
○ - surface-treated calcium carbonate grades

Particle size-dependent use of untreated and surface-treated dry calcium carbonate fillers.

tion have shifted. Whereas up to the Second World War chalk dominated the filler market for calcium carbonates, the dominance first shifted to limestone and then to marble, which today is the most important starting material for fillers. This trend mainly reflected the high requirements on whiteness.

But the trend in recent decades has not just been influenced by whiteness. The improvement in mineral dressing processes contributed to the steady growth in the product range of calcium carbonate fillers. For instance, only in recent years has it been possible to produce a dry-ground filler of high fineness and narrow particle size distribution (d50 = 0.9 µm and d98 = 5 µm: see Appendix "Definitions and Methods") in economic quantities. A similar situation applies to wet-ground products for the paper industry. By improving the grinding processes high opacity values were achieved in the slurries, without affecting the degree of whiteness.

Especially in the paper industry, marble products are today the preferred fillers and coating pigments. Their high degree of whiteness, however, also makes them interesting for the paint industry, because they can be used to reduce the proportion of expensive white pigments and titanium dioxide used in paints.

PCC has similar areas of application to the marble fillers and is additionally used in pharmaceutical products because of the possible purity of the product. The relatively high price of this artificially manufactured calcium carbonate is, however, an obstacle to its wide distribution, especially as it hardly differs from the natural carbonates.

Limestone products are frequently used in areas where the contribution of the fillers to the value-added of the finished product is comparatively low or raw materials of lower whiteness can be used with out compromising quality. Limestone is primarily used, for instance, in flue gas desulphurisation and forest fertilisation. There are, however, some high-quality grades of limestone which are used as fillers or pigments in the paper and paint industries.

The fillers made from chalk cannot be clearly assigned to one particular area of application either. There are high-grade types of chalk which in view of their positive characteristics are still used in the paper and paint industries, as well as in cable manufacture. And there are the other not so pure chalks which are used as raw materials for fertiliser production and similar areas of application involving relatively low value-added.

How long the current distribution of the calcium carbonate filler market will last cannot be predicted, because the main prerequisite for the use of calcium carbonate as a filler is that customers want to have a product with these specific characteristics or that a potential customer can recognise the advantages of a new product in detailed tests. This requires close contact and an intensive exchange of ideas and experience between manufacturer and customer.

This also means, however, that changing customer needs can impact strongly on the production processes for fillers. As, in addition, the competition from other products is considerable and the process technology is constantly developing, new improved calcium carbonates will come onto the market in future too. In the filler industry the need to innovate never ends.

3. From rock to filler

Many calcium carbonate deposits are directly usable as a source of raw materials, either as a feedstock for the production of cement or as crushed stone and aggregates, if the strength of the rock is high enough. Once the quality requirements for calcium carbonate filler are taken into account, however, the number of suitable deposits quickly reduces. Whether or not a deposit can be economically developed can only be determined by extensive prospecting.

Even if a raw material meets the quality requirements it does not mean that it can be extracted. Only if deposits of adequate size in high-grade quality are known in the right quantity and if the rock can be extracted without excessive technical investment is exploitation worthwhile.

But regardless of all the care with which a deposit has been selected, the properties of natural calcium carbonates only rarely match those required by industry and after extraction a mineral dressing process is usually necessary. And this must not just be technically feasible but must also fulfil economic criteria.

Not least, the extraction site and production facilities should be located close to potential customers so that the logistics costs are within reasonable bounds. Furthermore, it can sometimes be necessary to develop special product forms which make it possible to store and transport the material on favourable terms.

3.1 Prospecting

Before any extraction takes place the deposit must be closely explored. This applies to the mining of coal and ore just as much as to the extraction of industrial minerals. In all areas of mining, therefore, the same exploration methods are used to determine the quality and size of the deposit.

The work starts with the preparation of a precise map of the area being explored. The extent of the possible deposit is marked out. Rocks found on the surface give an initial indication of the quality of the deposit. By including the geological data of the area these results can be used to make an initial assessment of the richness of the deposit. In parallel, even at this early stage information on possible official ordinances and on the infrastructure of the surrounding region must be obtained, as this is of great significance for subsequent operation.

If the numerous individual recordings result in a positive overall picture for extraction, detailed surveys must start in order to test whether the deposit is worth exploiting. First, following a rough grid, boreholes are sunk which give a precise picture of the quality and size of the deposit. In complex intergrown deposits this drilling grid must be refined in order to exclude uncertainties in the assessment. In some cases it can even be useful to dig a prospecting trench or even an exploratory drift if underground mining is planned.

At the close of the explorations the following data must be available:

- Size of the deposit
- Quality and quality fluctuations
- Thickness and composition of the overburden to be removed
- Geographic extent and inclination of the deposit
- Foldings in the deposit
- Type and proportion of undesired foreign minerals
- Level of the groundwater
- Possible water channels
- Possible waste dump areas and volumes of waste

To avoid later conflicts and legal disputes from the outset, it is recommendable to carry out the necessary explorations in cooperation with the relevant authorities such as mining agencies and water authorities.

The results of the exploration permit an initial estimate to be made of the investment that will be necessary for extraction and subsequent processing. Together with the estimated investment expense for the infrastructure and the costs for product logistics this provides the basis for the decision as to whether the deposit is worth exploiting.

If the deposit is found to be workable the geological and mining exploration results are used to draft a model for the deposit which represents the course of extraction as follows:

- Removal and stockpiling of the overburden
- Opening of the deposit
- Development of the levels and ramps
- Sequence of different rock qualities over the course of extraction
- Estimate of selective extraction
- Requirements for mineral dressing
- Removal and stockpiling of waste materials
- Reclamation of the quarried site

This model not only represents the later sequence of quarrying over the course of extraction but also covers the necessary interventions in the environment and the recla-

Chalk extraction in Lägerdorf/Germany.

mation work after completion of quarrying. In view of the numerous regulations this is also urgently recommended because the costs for example of re-landscaping the site influence the total cost of extraction.

Before quarrying can start all official bodies involved in the approval process must accept this model. This is a long and drawn out process which can go on for several years. And only when all approvals have been granted is the prospecting regarded as completed because only then have all the prerequisites been met for developing the deposit.

3.2 Quarrying

The method applied to extracting a raw material primarily depends on the particular features of the deposit (underground or open-cast mining), but the characteristics of the selected raw material also influence the extraction technique. Calcareous rock appears in a wide range of different forms and different approaches are required for extracting chalk, which is a soft "earthy" material, and the hard rock of limestone and marble.

3.2.1 Chalk

Chalk is found in extensive deposits at depths of up to 80 metres and more. Because of its low compaction and strength, chalk can be extracted mechanically using bucket-wheel excavators or power shovels (see figure). This is highly favourable in terms of cost as the expensive drilling and blasting work normally necessary in quarrying is not required.

The object of chalk extraction is to obtain a raw material which leaves the quarry largely free of impurities. To develop a chalk deposit, therefore, the overburden of soil and sand is first removed by means of loaders or excavators in order to obtain a clean chalk surface for further extraction. This must be done very carefully because otherwise surface water and any contaminants in it can penetrate uncontrolled into the material to be quarried. If, in addition, large amounts of flint for example are embedded in the deposit (see figure) these are separately removed and stockpiled.

To achieve even extraction of the chalk, several levels are made in the quarry which are either worked simultaneously at a correspondingly high rate of production or consecutively. Depending on the distance, the chalk is transported from the individual extraction points to the dressing plant either by belt conveyor or truck.

3.2.2 Limestone and marble

As far as possible limestone and marble should be extracted by surface mining because underground working entails disproportionately higher costs. Only if the geological conditions require or the overburden is too difficult will the economic viability of underground mining have to be checked.

In open-cast mining the overburden, which in most cases consists of earth or clay material and which can have a thickness of several metres, is first removed. This overburden must be carefully stockpiled so that it can be used later when re-landscaping the mined deposit. It may be necessary to make a waste pile. Depending on the extraction strategy the overburden is removed continually, parallel to the extraction work or campaign-wise for opening up new areas of the deposit.

Chalk with typical flint intergrowth.

Limestone quarry in Burgberg/Germany.

The mineral is extracted in levels which can have a height of 10 metres and more. The exact height depends on the particular circumstances of the rock body. The level height is also influenced by faults in the rock and by inclusions of waste rock, which are frequently found in marble. If there are high proportions of waste rock, selective extraction is used: The quarry face is worked in such a way that waste rock and calcium carbonate are extracted separately as far as possible. The waste rock is taken to the waste pile and the marble enters the dressing process.

To carry out this specific type of extraction it is necessary to prepare a careful programme of drilling and blasting, with the boreholes being drilled according to the required size of broken rock. After blasting the raw rock must be of a size which permits economic loading and transport to the processing plant.

To create a clean working face and stable levels it is recommendable to incline the blast-holes up to 20° to the vertical. This results in favourable sizes of broken rock and clean breaks. The distance of the blast-holes to the quarry face and their distribution and diameter are set according to the characteristics of the rock but they also very significantly influence the composition of the broken material. The explosive used is also decisive for the configuration of the boring pattern and the broken material.

If all the broken material is not of the desired size, the excessively large pieces of rock have to be broken up by block-holing. A pneumatic drill is used to drive small holes into the lump of rock and small quantities of explosive are inserted to reduce it to a size which can be loaded and transported. At the quarry face a loader, bucket-wheel loader or excavator picks up the broken material and loads it onto a conveyor or a heavy-duty truck for transport to the primary crusher.

Blasting a marble quarry face.

Tracked drilling unit.

Wheel loaders are also widely used in mining operations.

Heavy duty trucks at work.

The chain of operations encompassing blasting, loading, truck transport and primary crushing must be carefully harmonised. It depends mainly on the extraction volume of the quarry, and all equipment must be configured in such way that optimal capacity utilisation is ensured in each phase. A large extraction volume requires large loaders and heavy-duty trucks as well as a large-capacity primary crusher. In this case the rock sizes resulting from blasting can be relatively big too. Smaller volumes of broken material require more-flexible operating equipment whose loading capacities must not be too high, in order to avoid idle time.

Standard equipment includes hydraulic excavators which exhibit high forces when picking up pieces of rock. As in most cases they are tracked these excavators are not affected by uneven rock that would otherwise cause severe wear. They can be flexibly deployed and can also handle work such as removal of the overburden or flattening of haul roads. Other standard units of equipment include bucket-wheel loaders (see figure). As their tear-out strength is not as high as that of excavators they can only handle smaller pieces of material. They are rubber-tyred which means that roadways have to be better prepared. On surfaces causing severe wear chains are used to protect the tyres. Wheel loaders can also be flexibly deployed. They have very short loading times and can serve several loading points at the same time.

Transport to the primary crusher normally takes place by heavy-duty trucks which can carry loads of 30 to 150 tonnes (see figure). As they are the most important component in the chain from the quarry face to the primary crusher, a careful selection of these trucks is essential.

For longer distances from the quarry to the dressing plant it can sometimes be recommendable to use a conveyor facility. So that the belts do not become too big the size of the material being conveyed must not be excessive. This requires primary crushing in a mobile or semi-mobile crushing unit located close to the quarry face upstream of the belt conveyor. In most cases an impact crusher or jaw crusher is used, which is fed by a wheel loader.

The primary crusher and conveyor unit should not be moved too frequently. A wheel loader is mobile enough to bridge even lengthy distances between the extraction point and the crusher. Nevertheless, careful calculation of the loader size and the distance is required in order to optimise the individual work steps.

If a shearer-loader is used the expensive primary crushing stage at the quarry face can be avoided. A shearer-loader produces a material with a diameter of less than 150 millimetres at the quarry face which can be loaded immediately onto a belt conveyor. These units can be deployed in marly limestone of low hardness, but in harder types of limestone the wear on the picks is so high that shearer-loaders are not yet viable.

Primary crushing

For primary or rough size-reduction crushers are used which apply pressure (jaw, gyratory crushers) or impact (impact crushers) to the material to reduce it to particle sizes of between 300 and 100 millimetres (see figure). Before the raw material enters the crushers it is often roughly screened using simple grizzlies. This removes clay and undesired constituents and also separates the pieces of limestone or marble which already have the required particle size and which would therefore only unnecessarily burden the crushers.

A **jaw** or **Blake crusher** consists of two crusher jaws arranged in v-formation. One of the two jaws is arranged to make oscillating movements against the non-moving jaw; the coarse material is crushed. The size of the feed material is determined by the width of the inlet opening and the crushing gap determines the product particle size.

The jaw crushers normally used in primary crushing can handle rocks of up to approx. 2 cubic metres at the rate of more than 1500 tonnes per hour. The level of throughput depends essentially on the size of the crushing gap; if a comparatively fine product is required the crushing gap must be narrow, which means a low rate of throughput. To attain a reasonable ratio for throughput and fineness, for primary crushers a size-reduction ratio of z = 5-9 is selected. If, however, jaw crushers are used as secondary crushers in mineral dressing the ratio is z = 3-6.

In a **gyratory crusher** a crusher cone moves eccentrically in a conical housing. As in the jaw crusher, pressure is applied to crush the material. Because of their large crushing chamber volume, however, gyratory crushers have a higher throughput. In addition, the rock can be fed from several sides. Nevertheless, the deployment of gyratory crushers is only cost-efficient at high throughputs, because their price is high owing to their elaborate design. Gyratory crushers are also used both in primary size-reduction (z = 10-12) and secondary size-reduction (z = 15). In the latter case product particle sizes of less than 4 millimetres can be achieved.

In an **impact crusher** a heavy rotor fitted with blow bars rotates around a horizontal axis inside a housing. Two movably suspended impact plates are fixed opposite the blow bars. The gap between the impact plates and blow bars determines the size of the product. The grinding chamber and the size of the inlet opening determine the size of the material that can be fed into the crusher.

The pieces of rock fed into the crusher are received by the rotor and crushed between the blow bars and impact plates. The size reduction by means of impact occurs mainly on the blow bars, where the energy is introduced into the material being crushed. The impact plates form the limit of the crushing chamber and return the material hit by the blow bars back to the rotor until it is small enough to pass through the gap between the blow bars and impact plate. Through variable design (shape of the crushing chamber, type and position of the impact plates, and shape and speed of the rotor) impact crushers can be built as primary crushers (z = 20-30) handling feed material sizes of 2 cubic metres and throughputs of 2,000 t/h and also as secondary crushers delivering products with particle sizes of less than 3 millimetres at throughputs of 20 t/h. Owing to the relatively high crushing speed, particular attention must be paid to wear because in the case of highly abrasive material the use of an impact crusher can very quickly cease to be viable.

The crushed raw material is then screened to a size of approx. 10 millimetres, according to further processing separated into wet and dry product and passed on to further size-reduction. The finer constituents, howe-

Machines for primary and secondary crushing (taken from Kellerwessel, p. 33 ff.):

a) Jaw crusher
b) Cone crusher
c) Impact crusher
d) Roll-type crusher
e) Hammer crusher

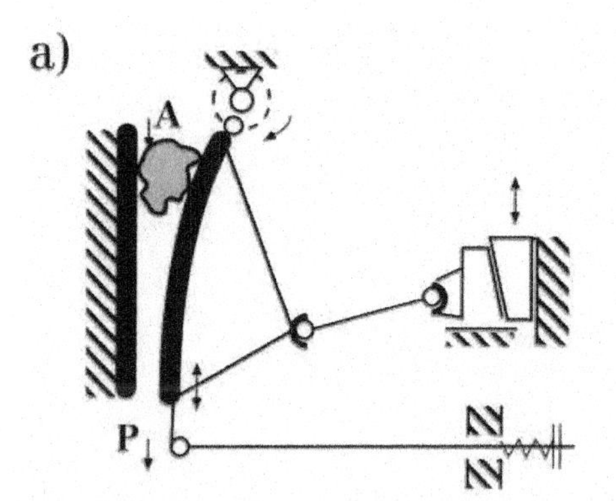

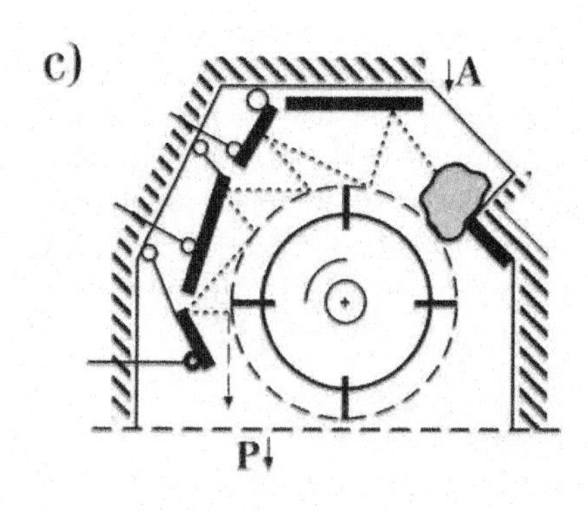

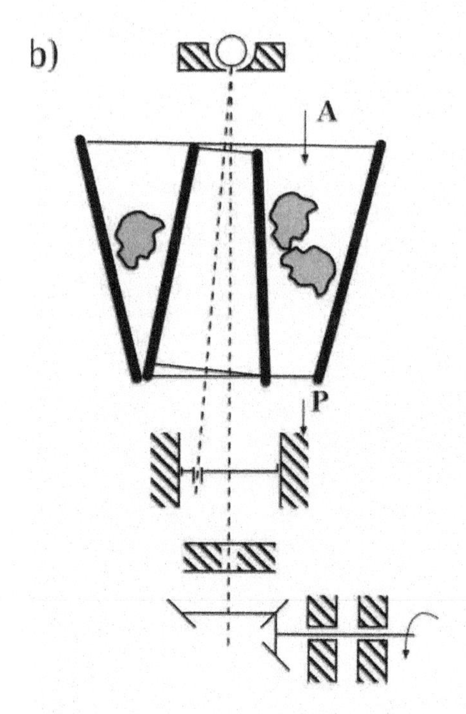

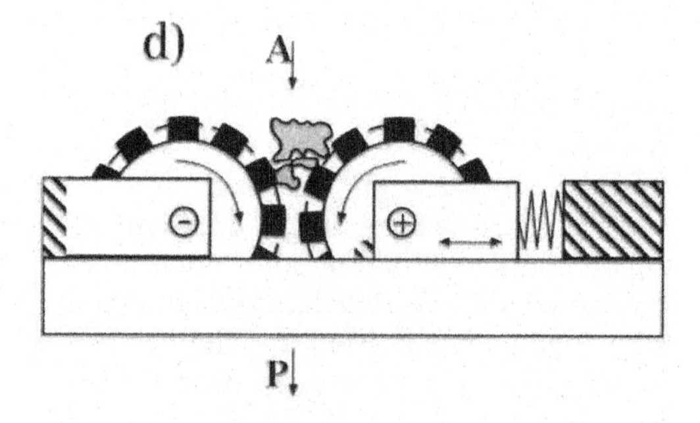

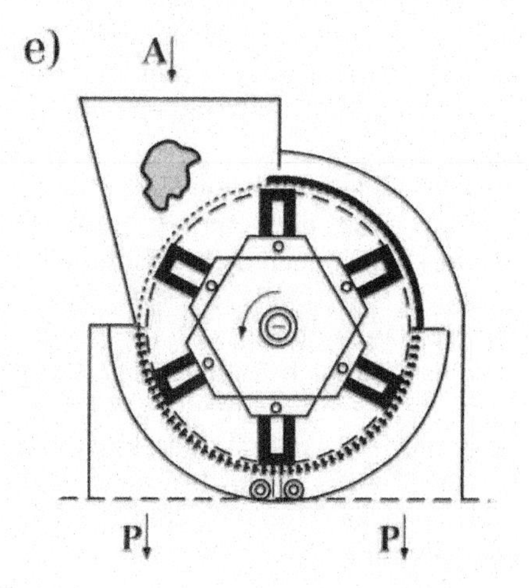

ver, jump one or several processing stages and at a suitable point are fed to the grinding process.

3.3 Mineral dressing

Mineral dressing has the purpose of converting an extracted raw material into a product tailored to a specific application. For this, several processing stages are generally necessary, which can be divided into the following main sections:

- Crushing, grinding and classifying
- Sorting
- Dewatering including drying
- Storage, packing, transporting

The individual stages do not necessarily take place consecutively. Frequently, for example, there is a sorting stage between two crushing or grinding processes and dewatering and drying operations occur at different points within the general mineral dressing process.

If we compare the dressing of calcium carbonate fillers with that of other minerals or ores we see that the individual stages do not differ fundamentally from the familiar processes – the crushing and grinding of limestone and marble are carried out using the generally deployed crushers and mills – but certain process stages have to be adapted to the particular characteristics of calcium carbonate fillers. For example, dry and wet fillers are ground using special mills, because the required fineness cannot be achieved with conventional mills and classifiers.

The process steps, technical equipment and machinery that are required in the specific case are determined by two factors:

- by the raw material with its specific characteristics such as hardness, lump size and wetness of the feed material, as well as impurities caused by accessory minerals
- by the product and its desired characteristics such as whiteness and particle size distribution.

Processing steps - Definitions -

The first size-reduction steps to which the quarried material is subjected are normally **crushing** steps effected by various means. Crushing may involve several stages before material is obtained that, for example, can be fed into a ball mill.

Grinding is carried out to reduce the grain to the particle size required for filler applications. Fine and ultrafine grinding are distinguished.

There is no definite particle size separating the terms crushing and grinding. Instead, the use of these terms is governed by the machinery employed.

The efficiency of comminution machinery is characterized by the ratio of reduction that compares the mean feed particle size with the average particle size of the product. In addition to this parameter, the particle size distribution in the product is also an important performance indicator. A narrow particle size distribution corresponds with a steep "grain curve", whereas a broad distribution leads to a flat "grain curve".

Classifying is defined as the separation of a collective of particles into different size classes. In dry processing, screens and/or air classifiers are used for classifying, whereas in wet processes, classifying involves the use of hydrocyclones or centrifuges. Classifying separates only on the basis of size and does not imply chemical or mineralogical changes.

Sorting is the separation of a collective of particles into components with different chemical and/or mineralogical make-up. An example would be the separation of a mixture of two minerals into components predominantly made up by either one of the two minerals. Common sorting processes are floatation and magnetic separation.

If this is transferred to the processing of calcium carbonate fillers the entire process can be divided into two parts: Firstly the processing is determined by the raw material side before, once a 'pre-ground product' has been attained, the required properties of the fillers move into the forefront.

Such a boundary cannot be precisely fixed and , inevitably, more of an educated choice takes place, because at the start of the dressing process the later end product is always considered. Finally, of course, it makes a difference whether a filler is made of chalk, limestone or marble. Nevertheless, in the first dressing stages up to the pre-ground product it is useful to make a distinction according to raw material, because each of the calcium carbonate varieties behaves differently, and even within the chalks, limestones and marbles there are differences between individual deposits. By contrast, in fine and ultrafine grinding a distinction has to be made primarily between wet and dry processes.

Extraction and processing of $CaCO_3$.

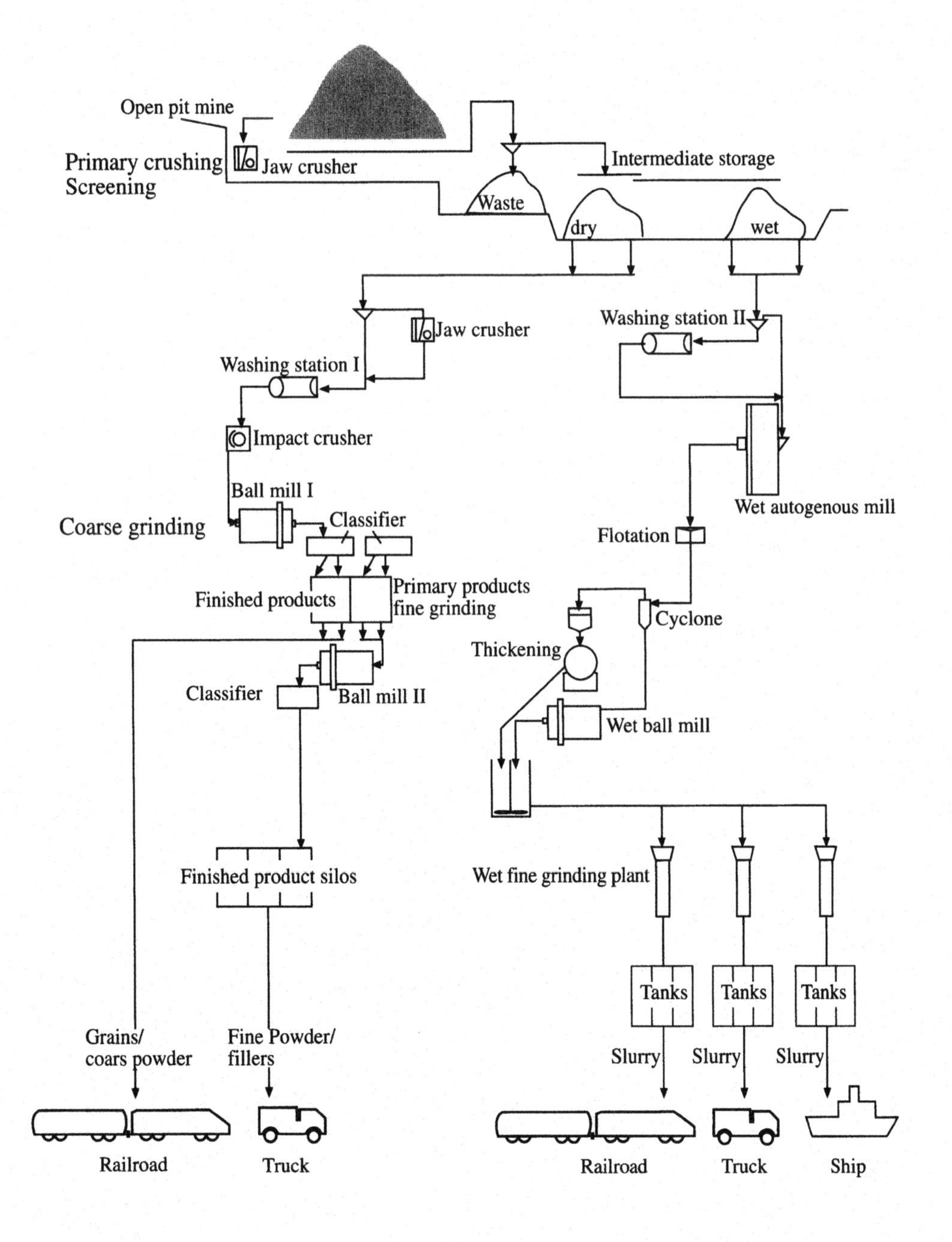

3.3.1 Production of the pre-ground product

Dressing begins with the clean extraction of the rock or its separation from coarse clay or from soil. However, not all raw minerals need to be cleaned, sometimes they are present in such a clean form that only crushing and grinding to the desired fineness are needed to attain products of high quality.

Chalk

Freshly quarried chalk in most cases has a water content of 20 percent and more when it enters the dressing process. Depending on the cleanness of the chalk, different process routes can be deployed.

Very clean, best-quality chalk, as is found in Champagne, is processed separately for the wet and dry product line. On the dry side the chalk is simultaneously ground and dried in an autogenous mill (Aerofall mill). The material leaving the mill is transferred to filler production. On the wet side a wet autogenous mill is used (see figure) which grinds or slurries the chalk to the concentration required for extremely fine grinding.

Usually, however, the chalk is interspersed to a lesser or greater extent with flint. It is therefore first slurried to separate flint and grid as far as possible. Then the slurried suspension is filtered, dried and transferred to further processing. Dry products are processed on separators, wet products are ground in special mills to attain the required fineness.

Limestone

The limestone pre-crushed and pre-cleaned in the quarry is successively size-reduced in the plant, where in addition to the crusher types already mentioned roll crushers and roller mills are used.

Roll crushers consist of two contra-rotating rolls working in parallel. The material to be crushed is fed from above, drawn into the roll gap and crushed by means of pressure (see figure, page 177). To improve the material draw-in conditions in the case of large lump sizes the rolls are fitted with cams or spikes. Such a roll crusher permits size-reduction ratios of up to z = 10. With a size-reduction ratio of z = 4-5 smooth rolls permit much finer product sizes. Gap widths and therefore particle sizes of less than 1 millimetre should, however, be avoided in fine crushing.

In a **roller mill** two geometrically defined bodies roll against each other. The material to be crushed is fed between these bodies, which are pressed against each other by gravitational, centrifugal or spring forces. A bed of several different layers of particles quickly forms. Crushing mainly takes place through the pressure exerted by the roller on the bed of material; to a small extent the friction between the particles also plays a role.

The oldest type of this mill is the edge runner, which has been known for several thousand years. For dressing calcareous rocks in particular the simply constructed ring-roll mills with throughputs of up to 100 t/h and the roller mill which can handle a throughput of 200 t/h and which also permits grin-

Wet autogenous mills of huge dimensions are used in the processing of limestone and marble.

ding/drying are important, as well as the high-pressure grinding mill which is like a roll crusher but which operates at distinctly higher pressures.

The machine most commonly used in primary grinding is the ball mill. Both the pre-ground products for dry and for wet fine grinding are first size-reduced in ball mills. For dry fillers the ball mill is also used in the fine grinding to product particle size (see section 3.3.2 "Production of fillers").

After primary size-reduction, secondary size-reduction and one or several grinding operations the limestone has the requisite size to be fed to the fine and ultrafine mills. The pre-ground product for the production of wet fillers has a top cut of d98 < 45 µm, for dry fillers it is d98 < 100 µm or finer.

Marble

In most cases the dressing of marble is more complex than the processing of limestone. This applies less to the size-reduction machinery – which is virtually identical for limestone and marble – and much more to sorting, because the requirements on whiteness for calcium carbonates derived from marble are high. Only rarely can white marbles such as Carrara be ground to make highly white fillers without having been sorted. In most cases they contain a series of dark accessory minerals, which are referred to in the rock analyses generally as ‚Acid Insoluble Residue' (AIR). The range of these minerals extends from graphite through sulphides such as pyrite to the dark silicates such as hornblende and amphibolite.

These minerals not only cause a deterioration in whiteness, they also have higher abrasivity than marble. The hard silicates are thus far less amenable to grinding than calcium carbonate, and they therefore become concentrated in the coarser fractions of the finished fillers, increasing the wear effect of these products.

For this reason marble, and limestone too, almost always undergoes a sorting process between the individual size-reduction steps (see box "Processing steps", box, page 178) in which impurities in and on the mineral are removed from the crushed material.

Sorting

Sorting is required for the following reasons:

- The raw mineral is contaminated on the surface with sand, clay or similar. This dirt can be removed by intensive washing.
- The rock can be badly embedded with dark foreign minerals which would reduce the whiteness of a ground calcium carbonate. These foreign materials can be removed by hand picking or, much more effectively, by optical sorting.
- The foreign minerals in the calcium carbonate are finely intergrown and would reduce the degree of whiteness on grinding. Fine intergrowth only occurs in marble; depending on the material, flotation or magnetic separation must be carried out.

Mineral washing and optical sorting are typical processes both for limestone and for marble and are carried out mainly at the beginning of mineral dressing.

Mineral washing is carried out in many dressing processes to clean sand or clay off the raw rock. In most cases drum or vibratory washers are used. In the washing process the important thing is to scour the surface of the rock so intensively that as far as possible all coatings of foreign material become suspended in the water and can be separated by subsequent screening. However, the washing process must not be so severe that the rock breaks and good white material passes through the screens along with the dirt.

If such water-intensive cleaning is not possible, "dry washing" is applied. In a rotating dry drum the dirty rock is dried by a stream of air and the clay constituents are comminuted to a fine powder. The contents of the drum are then passed to a screen whose lining is such that most of the fine clay can be separated without losing too much fine limestone.

View of flotation hall.

Optical sorting is based on the technique of picking in which undesired mineral constituents are removed by hand from a flow of material. Since TV cameras and other optical detectors have been able to distinguish different materials by shape or colour, attempts have been made to mechanise the picking process. Using the CCD chips and rapid computers available today it is now possible to distinguish approx. 4,000 items per second on the basis of their shape. Mechanical implementation of the signals is not yet quite as quick. Separation according to colour is now also possible, permitting a precise differentiation of the undesired constituents. Optical sorters have now reached a stage where they can be reliably deployed in operating practice.

The feed volume, however, is strongly dependent on the particle size. Whereas around 80 t/h of material with a particle size of 50 to 150 millimetres can be classified, the throughput drops to just a few tonnes with a particle size of 4 to 20 millimetres.

The cleaning process is completed for limestone after washing and/or optical sorting. The whiteness of the product cannot be increased any further. For marble, however, a flotation process is also often necessary in order to separate finely embedded minerals.

Flotation is a process in which a mineral phase of an ore is separated by floating to the surface of a pulp while the other mineral constituents remain in the pulp (see figure).

Mineral phases always have a higher density than the water or the pulp. A medium therefore has to be found which transports the desired particles to the surface of the pulp. This medium is finely dispersed air, to which the solid particles attach themselves.

If a solid is hydrophobic (water-repellent) the air displaces the water wetting the surface of the solid in a three-phase mixture of water, air and solid. The force binding the solid particles and the air bubble is determined by the strength of the hydrophobicity and the stability of the bubbles. As calcium carbonate does not naturally exhibit hydrophobicity, this has to be induced by adding reagents.

In the same way that the stability of a foam can be influenced by adding a foaming agent or frother the surface properties of the solid can be influenced by adding reagents. Collecting agents make the surfaces of a mineral hydrophobic while depressing agents increase hydrophilicity. The development of these reagents is now so advanced that virtually any mineral can be induced to float (direct flotation) or to sink (indirect flotation).

Because in marble the proportion of accessory minerals is smaller than the proportion of calcium carbonate, as a rule indirect flotation is applied, causing the accessory minerals to float to the surface. If direct flotation were applied the probability of undesirable material being entrained to the surface would be too high.

The reagents used in calcium carbonate flotation can either be mixtures which are matched to graphite, sulphides and silicates, or mineral-specific collectors are used consecutively. With them each mineral can be specifically induced to float. The strategy applied depends on the volume and composition of the accessory materials to be floated and the price and quality of the reagents.

Flotation is determined both by chemical-physical factors (mineral surface – reagent – pH value of the pulp – air) and by mechanical-hydrodynamic variables (flotation cell – dispersion – phase separation – foam formation). It can be divided into the following steps:

- Treatment of the mineral phase with reagents
- Suspension of the pulp
- Admixture and distribution of air
- Formation of a laden foam on the slurry surface
- Removal of the foam

The first three tasks stand in contradiction to the last two. First it is necessary to create high turbulence in order to obtain a good distribution of solid material and air in the water and as high a frequency of contact as possible between the solid material and the air. This is required to give all particles the opportunity to adhere to a bubble. Then the particle/bubble complex must be able as far as possible to rise unimpeded to the foam bed. As far as possible unimpeded because any mechanical loading can lead to the destruction of the complex. In a flotation unit, therefore, the mixing zone in the lower part is separated from the contact zone in the upper part of the cell.

In order to float an ore the mineral constituents must be in a liberated or "unlocked" condition. Only if they are available liberated the collecting agent can attach to the desired mineral surface and effect selective separation. The particle size to which an ore has to be ground depends on the degree of intergrowth – i.e. the size of the embedded particles in an ore which have to be separated. In the case of marble this is the size of the embedded silicate particles and the other ancillary constituents.

In most cases grinding to an upper particle size of 100 to 200 micrometres is enough. Grinding takes place in a closed circuit on a wet autogenous, rod or ball mill. The aim should be to attain as steep a particle size distribution as possible of the floating material, because the finest components consume a very high quantity of reagents owing to their large specific surface. Moreover, the finest components cause an increase in misplaced material as the fine and therefore light particles are very easily entrained with the foam and can thus enter the silicate phase for separation.

If there is a very varied degree of intergrowth, flotation can be subdivided into several stages. In this case the coarse material from a classifier is initially treated in a coarse grain flotation process before being returned to the mill. The coarse silicate constituents are floated off and thus taken out of the grinding process. This is because the hard silicate particles can increase the wear on the mill and should therefore be removed as quickly as possible. The fine material from the classifier is then transferred to the main flotation process.

The decision to implement primary flotation and a joint or separate flotation process depends on the intergrowth of the ore and the content of foreign materials. Each rock must be closely examined to determine the type and quantity of accessory minerals and laboratory flotation must be carried out to draw up a flow sheet for the flotation operation.

Flotation has evolved as the most important process step in the sorting of marble. It can be used for processing a large proportion of extracted marble. Sorting will not succeed, however, if there is fine intergrowth of iron hydroxide or graphite.

In such cases recourse can be had to **magnetic separation**. In marble processing it is used before flotation to separate abraded iron particles and ferromagnetic minerals. HGMS separators also enable some paramagnetic minerals to be separated.

Modern separators based on supraconduction can develop field strengths of up to 5 teslas. These separators make it possible to separate ultrafine particles from slurries for the paper industry and thus to increase the degree of whiteness further.

While flotation as discussed above is able to economically process suspensions with a top cut of 100 micrometres, the new magnetic separators can handle much finer products if they contain magnetic minerals. A further advantage of magnetic separation is that it is not necessary to deploy chemicals and the burden on the environment is therefore considerably reduced. As the use of magnetic technology in processing marble is still very new, intensive studies and machine developments are required in order to optimise these processes. The machines available on the market have already attained a size which permits industrial deployment.

After flotation and/or magnetic separation marble too exhibits the necessary whiteness and adequate cleanness to be processed into fillers. Depending on whether further processing will be dry or wet, the marble has to be ground to the requisite particle size for the particular pre-ground product.

3.3.2 Production of fillers

Fineness, or more precisely the particle size distribution, is the most important criterion for the differing use of fillers. With the particle size distributions achievable today of $d_{98}<1.0$ µm areas are being entered which would have seemed inconceivable just a few years ago. An enormously increased energy requirement is still required, however, because the energy input increases with the fineness of the product to be ground. And grinding has to take place in a wet process because dry mills cannot achieve these levels of fineness at reasonable expense. But the particle sizes of dry-ground fillers still meet the requirements for many purposes, and 60 percent of calcium carbonate fillers are produced in a dry process.

Dry fillers

Dry fillers are produced in almost all production plants: of chalk, limestone and marble. Depending on the quality of the raw metals the pre-ground products are in dry or wet condition. If further grinding takes place in dry mills the wet primary product first has to be dried. Depending on the fineness of the material, flash or drum dryers are used. If the wetness of the feed material is not too high, grinding/drying can also take place in the mill. For this the heat from the grinding process is utilised, because the energy input for grinding is converted almost entirely into heat.

By far the most common type of mill used in dry processing is the ball mill. A **ball mill** is a drum mill which uses steel balls as grinding media (see box). It is a universally deployable, sturdy and low-maintenance grinding machine which can operate wet or dry, although a dry mill normally cannot

Drum mills

A drum mill is basically a cylinder with conical faces which is filled with the material to be ground and a grinding medium and which rotates around the longitudinal axis. Size reduction is achieved by impact and friction of the falling grinding media between which the material being ground is located.

The grinding capacity is a factor of the mill's volume, the ratio of length and diameter (L/D ratio) and the mass of the grinding media employed. The size of the grinding media is determined by the feed's grain size. Fine grinding in drum mills requires small grinding media that ensure a sufficient number of contact points with the material. However, a minimum size of the grinding media of some 10mm to 12mm applies generally since media of smaller size do not carry enough kinetic energy for efficient size reduction. In coarse grinding, rods are usually used as grinding medium, whereas spherical media or so-called cylpebs (cylinders with a quadratic cross-section) are employed for fine grinding. Energy input into the material being ground is a function of the specific weight of the grinding medium, which should therefore be as high as possible. Cast-alloy steel balls are widely used for this reason. In applications where metallic contamination of the product is to be avoided, media made of aluminium oxide or ceramic materials are used. Quartz pebbles are a very cost-efficient grinding medium for highly abrasive materials.

Apart from the type of grinding medium used, the mill speed is crucial for grinding performance. An important limiting factor in this respect is the "critical" speed (n_{crit}). At the speed n_{crit}, the centrifugal forces acting upon the grinding media are equal to the gravitational forces. As a result the grinding media adhere to the mill wall, which causes the grinding efficiency to drop to zero. At speeds below n_{crit}, the grinding media form a cataract coming down right before the mill's peak when the centrifugal force and friction are no longer able to hold them on the wall. The grinding media fall in a parabolic trajectory and impact on the material resulting in size reduction. Further lowering of the speed causes the grinding media to roll over each other in the form of a cascade, resulting in low grinding efficiency.

The actual behaviour of the material being ground depends on numerous factors including the size distribution of the grinding medium, the mill's degree of filling and the coefficient of friction between the material and the grinding medium. The critical speed therefore has to be determined empirically.

The high mechanical stresses caused by impact and friction make it necessary to clad the drum's wall to protect against wear. Usually, the wall is lined with steel or ceramic panels or protected by means of rubber elements.

Drum mills are among the most important processing machines. Their uncomplicated design is a main advantage providing ease of maintenance and ruggedness. Also, they can be tailored to the requirements of practically any application by variation of the L/D ratio and selection of the appropriate grinding medium. An additional advantage is this type of mill's suitability for either dry or wet processing. The autogenous mills and ball mills used in the production of calcium carbonate fillers fall under this class.

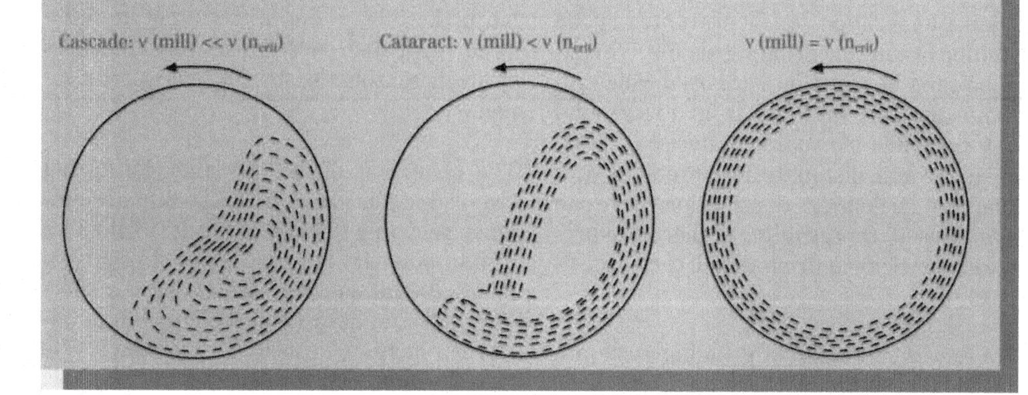

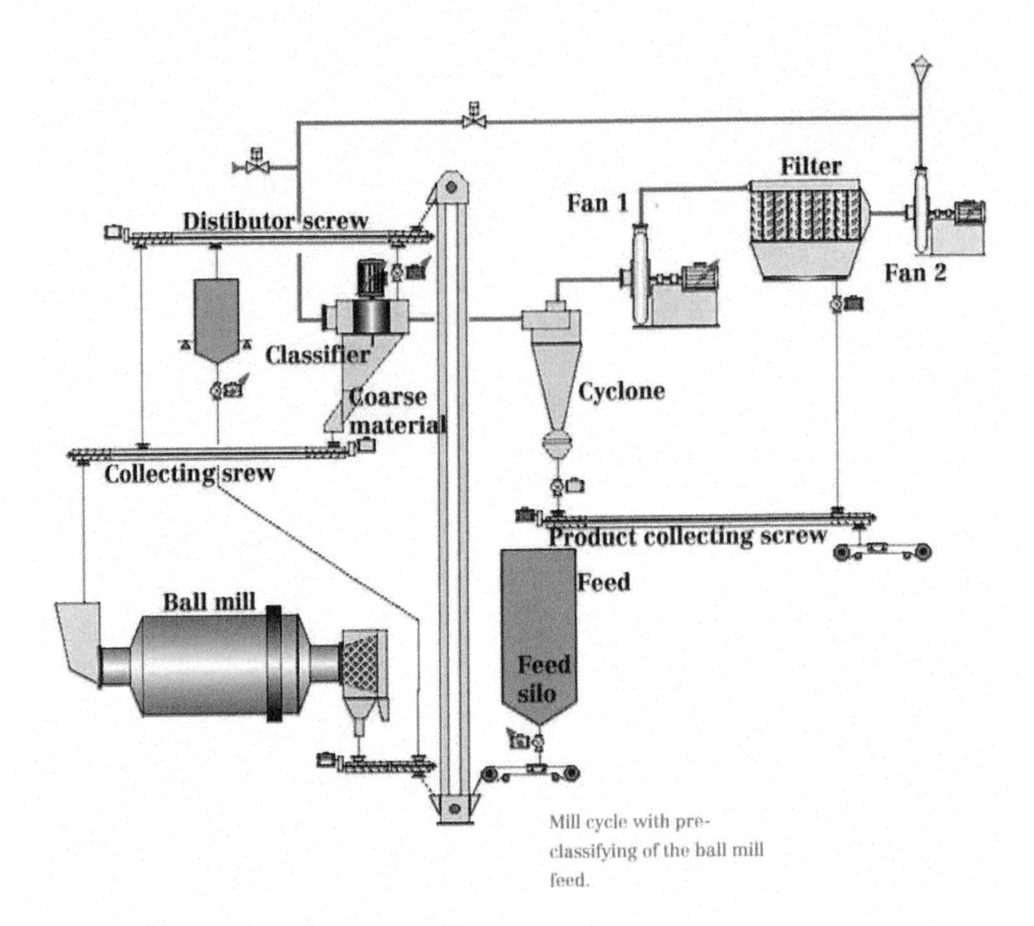

Mill cycle with pre-classifying of the ball mill feed.

produce products which are much finer than d98= 100 µm – because if the proportion of fine material in the mill is too high the grinding media 'fur up' i.e. agglomerates from the ground material adhere to the grinding media, reducing the impact between the material being ground and the grinding media and hindering further size-reduction.

This agglomeration can be prevented by the addition of suitable grinding agents. If in addition suitable classifiers are used which separate the fine material from the ground material while the coarse material is returned to the mill, distinctly finer products with a narrow particle size distribution curve can be produced. Dry grinding is therefore preferably carried out in a ball mill cycle (see figure).

The requirements for a grinding cycle are simple: The ground material must leave the mill as quickly as possible and in the classifier the already finished product must be separated as completely as possible. The longer the ground material dwells in the mill the greater is the possibility that material which has already been finish-ground will be further size-reduced, resulting in undesired ultra fine constituents and unnecessarily increasing the specific energy requirement.

The behaviour of the raw material during size-reduction has to be taken into account when selecting the mill: A product with a lot of fine material is produced in a relatively long ball mill. For less fine material a short ball mill attaining favourable grinding behaviour needs to be used. The grinding cycle must be equipped with adequate classifying capacity. If the size-reduction behaviour of the calcium carbonate is very favourable a roller mill can even be used, requiring a lower energy requirement than a ball mill.

Just as important as selecting the suitable mill type is operating the mills in a close circuit with a classifier because the energy deployed for grinding can only be minimised with an effectively separating classifier.

Screens are not suitable for classifying particles of this particle size. The finest screens used in the industrial minerals sector have a mesh width of approx. 100 µm, and wet screening can even be deployed at approx. 40 µm. Finer screens are conceivable – they are used in the laboratory – but the requisite expense of investment and maintenance is excessive.

Bladed rotor separators

Exact separation with a minimum of misplaced particles in the fines and in the coarse material makes high demands on the machinery that currently are only met by advanced bladed rotor separators (see figure). One face of the bladed rotor revolving in the spiral housing of such machines is shut whereas the other face features a circular opening for withdrawal of the product. Air is conducted into the spiral housing from below to flow through the rotor and exit the machine through the rotor's opening in the upper face. The separation conditions and thus the separation grain size can be adjusted directly at the periphery of the rotor.

When the machine is running, a ring of particles is formed around the rotor whose size is close to the separation grain size. The feed material penetrates this ring and the fines pass through it whereas the coarse grains migrate back into the coarse material chamber. As the ring of material around the rotor is a dynamic phenomenon a certain number of coarse particles can be entrained into the rotor and some fine particles will not be able to enter the rotor. Bladed rotor separators must therefore be designed very carefully to ensure the kind of precise, even and vortex-free flow which enables the separation conditions to be met as closely as possible.

For flow behaviour reasons, most classifiers of this type have rotors with l/D = 1. However, variants with elongated rotors (l/D = 2) are also possible. Such types feature extraction of the fines on both faces of the rotor. Advanced design concepts of this kind were the basis for compact machines with an air throughput of over 120,000 m^3/h and excellent yield in products with d_{40} = 2 µm and steep separation characteristics. Such machines are the biggest classifiers built hitherto for products of this fineness.

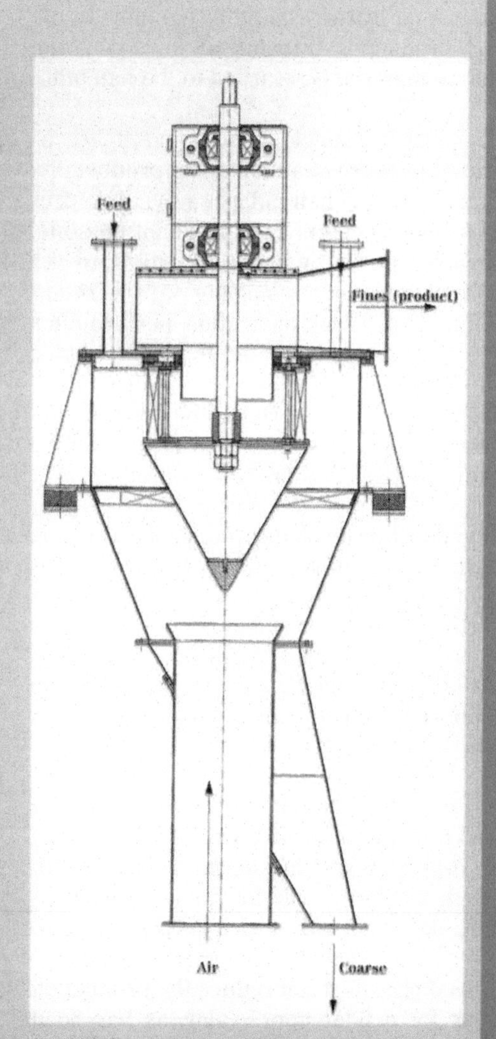

Air classifier.

Air classifiers function according to the elutriation process: If an air stream of specific speed is directed through a comminuted rock consisting of particles of different particle size it entrains all the particles whose rate of fall is less than the up-current speed of the air. If, however, the rate of fall is equal to the up-current speed the particle hovers; if it is larger it falls. As the rate of fall for particles of the same density depends on the diameter, the material can be classified according to particle size (see box). The entrained particles are kept separate from each other by corresponding equipment, the end product is transferred to a silo, the coarse material is returned to the ball mill.

The classifier should as far as possible separate the entire quantity of end product discharged by the ball mill. In a dynamic process such as classifying this is not possible, however, as the material returning to the mill always contains some finish-ground components. The separation performance of a classifier is indicated by the separator efficiency. The efficiency (η) indicates the proportion of a material in the desired particle size which is separated from the feed material by a classifier.

As separation involving finer particle sizes is always more difficult, the separator efficiency of a classifier decreases as the separation particle size gets smaller. Good classifiers achieve efficiencies of $\eta > 80$ percent at a top cut of 24 µm and $d_{50} = 4$ µm, but the value drops to $\eta = 60$ percent at a top cut of 10 µm and d50 = 1.8 µm.

The values stated relate to classifiers with a feed rate of approx. 40 t/h and a product rate of approx. 6 t/h. In smaller classifiers, in which the flow conditions are easier to optimise, separator efficiencies can be increased.

A good classifier can reduce the production costs for a filler appreciably, as has been shown in numerous tests and the experience of many years of operating practice. A grinding circuit consisting of a mill and classifier always saves grinding energy, because the additional expense for classifying is considerably less than the amount of energy saved in the grinding process as the mill is not unnecessarily burdened with product which has already been finish-ground.

Grinding to product particle size without classifying is therefore the least favoured way of grinding. Without classifying the mill can be given only a very low loading to ensure that the entire material is ground to the desired product particle size. Through the repeated grinding of the individual particles a lot of ultrafine material is produced which is not desired. Short retention times in the mill in connection with a good classifier of adequate capacity are necessary to cost-efficiently attain products with a steep particle size distribution.

If the intention is to produce several products at the same time in the grinding process a ball mill has to be combined with several classifiers in a cycle. Depending on the number of classifiers deployed various possible combinations arise (see figure)

The production of combined products reduces production costs considerably, because in a ball mill only the coarsest particle sizes have to be produced, the finer products are obtained as an 'extra'. In order to obtain enough fine material in the output from the ball mill, however, the entire cycle has to be carefully configured.

The ball mill cycle with air classifiers is also subject to limits: The economic upper limit for bulk fillers is currently a fine product with $d_{90} = 2$ µm. If finer products are required small grinding media have to be used and at the same time enough energy must be introduced into the grinding media/ground material mixture. This can only be accomplished with agitated ball mills.

In order to introduce enough energy into a charge of grinding media they have to agitated. **Agitated ball mills** consist of a vertical cylinder with a large L/D ratio in which a shaft with agitator elements rotates. These agitator elements are distributed over the entire length of the shaft and agitate the grinding media – high-strength ceramic balls with a diameter of approx. 2 millimetres. These balls have enough contact surfaces with the material being ground to make products with d_{98}=10 μm. Finer products can be attained in a mill cycle with a classifier.

Material is fed into the mill from above. The material being ground is moved from top to bottom with the grinding media. Both, the material being ground and the grinding media, are discharged from the bottom of the mill and separated by screening. The ground material is then fed into a classifier. The fine material enters a silo, the coarse material is returned to the mill with fresh material and grinding media.

The advantage of these mills is the high energy input into a compact mill volume. This energy input is determined by the rotational speed of the mill shaft and the density of the grinding media. The mill can be provided with a ceramic lining to prevent the ground material being contaminated by abrasion. The grinding elements of the shaft can also be protected by a lining of ceramic or wear-resistant material. To prevent agglomeration grinding aids are used in ball mills.

Wet (suspended) fillers

Wet fillers are used in the paper industry and they must therefore be matched to the requirements of the paper and the paper machine. This can happen in three different ways.

- Dry grinding followed by mixing to form a slurry of the required solid concentration.
- Wet grinding at the solid concentration required for fillers or coating pigments.
- Wet grinding at low solid concentration followed by increased concentration.

Dry grinding followed by mixing into a slurry in the paper plant is now only carried out in exceptional cases, for example when the transport or storage situation requires. The majority of fillers and pigments are wet-ground, in the final concentration.

Firstly a suspension with a solids content of over 70 percent is produced from the primary product. Normally a mixture of 70 percent calcium carbonate and 30 percent

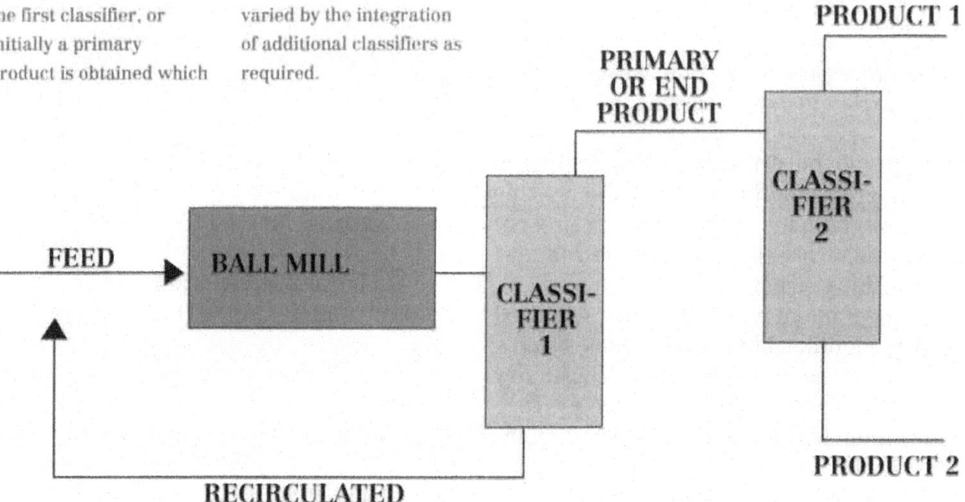

Classifier/ball mill cycle. In this cycle a product can be obtained directly after the first classifier, or initially a primary product is obtained which is then separated into 2 end products by classifier 2. The pattern can be varied by the integration of additional classifiers as required.

water is unable to flow. But the filler manufacturers have developed highly active dispersion agents specially for the use of calcium carbonate slurries in the paper industry which permit such concentrations without affecting the complex chemistry of a paper machine.

This highly concentrated suspension is ground in wet stirred ball mills because only such a mill permits the introduction into the slurry of the energy required for ultrafine grinding – depending on the fineness 60 to 200 kilowatt-hours per tonne of product.

Wet stirred ball mills are similar in construction to the dry variants, only the configuration of the grinding elements is different. Whereas in dry grinding rods are mainly used for the agitation, in wet grinding discs of diverse configuration are deployed along with rods in order to attain as smooth and effective an energy input as possible.

3.3.3 Other processes

Once the fillers leave the fine-grinding mills their processing has in most cases been completed, but for some special applications a post-treatment stage is required. Final drying of the fillers and surface treatment with stearates in particular are frequently encountered.

Drying

In the processing of calcium carbonate drying is carried in two areas:

- Drying of the primary product which is being processed to make dry fillers. This removes both the natural wetness present in the mineral and the wetness introduced by washing or flotation.
- Drying of fillers which are to be deployed in dry condition but which can only be produced by a wet process because dry methods cannot cost-efficiently produce the product quality or the desired volumes.

As drying is a highly energy-intensive process, prior mechanical dewatering should as far as possible be carried out. In the case of coarse material screens are suitable for this, while filters or centrifuges can be used in the case of fine and ultrafine products.

3.3.4 Production of PCC

PCC can be produced in different ways. Familiar processes are precipitation with carbon dioxide, the lime soda process and the Solvay process in which PCC is a by-product of ammonia production.

Precipitation with carbon dioxide is the most frequently deployed process, especially in the on-site facilities of the paper industry. As such facilities are located directly on the user's site the PCC slurry produced can be fed immediately into paper production. In off-site facilities, which produce both slurries and dry fillers for different consumers, precipitation with carbon dioxide is also usually applied.

The prerequisite for a PCC of high quality is a clean limestone or lime; any traces of iron or manganese can reduce the whiteness of the product sensitively. The lime is initially slaked to the calcium hydroxide (milk of lime) and then added to the reaction vessel as a thin suspension. There carbon dioxide is added until the calcium hydroxide has been converted completely into calcium carbonate. The duration of the reaction can be assessed and controlled by the course of the pH value. Overall the following reactions take place in the production of PCC:

$$CaCO_3 \quad CaO + CO_2\nearrow$$
(Calcination, $\Delta H = \angle\, 3130$ kJ/kg CaO)

$$CaO + H_2O \quad Ca(OH)_2$$
(Slaking, $\Delta H = +\, 1134$ kJ/kg CaO)

$$Ca(OH)_2 + CO_2 \quad CaCO_3 + H_2O$$
(Precipitating, $\Delta H = +\, 1996$ kJ/kg CaO)

By adjusting the reaction conditions such as pressure, temperature and time and by adding chemicals the crystal form and particle size distribution of the PCC produced can be influenced. To keep the PCC free of any agglomerates or insoluble constituents both the milk of lime and the finished product are classified.

In an on-site facility the finished product passes directly to the paper machine. In an off-site facility, however, it has to be thickened to similar concentrations as the GCC slurries in order to obtain favourable transport conditions. If dry fillers are required, mechanical dewatering and perhaps also drying have to be carried out. Like GCC products PCC products can be surface-treated in order to attain particular characteristics for use in plastics or rubber.

Bagging machines in action.

3.3.5 Storage and packing

Storage and packing form an important part of the production process, even though both are frequently regarded as necessary evils because they add nothing to the product. But only with adequate storage in sizes of packaging which the customer accepts can a logistics system be built up that permits a reliable supply to the customer. In some cases this requires considerable expenditures within the production plant.

Liquid calcium carbonate products are transported in trucks, railway wagons, river vessels or sea vessels. Post-production storage and intermediate storage is in tanks of up to 3,000 cubic metres. Dry products are stored in slender silos of up to 1,000 cubic metres. In designing the silos the flow characteristics of the products have to be taken into account. For example, products which do not flow very well require a steep cone to guarantee even discharge. In addition the flow properties can be improved by installing aeration floors. Rotary airlock feeders or screws adapted to the product properties and the quantities to be discharged serve as discharge units on the silos. Apart from these standard devices there are numerous special discharge units which are adapted to difficult flow properties.

Very fine products of low bulk weight can exhibit flow properties which are similar to those of a liquid. For these special attention must be paid to the sealing devices on the silo discharge unit and to the transfer points in the transport sequence. If the sealing devices are not carefully selected uncontrolled flowing of the product can occur.

Dry fillers are shipped mainly in two forms:

- Loose in silo cars or trucks which are filled directly from the silo.
- Packed in bags of 25-40 kilograms which are stacked on pallets and loaded onto transport vehicles.

A further possibility is packing in big bags with weights of 0.5-1 tonne according to the bulk density of the product. This type of shipment is not the most popular because the packing, storage and handling of the big bags is far less convenient than with normal bags.

The bags are filled on packing machines (see figure). The valve bag with only one opening (valve) on one corner is matched to the requirements of the product being packed. It can be made of multi-layer paper in order to obtain increased strength. The bags may have small holes to ensure ventilation of the contents, which makes for stable pallet layers. The suitable bag size and type for a specific product is determined with the bag manufacturer and the supplier of the packing unit.

There are two designs of packing facility: In-line packing units for small and medium-size filling quantities and roto-packers for medium-size to large filling quantities.

A high-performance roto-packer can fill more than 2,000 bags per hour if the flow characteristics of the product permit. For small throughputs the empty bag can be placed by hand, while for larger outputs bag placers are used which take the bag from a roll or a magazine and automatically guide it to the filling nozzle.

The filled bags are then conveyed to the palletising unit and packed to form pallets weighing around 1 tonne. To protect the bags the finished pallets can be covered with shrink wrapping or sheeting. The capacity of the palletising units is such that they can be deployed downstream of a roto-packer without the need for intermediate stacking. After

Just-in-time delivery made possible by round-the-clock loading of silo trucks with dry calcium carbonate powders.

palletising the pallets enter an intermediate store from where they are loaded onto the transport vehicles.

Bagging and palletising units also form a complex and expensive part of production. Not least because of the high investment and maintenance cost their installation requires precise planning and coordination with upstream production.

3.4 Logistics – the route to the customer

An important "property" of a product for a processor is the assured availability of the material at any time in the required quantity. To ensure that the desired quantities are available, production has to be built up in the required magnitude including a reserve capacity which can be deployed for production in the event of technical difficulties. Additional security can be achieved by making the products of several production sites interchangeable.

The latter requirement can be easily met in other industries by means of uniform, standardised primary products. In the case of a natural primary product, however, disproportionate efforts are required: even just a low level of impurities can change the whiteness of a calcium carbonate product considerably. Being able to make products of the same whiteness and yellowness index and of the same particle size distribution is therefore an essential prerequirement for the interchangeability of products from different deposits.

Interchangeability can reduce the necessary reserve capacities at the individual production sites. To ensure supply reliability, however, a logistics system has to be set up which permits rapid switching of supply from one production site to the other. At the same time the transport costs must be kept within economically acceptable bounds.

Even if the logistics system meets all these requirements it will still be expensive. The logistics costs for calcium carbonate are on average 25% of the selling price, but they can vary considerably. In some cases it is therefore worthwhile developing completely new forms of shipment, as reflected in the example of the supply of calcium carbonate to the paper industry.

Cost-efficient transport requires a cost-efficient form of shipment for the product and adequate buffer capacity in order to be able to balance out bottlenecks, as well as a restricted number of products which cover as wide a range of uses as possible so that storage capacities can be minimised. Dry powder is one shipment form of calcium carbonate fillers. This is stored in silos, transported to the customer in special trucks, blown into silos again, mixed upstream of the paper machine and finally supplied to the process. Powder is therefore one possible shipment form, but is it the cheapest?

Owing to their low bulk density of 0.4-0.8 tonnes per cubic metre, dry fillers need a large storage and transport volume. A relatively energy-expensive pneumatic conveying system has to be deployed and finally the customers must suspend and disperse the dry powder to make it ready for the paper machine to use.

It therefore seemed appropriate to "liquefy" the filler in order to improve handling, and the idea of calcium carbonate slurry was born. A calcium carbonate slurry is a suspension of water and calcium carbonate. The requirements for a slurry are:

- high solids content
- good rheological behaviour
- high stability
- little influence on the downstream processes
- good dispersibility

The high solids content minimises the storage and transport volume, and stability makes storage easier. The rheological behaviour influences the pumpability of the slurry and its behaviour in the paper machine. Dispersibility is necessary to dilute a slurry to the solid concentration needed on the paper machine. The chemistry of a slurry must be compatible with the paper machine.

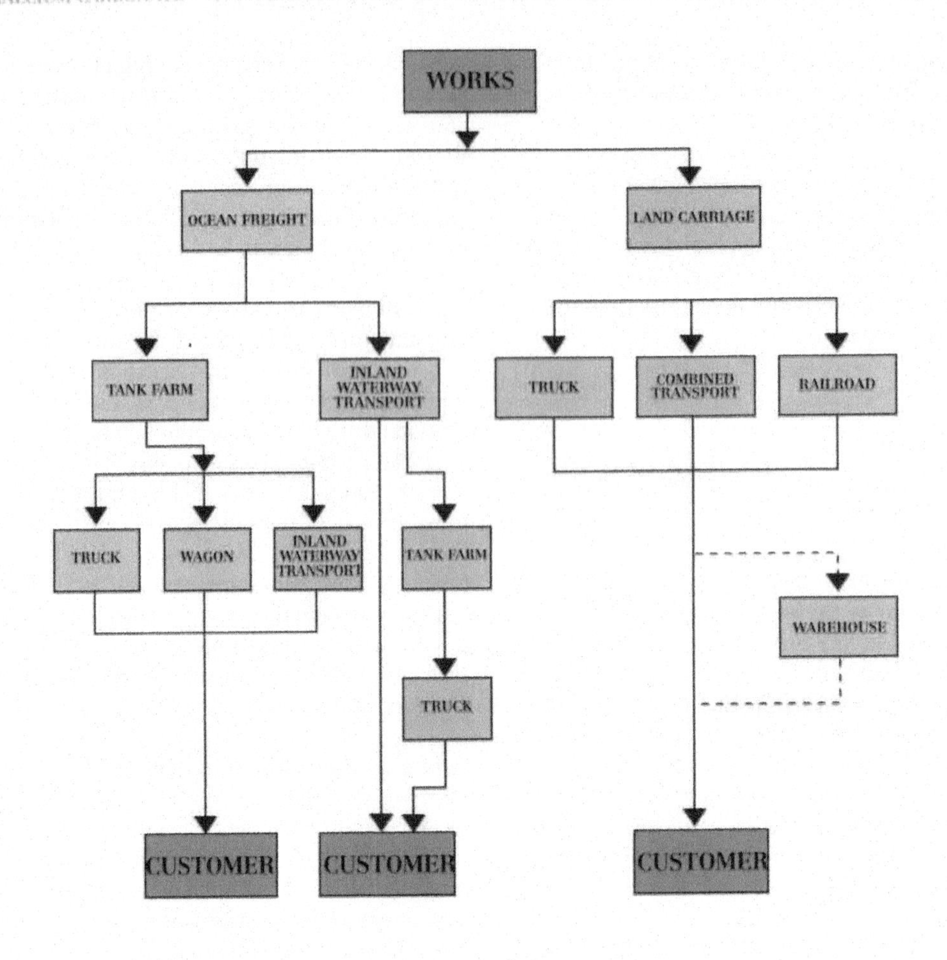

Logistics – the route to the customer.

To understand the chemistry of the slurry it is necessary to know some slurry properties. A slurry is produced with solid contents of up to 78 percent calcium carbonate. Without the admixture of dispersion agents this would be a solid sludge – moist but breakable and definitely not liquid. By contrast, a slurry has a viscosity of approx. 300-500 millipascal seconds (mPa*s); it flows easily, is pumpable and amenable to storage. Only with the development of suitable grinding processes and in particular suitable dispersion agents did it become possible to produce such slurries. Developments proceeded hand in hand. New dispersion agents facilitated more highly concentrated grinding, new findings in grinding technology set higher requirements for the dispersion agents.

Today transport densities of approx. 1.5 tonnes of calcium carbonate per cubic metre of slurry are attained with a 78-percent slurry, which is 2-4 times the transport density with dry calcium carbonate. If the viscosity of the slurry and its stability are right, the slurry will meet all the requirements for shipment: it can be cost-efficiently stored, is easy to handle and cheap to transport. If in addition the requirements in respect of dispersibility and compatibility with the chemistry of a paper machine are fulfilled, an efficient logistics network can be built up.

Slurry tanker – a highly efficient and at the same time ecologically beneficial mode of transport.

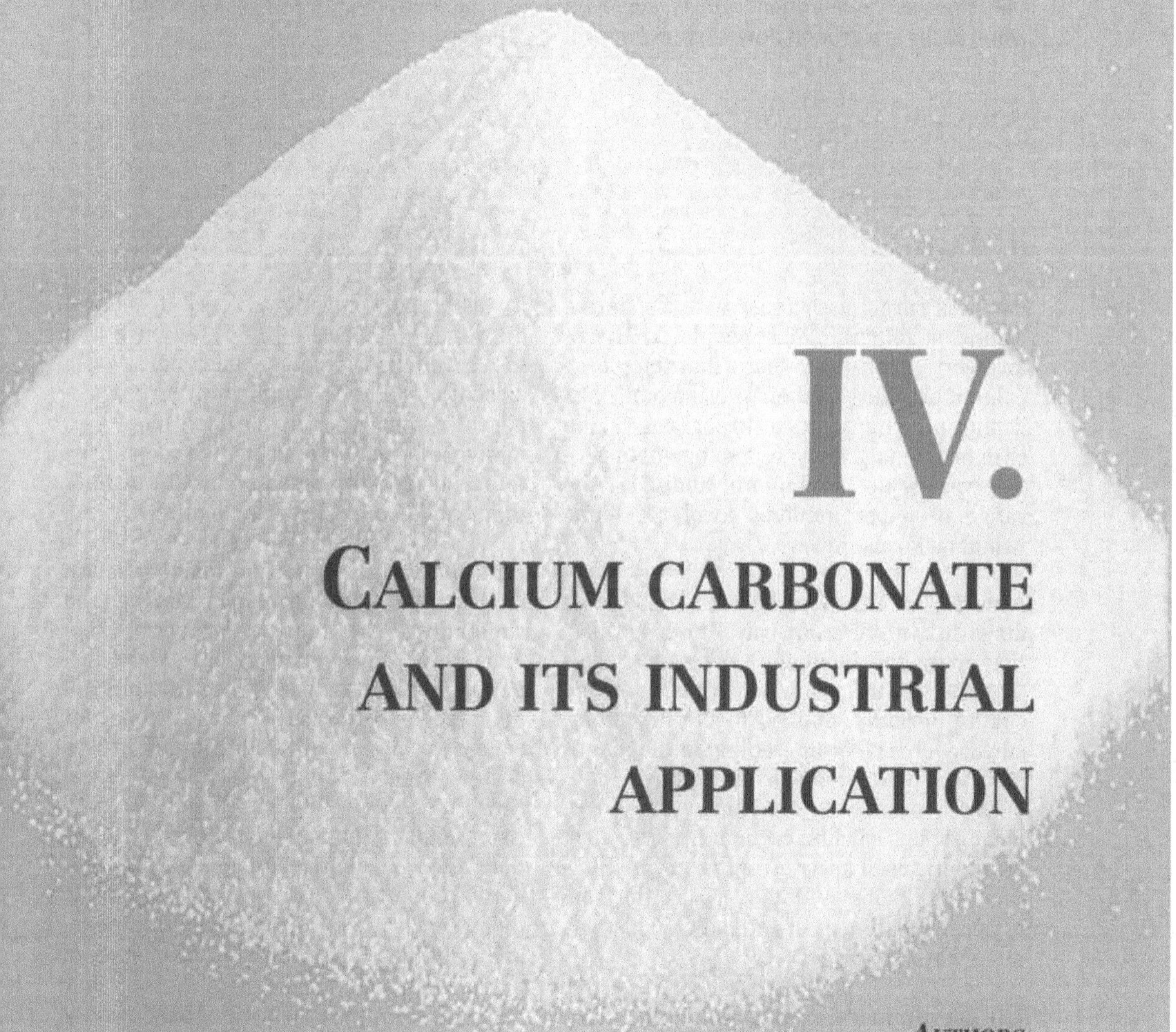

IV.

Calcium carbonate and its industrial application

Authors:
Christian Naydowski
Peter Hess
Dieter Strauch
Ralph Kuhlmann
and Johannes Rohleder

There have been mineral raw materials that have had such a formative influence on entire epoches that they have been named after them. Bronze and iron were such raw materials, and the same also applied to stones. However, as the importance of metals continuously declined, "stones and earths" gradually made a come back. In fact their consumption rate even became an indicator of the prosperity of a country.

The highest value-added rate is achieved when rocks are ground down to fine mineral flour and powder. Many present-day materials and consumer goods would be unthinkable without mineral fillers. This applies particularly to the paper, plastics, paints and varnishes industries, all of which consume vast quantities of different industrial minerals. Calcium carbonate ranks right at the top within these industries and in other areas. There is no mineral that is so diverse in its uses as calcium carbonate, either in fertilisers or medicines, as an additive in foods, as a pigment in paints or as a filler in other areas.

1. Paper

Paper is rarely associated with the terms 'stone' or 'mineral'. Most people are therefore very surprised to learn that the glossy printed brochure in their travel agency has a mineral content of up to 40 per cent – in the case of TV magazines, it is somewhat less – but were it not for mineral additives, the range of paper qualities available today would be far narrower.

Minerals are incorporated within paper products in two different ways: both as filler that enters the pulp before the web is formed, and as the main constituent of the coating formula, applied to the dry paper surface for further upgrading. In the manufacture of high-quality printing papers, both of these processes are employed, and in most regions calcium carbonate is the most frequently used mineral, either in combination with kaolin and talc, or as the sole mineral constituent of the paper. This was not always the case.

The use of minerals in paper manufacturing did not begin on a larger scale until the 20th century. To be more precise, it was not until the 1950s that the application of mineral slurries, which had originally been used in special processes for exclusive qualities, became a common procedure in upgrading. This led indirectly to an increase in the amount of filler in papers, since, as it is today, it was already common practice to reintroduce the paper waste generated back into the production process. Since this waste already contained a significantly larger proportion of minerals from the fillers employed, many paper manufacturers had to get used to a considerably higher quantity of filler than had been incorporated before.

Until then, fillers had been mainly used in modest quantities for paper glazing and whitening. The quantities were kept 'moderate', since many people were disparaging about the use of large amounts of filler. The substitution of cheap 'stone' for expensive, high-quality, fibre material was even considered to be a deceptive practice. Not until the acceptance of minerals had risen, and the paper industry had studied their effects in depth, were entirely new mineral properties discovered. This lead to their wide- spread introduction in increasing quantities. The functional properties of pulverised calcium carbonate in particular have subsequently revolutionised paper manufacturing.

The decisive breakthrough for calcium carbonate occurred in the 1970s. Properties such as high whiteness and solubility, together with alkaline pH value, and also the rhombohedral pigment form of the mineral, fostered the development of hitherto unknown processes and technologies. Examples of this are the neutral process, multiple coatings with up to four separate, successively dried, mineral films per sheet, and a heat-set offset printing process that now has high resolution, creating an unprecedented printing gloss.

The use of minerals – and particularly of calcium carbonate – in paper manufacturing has expanded exponentially over recent years, as freshly precipitated calcium carbonate (PCC) has supplemented natural, ground calcium carbonate.

At the turn of the millennium, 30 years to the day after calcium carbonate first found widespread acceptance in the paper industry, Europe is already using more calcium carbonate than all other mineral additives combined. In the Asian Pacific Rim, calcium carbonate has become the leading mineral for paper manufacturing. Solely in the United States of America is the situation different.

1.1 Calcium carbonate as filler

Mineral paper fillers were long regarded – and employed – as an inferior substitute for the prestigious raw fibre material. In this vein, an English author recommended in the "Annals of Philosophy" in 1823 that paper manufacturers who used gypsum in paper manufacturing should be denounced:

"To increase the weight of printing papers, certain paper manufacturers have introduced monstrous quantities of gypsum to the rags. The fraud may easily be identified

SEM shots of various minerals (scanning electron microscop). (a) GCC (b) PCC (c) Kaolin (d) Talc

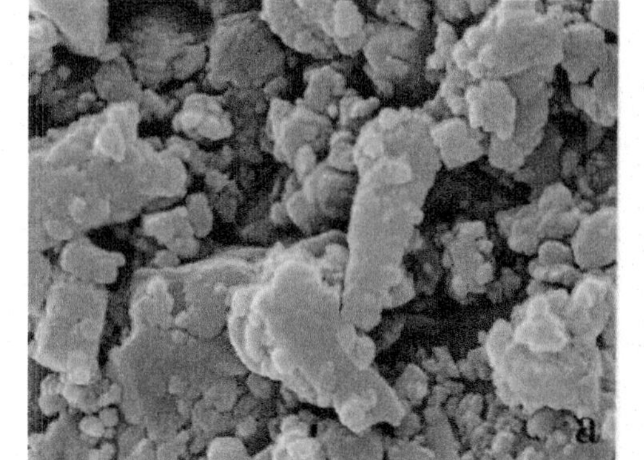

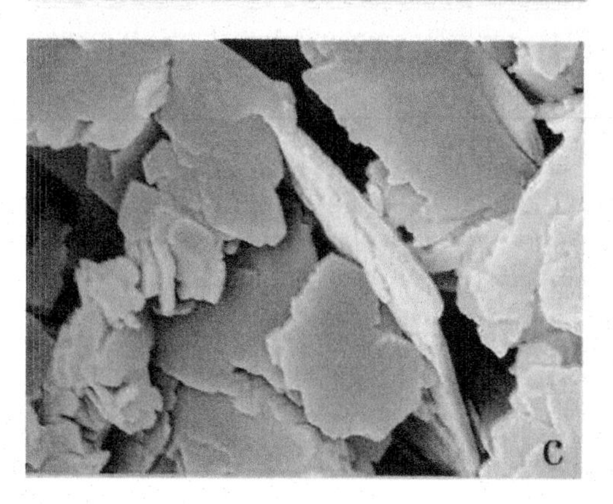

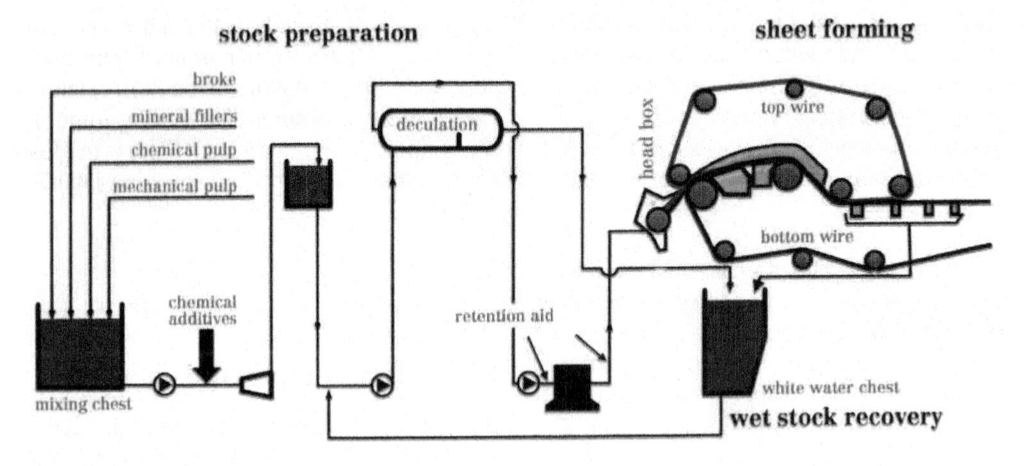

Process diagram for the pulp circulation system of a modern paper machine.

by burning a small quantity of this paper in a crucible and testing the resulting ash for gypsum ..."

However, this attitude quickly changed. As early as 1871, Rudolf Wagner, in his "Handbook of Chemical Technology", wrote on the theme of paper manufacturing:

"The moderate addition of a suitable mineral body to paper pulp is by no means disadvantageous, and is useful in several ways. By this means, medium fine papers are given improved whiteness, the untoward translucence with very thin papers is to some extent mitigated, strength is not compromised, and, finally, through the addition of organic material, less expensive paper may be produced ..."

It may be safely assumed that at the time, calcium carbonate played no role as a paper filler. Rather, gypsum or kaolin was employed. Gypsum was also traded under the names of aniline, pearl hardening and unburnt plaster, while kaolin was known by the names of Lenzin, clay and China clay. Together with the dazzling white barium sulphate, they were known collectively as 'rag surrogate'.

Calcium carbonate was not introduced into modern printing papers until around the mid 20th century, and has subsequently transformed paper manufacturing. The reason for this is that the introduction of calcium carbonate prompted the switch from acid paper manufacturing.

1.1.1 Paper manufacturing

Paper has been produced to the present day from an aqueous fibre pulp by filtering out the solid constituents on a wire. The proportion of fibres and other solid constituents in this pulp is less than 1 per cent, the remainder being water. The solid constituents, i.e. fibre materials, fillers and chemical additives, are introduced to the mixing chest (see figure) and pumped to the so-called 'head box' that distributes the mixture to the moving wire. In the course of dewatering, the fibres form a fibrous mat, or, in papermakers' jargon, 'web'.

The mixing chest is the principal vessel in the manufacturing process, in which all the constituents that have been carefully prepared are brought together from the various process stages. These include the fillers.

The make-up of this mixture is decisive for paper quality. The fibre materials may consist of chemical pulp, 'mechanical' wood containing pulp (ground wood), or redispersed paper deriving from broke or used paper. Calcium carbonate, kaolin and talc are now used worldwide as fillers in paper pulp. The use of gypsum has practically ceased. Other mineral substances, such as titanium dioxide, barium sulphate, as well as precipitated silicic acids with pronounced functional properties or high specific area are applied in small quantities in addition to the classical fillers. The object is to improve wet opacity, whiteness and smoothness. Their price often exceeds that of the fibre materials.

The mixing chest contains a viscous mixture of 4 per cent solids and 96 per cent water that must be further diluted before it is introduced to the head box of the paper machine wire. The precise composition of the contents of the mixing chest remains, to this day, the well-guarded secret of each and every paper manufacturer.

The pulp is separated into its solid and liquid components on the paper machine wire (web formation). The object is to distribute the solid substances over the wire as completely and homogeneously as possible. The term retention (latin: retendere) designates the relative percentage of solid substances retained by the wire to the solid substances washed through the wire. Experts distinguish here between fibre retention and filler retention, as a function of the total quantity of fibres or fillers used, respectively.

Irrespective of all process variations, the contents of the mixing chests in the early 20th century had one thing in common – an acid pH value. In the early years of paper production, the acid pH value was caused by microbial decomposition (fermentation) of the rag fibres, thereby improving the binding capacity of the fibre surface and increasing paper strength. Even though the unpleasant smell arising from microbial decomposition could be quelled by introducing slaked lime, or by using modern, and sometimes even alkaline, fibre separation techniques, the paper still remained acid. The reason for this was that alum salts were added to the paper in the mixing chest, reducing the pH value to 3-4. The alum salts were necessary in order to precipitate the binding substance consisting of animal glue or rosin acid onto the fibre surface, thereby making the paper writable and ink resistant.

1.1.2 The role of fillers in paper

As a means of achieving not only qualitative, but also economic added value in paper production, mineral substances only gradually attained acceptance and recognition. It was not until the advent of automation in paper production in the 19th century that the principal technological advantages of filler substances became apparent. The mineral substances, like the fibrous materials, have an effect on web formation and drying, and also on the development of optical and mechanical properties, such as strength, smoothness and porosity. These properties have a decisive influence on the printing of paper and its writability, irrespective of whether it is a filled base paper or an uncoated natural paper. Whereas mineral substances are only occasionally used for tissue papers (toilet paper, paper serviettes, etc.), they considerably improve the functional properties of other paper grades and cardboard, for example their printability.

Almost all the process stages affected by the use of minerals in paper production change significantly with production speed. This may be clearly illustrated by taking filler retention as an example. The object here is to distribute the filler homogeneously between the fibres, thereby indirectly improving the printability of the finished paper. If the filler level and dispersion of the mineral filler are not adjusted to the production speed during web formation, the fibre/filler distribution remains heterogeneous, leading to an imperfect printed page.

As production speeds have risen continuously since the invention of the first

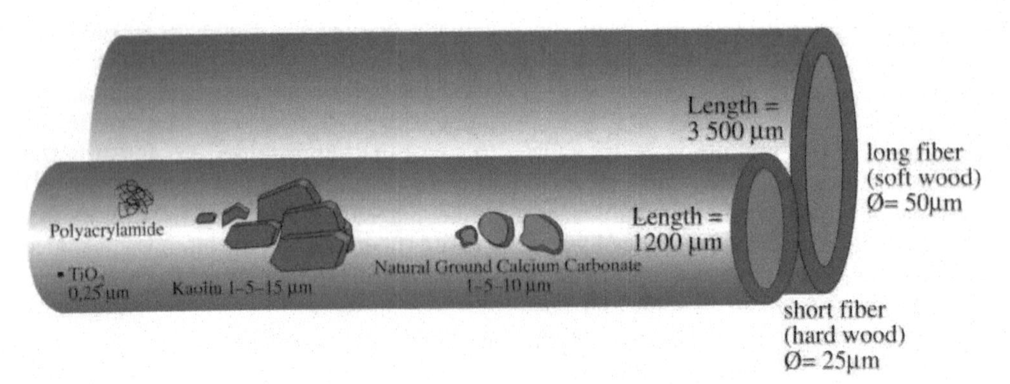

Relative proportions of typical additives.

automatic paper machine, and will continue to rise in the future, all raw materials must be adjusted to suit. This is particularly true of mineral fillers, which have been a constituent of printing and packaging papers for a very long time. Not only have the speeds of paper machines risen, but those of printing machines have also.

A comparison of the relative proportions of filler to other mineral components, particularly fibre materials shows clearly that via the addition of retention agents, the fibre/filler distribution has not only chemical, but also mechanical and physical aspects (see figure).

Filler properties and functions

"The main characteristics that a rag surrogate should have are low price, white colour, insolubility in water and extremely fine granulation."

More than a hundred years have passed since Rudolf Wagner stipulated these requirements for fillers in his "Chemical Technology", and they remain valid to the present day. Owing to the high speeds used in paper manufacturing, new requirements have, however, arisen for modern fillers (see figure).

First, the **abrasion profile** of a filler is decisive for the life expectancy of production equipment subject to wear. The reason for this is that, during paper production, fillers come into contact with a variety of very different materials, such as felt, stone, Teflon and cast steel, at high production speeds of 1400 meter per minute and more.

Functional properties of paper fillers.

	Calcium Carbonate	Kaolin	Talc
ISO-Whiteness [%]	85-97	75-85	70-90
refractive index	1.65	1.55	1.57
aspect ratio	1	5-15	5-100
pH	8.6	3-5	7
abrasivity AT 2000 [mg]	3-20	10-20	10

The so-called **form factor** (aspect ratio) of a filler is of significance in dewatering, retention and abrasion, while the pH value determines whether the papermaking process must take place under acid or neutral conditions. The technologies involved vary significantly.

In addition to the high **degree of whiteness** of fillers, certain applications also require a high refractive index of refraction, which is characteristic of satisfactory paper opacity. This is particularly significant for thin, lightweight papers that could not be produced at the low speeds common at the time.

The **surface energy** of a filler is significant for the hydrophobic and hydrophilic interactions within a paper machine circulation system, and also for the paper's printing properties. The interactions have increased enormously with the increasing numbers of process chemicals in paper manufacturing, e.g. retention aids, biocides, dispersant agents, defoamers, bleaching chemicals and soaps.

As far as the 'extremely fine granulation' of a filler is concerned, the present technological status of wet grinding goes far beyond any reasonable demands on fineness in paper manufacturing. This is particularly true of calcium carbonate, which retains its rhombohedral form (form factor = 1) during grinding. Conversely, materials such as kaolin and talc, which have a laminated structure, change their form factor during fine grinding. Rupture can occur either in the x-y-plane or in the z-plane, resulting in different form factors. For the former, the form factor is increased (so-called delamination), whereas for rupture along the z-axis, it is reduced. In fine grinding, mixtures therefore arise with very different form factors.

While in the paper coating process the paper quality increases with the **fineness** of the coating pigments, the fineness of fillers is limited, since with increased fineness their retention capacity is reduced. The art of modern filler manufacturing consists in producing particle sizes that combine optimum retention with the highest possible optical properties (see figure).

Typical fineness of calcium carbonate products for paper manufacturing.

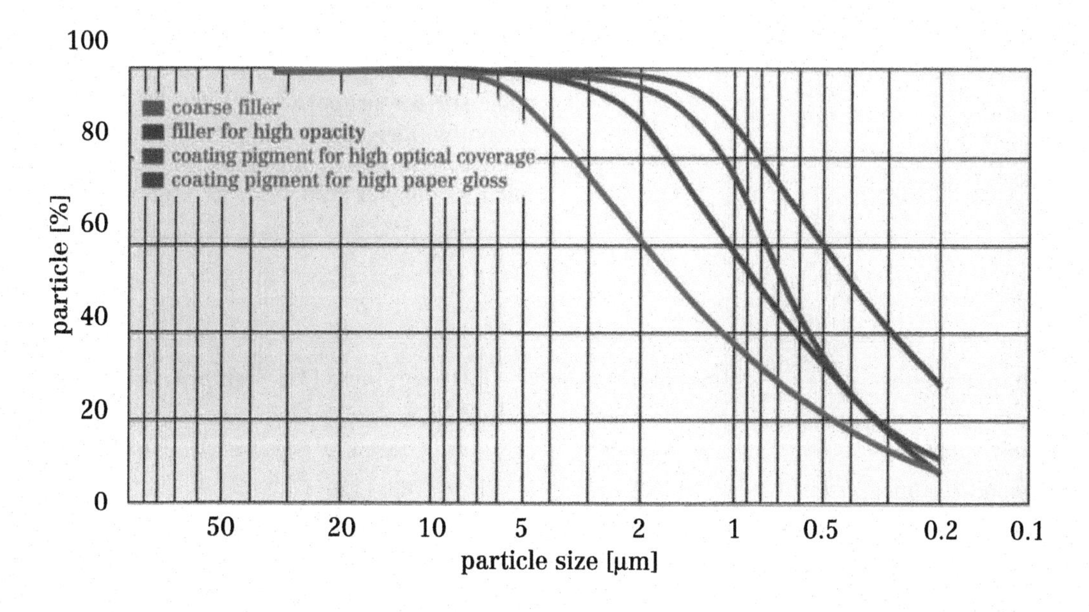

This consideration led in the 1980s to the development of cationically dispersed fillers, and among these, cationic calcium carbonate alone could become established as filler.

Today, the **purity** of the filler must satisfy much higher requirements than in earlier times. Natural fillers, such as calcium carbonate, kaolin and talc, may be contaminated with escort substances, such as metal oxides, organic humic acids or graphite, which considerably detract from the optical properties of a filler. On the whole, natural filler substances are only used after they have been subjected to chemical separation and passed through a bleaching process to improve their optical properties.

Conversely, precipitated calcium carbonate is largely free from organic impurities and some metal oxides, since they may be quantitatively separated during the chemical production process. PCC may nevertheless be contaminated with residues of calcium hydroxide that has not reacted quantitatively during precipitation. In such cases, the pH value at the surface of the filler particle may be as high as 12 and above, and this can lead to undesired chemical reactions (yellowing) with other constituents in the paper.

Numerous material properties beyond those mentioned above are undoubtedly present, but only a few of these are important, or could be attributed to a specific function in paper manufacturing.

Improved added value through fillers

It is evident that in order to substitute fibre materials economically, a filler must be cheap. Kaolin, which was already in use as a paper filler a hundred years ago, may be used to illustrate the relative prices of filler and printing paper over this period.

A hundred years ago the price per tonne of bleached cellulose amounted to between 40 and 50 per cent of the final price of a tonne of printing paper. The price of one tonne of kaolin amounted to approximately 10 per cent of the price of one tonne of printing paper (subject to variations for technical reasons). This relationship has remained constant as an average, despite the fact that pulp and paper prices have fluctuated dramatically during recent decades.

The added value resulting from the use of fillers has increased substantially over time. Whereas a hundred years ago, fillers only constituted 5 to 10 per cent of a tonne of fibre, today the filler level in high-quality printing and writing papers may be as high as 38 per cent. As far as the quantitatively important woodfree office papers are concerned, even a filler level of 29 per cent is now common, and calcium carbonate is the most important, and often the only, filler.

1.1.3 Uncoated filled papers

All uncoated papers share one attribute: their surface properties are strongly influenced by fibre characteristics. This applies both to typical writing and office papers, such as copying and laser printer papers,

Paper grades for natural papers.

paper grades	basis weight [g/m^2]	ISO-Whiteness [%]	filler level [%]
newspaper	43-47	58-66	0-15
woodfree uncoated	75-80	82-112	5-29
wood containing uncoated	53-59	55-68	12-38

region (estimated)	filler level [%]	relative market share of filler
Europe	15-29	GCC, PCC, Kaolin
North America	8-13	PCC, Kaolin
South America	9-14	GCC, PCC, Kaolin
South Africa	15-23	GCC, Kaolin
South East Asia	5-18	GCC, PCC, Kaolin, Talc

Fillers and filler level for copying and office papers worldwide (1996).

and to magazine papers based on SC papers made from wood fibres or recycled paper. Although newsprint also belongs to this category, it is often treated as a separate class of paper (see figure).

Understandably, calcium carbonate was the first pigment to be incorporated in woodfree, high-bright white papers. In this grade, the whiteness of the filling substance complemented the objectives of the more expensive whiter fibre materials. An international study on the use of fillers in the copying and office paper sectors carried out in the mid 1990s showed that calcium carbonate had already become the leading mineral in this market segment (see figure). Either natural, ground, calcium carbonate (GCC) or PCC is employed, depending on the region.

Developments here were rapid: calcium carbonate had been used for the first time in an experimental context in the mid 1960s as a filler in uncoated papers; 20 years later, fifty per cent of all West European paper mills were using calcium carbonate for the production of neutral uncoated paper. Today, not only are woodfree papers the most important uncoated papers in terms of volume, but also possess an above-average growth potential, since the market for office and home printers is continuing to expand (see figure, next page).

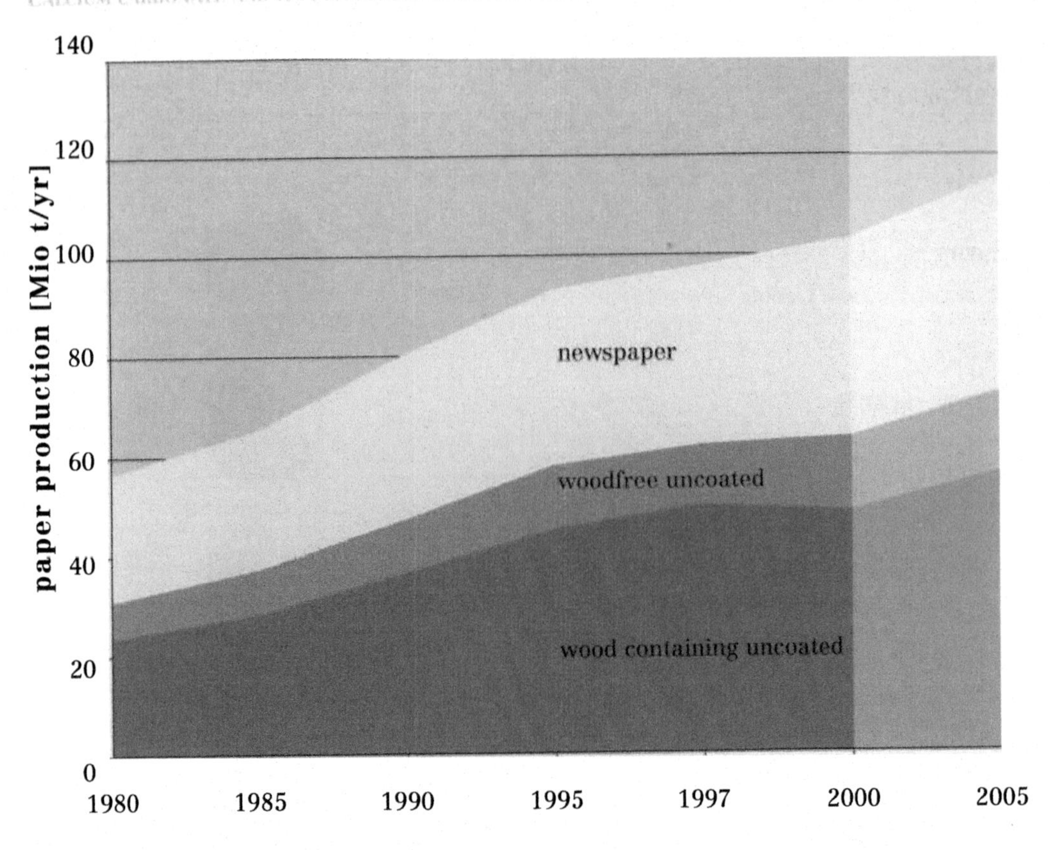

Fillers and filler level for copying and office papers worldwide (1996).

At the beginning of the 1990s, newsprint was a classical example of unfilled paper. Especially in Germany, when recycling of used paper began, significant quantities of filler were introduced into newspapers, and this particularly concerned calcium carbonate. For typical specific weights of 45 grams to 53 grams per square meter (g/m^2), filler levels of up to 15 per cent can be detected. In addition, the neutral paper process had to be introduced for this type of paper.

This development opened the way for the incorporation of fresh calcium carbonate in uncoated, wood containing papers. In the meantime, certain newspaper producers have pioneered the inclusion of up to 10 per cent fresh calcium carbonate as filler even in the absence of recycled paper, since the level of brightness for filled newsprint is appreciably higher. Special qualities, based on used paper with a high proportion of chemical pulp, today achieve brightness levels of up to 82 per cent.

Although super-calendered (glossy) SC papers only represent a small proportion of uncoated papers on the world market, they permit the highest filler level of any uncoated paper. In Europe, the average filler level in the 1990s was around 30 per cent, with peak values of up to 38 per cent. Until the middle of the 1990s, kaolin was the principal filler for this type of paper owing to the improved gloss achievable by calendering and the ease of rotogravure printing, while in Finland, talc had a role to play.

As late as 1998, calcium carbonate filled SC paper entered the market from Finland, this being filled with 20-25 per cent calcium carbonate, so that the brightness level in this segment was raised in a single step by five per cent. Using new calendering techniques, the level of gloss could be increased to satisfy customary standards. Today, the proportion of calcium carbonate (approximately 15-20 per cent) represents about half the total filler used, the rest being kaolin.

Although base papers for a subsequent coating do not count as uncoated papers, they are nevertheless filled. The filler level can be as high as 17 per cent for high-weight, coated papers (> $100 g/m^2$). For light-weight, coated base papers, small quantities of 'primary filler' are added, since the filler level is achieved mostly through the reintroduction of coating constituents from the broke or from the coating process. At times, the entire filler requirement can be met from this source. Unlike uncoated papers, individual properties of broke-filled papers, such as brightness, opacity and gloss, are much less amenable to specific control than via primary filler. Often, broke management has become so complex that the targeted improvement of single properties is no longer possible, and the filler is left to 'fill' as best it can.

1.1.4 Neutral paper manufacturing with calcium carbonate

Neutral paper manufacturing is a processing procedure in which the pH value in the mixing chest, and in all subsequent aqueous process stages, is maintained in the range 6.5 to 7.5 by virtue of calcium carbonate. By comparison, the pH value for acid processes lies between pH 3 to 6, this being maintained by continuous injection of up to 3 per cent aluminium sulphate.

There also exists an alkaline paper process in which the pH values remain above 7.5. These are often achieved by the use of additional, non-buffered alkaline sources in the presence of calcium carbonate. Sodium hydroxide and sodium hydrogen carbonate from fibre production may be used as alkali sources in the circulation system of a paper mill, while calcium hydroxide waste from PCC manufacturing has also featured.

In papermaker's jargon, the imprecise term pseudo-neutral paper manufacturing was coined spontaneously in the mid 1980s. This meant a neutral paper process but with the addition of approximately 0.1-0.5 per cent aluminium sulphate. This procedure in the presence of aluminium sulphate has become so entrenched as to represent the standard, particularly in the manufacturing of wood containing papers.

The terms neutral and alkaline paper processes are synonymous with the use of calcium carbonate in the circulation system of a paper machine.

Calcium carbonate – a filler changes paper technology

Through the use of calcium carbonate, chemical paper manufacturing technology has changed significantly since the end of the 1960s. The desire for higher whiteness has led papermakers to adopt calcium carbonate, and forced them to reject the acid paper process. This was because the solubility profile (see figure. next page) of this mineral is pH- dependent, necessitating either a neutral or an alkaline process. Calcium carbonate had proved unstable under acid conditions with pH values of 3 to 6 in the material and water circulation systems of a paper machine.

Attempts to employ calcium carbonate under acid conditions using the surplus aluminium sulphate invariably caused insurmountable problems:

- foam generation on the surface of the process water, due to copious carbon dioxide generation
- fibre flotation in the process water due to rising carbon dioxide gas
- loss of retention on the paper machine wire and in fibre recovery from the process water
- deposits and holes in the paper caused by the collection and accumulation of hydro-

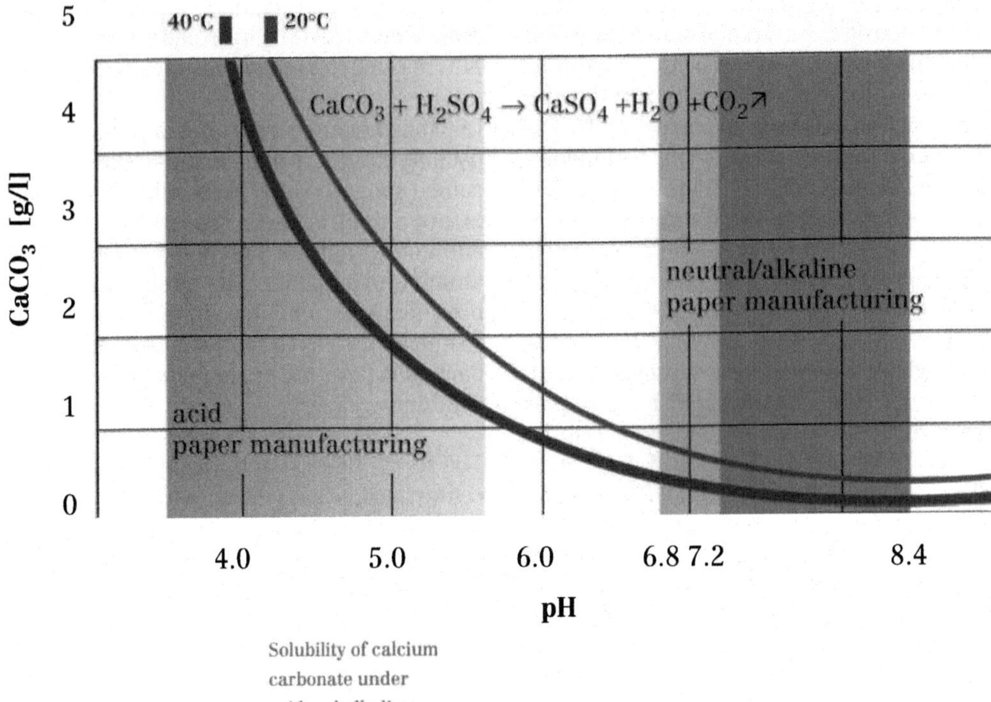

Solubility of calcium carbonate under acid and alkaline pH conditions.

phobic substances in the interfaces between the carbon dioxide gas and water
- salting-up of all process and waste waters
- loss of sizing efficiency

The disadvantages relative to production costs and paper quality were so great that a consistently neutral or alkaline paper process had to be introduced in order to use calcium carbonate as a filler. The transfer from an acid to a neutral paper process does, however, make it necessary to substitute other materials for aluminium sulphate in the circulation system of a paper machine.

The chemical mechanisms of the aluminium ion and their technological ramifications may be understood from pH-dependent chemistry (see figure). Under acid conditions, aluminium is present as a positively charged cation, and this can react with the anionic groups on the surface of the cellulose and wood fibres, thereby neutralising their charge. This results in fibre and particle aggregation, kaolin retention and fixation of resin size on the fibres.

Under neutral and alkaline conditions, aluminium sulphate no longer possesses cationic charge carriers, enabling simplex generation on the basis of charge exchange alone. Nevertheless, an auxiliary mechanism of aluminium cations may be used to advantage under neutral or alkaline conditions in paper manufacturing. To do so, the fact that acid (cationic) aluminium sulphate does not lose its charge characteristics instantly, but over a period of 30 to 45 minutes may be exploited. This is referred to as kinetic equilibration. Over this period, the cationic charge steadily diminishes until the aluminium has reached a thermodynamically stable state.

This knowledge has led to the practice of adding small quantities of aluminium sulphate (0.1-0.5 per cent) to the pulp between the mixing chest and the head box in neutral and alkaline processes as well. In the few seconds' delay between injection and final distribution in the paper, the aluminium complex may exert its beneficial effect.

Substitutes for aluminium

However, the necessary cationic and colloidal polymeric character of aluminium is absent in the water circulation systems following de-watering as a consequence of neutral processes. For this reason, new polymeric substances were developed on the basis of acrylamid, ethyleneimine, Dadmac or natural starch as substitution materials, which form cationic groups in aqueous solution at pH 7 and above. Thanks to these substitute substances, aluminium sulphate may be replaced entirely in neutral processes.

It should not be forgotten, however, that the neutral process is a meta-stable system that reacts sensitively even to minute quantities of acid. Thus the above-mentioned problems may occur locally at any time whenever an acid is added to the circulation system. This is of significance whenever it is necessary to introduce acid process chemicals. To avoid local pH perturbations, rapid and massive dilution must be provided.

Biological equilibrium in the presence of calcium carbonate

Microorganisms are ever present in the living world. This also applies to the various circulation systems in paper manufacturing and, of course, to calcium carbonate itself. By changing the pH value from acid to neutral, the living conditions of the micro- organisms in the paper circulation system also change, and a new state of equilibrium arises.

Whereas when using kaolin as filler under acid conditions, mainly cocci and endospore-forming bacteria of the bacillus type are the most frequently occurring strain of micro-organism, various species of pseudomonad develop on natural calcium carbonate. In commercial calcium carbonate products, however, yeasts and fungi are seldom found.

Although the concentration of micro-organisms in calcium carbonate slurry amounts to 10^2 to 10^3 per millilitre (the permissible concentration in drinking water is 10^2 micro-organisms per millilitre), the slime counter-acting biocides must be adjusted to suit the neutral process. Otherwise there is a risk of uncontrolled deposits being formed that may cause holes and ruptures to form in the paper.

Behaviour of aluminium ions as a function of pH in aqueos solution.

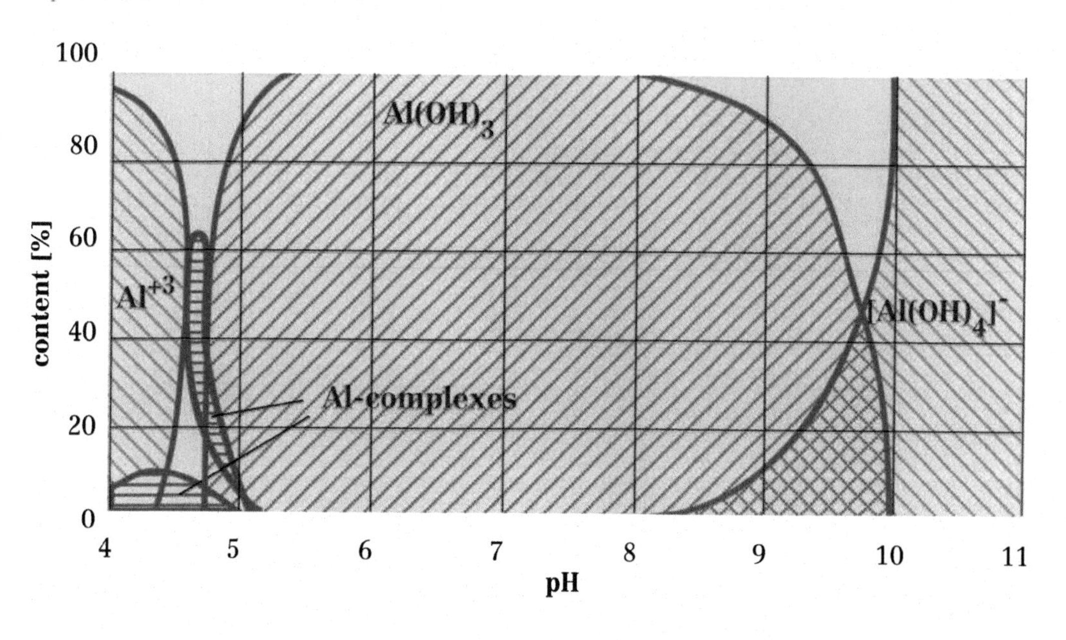

Retention – the challenge for calcium carbonate

Fibrous materials are easier than mineral fillers to retain on a paper machine wire. The problem with mineral fillers has intensified owing to the use of modern, high-speed paper machine wires, since the flow rate through the web is now several factors of ten higher than for hand couching. Whereas a hundred years ago, the filler retention rate for hand couching was around 50 per cent, the rate today would be less than 10 per cent if retention aids were not used.

Owing to its characteristic surface, kaolin powder may be caused to aggregate and be retained at a relatively low level of cationically charged polymers. Thus the aluminium ion itself may exert a retention effect under acid conditions. Even for high-speed paper machines, a relatively weak cationically charged polymer such as polyacrylamide was found to suffice under these conditions.

Unlike polyacrylamide, natural calcium carbonate in powder form carries a negligible charge. Slurry systems are anionically dispersed. In introducing calcium carbonate and the neutral paper process, retention agents had to be developed that took account of the charging characteristics and reactivity of calcium carbonate. A decisive breakthrough was BASF's development of retention agents on the basis of ethyleneimines in the 1970s. Numerous new polymers have been developed subsequently (see figure), and the filler retention for neutral paper manufacturing today lies between 30 and 60 per cent.

Polymer retention systems.

single component systems	
polymer	
• cationic[1] polyethylenimine (PEI), polyvinylamine (PVAM) • cationic polyacrylamide (PAM) • non-ionic polyethylenoxide (PEO)	
dual component systems	
polymer component 1	polymer component 2
• cationic PAM • non-ionic phenol resin	• anionic PAM • non-ionic PEO
micro particle systems	
cationic polymer	micro particle
• cationic polyacrylamide (PAA) • cationic starch • cationic PAM • cationic starch + anionic PAM	• organic micropolymer • silicasole • bentonite • silicasole
[1] at pH 7.	

Interactions with other substances in the presence of calcium carbonate

The employment of calcium carbonate as a paper filler has fuelled a heated debate on the undesired interactions within paper processes and the most appropriate ways of combating them. The initial search for so-called 'problem substances' did not, however, lead to satisfactory results. The problems were simply attributed to individual substances or to groups of substances, and a new terminology was developed. Yet no really effective proposals could be advanced to create a stable, neutral, paper process.

In fact, the discussion relating to problem substances has hampered the introduction of urgently needed neutral process techniques and fundamental production modifications. Indeed, for the simple reason that these interacted under neutral process conditions and appeared in the paper in the form of deposits, substances that had always been a part of paper manufacturing were unexpectedly dismissed as to 'problem substances'.

In fact, there is a fine dividing line between useful and problem substances. Although rosin sizes, defoamers, fillers, starches, binders, fixatives and dispersants, as well as the whole range of wood extracts and even fibres, are indispensable auxiliary substances, they are nevertheless present in almost all types of deposit.

However, all these discussions and research on the theme of 'problem substances' have led to one fundamental conclusion. That is that the neutral process is a thermodynamically unstable process, and, contrary to the acid process, its chemical equilibrium can easily be disturbed. Even quite small variations in pH value, temperature, conductivity, charge state, air content and content of dissolved colloidal organic substances, can alter the fibre and filler retention factors, thereby affecting paper quality. At the same time, agglomerations of substances that are incompatible with the paper may arise and find their way into the paper, where they lead to holes and ruptures.

In consequence, high concentration gradients of particular substances in the primary circulation system must be avoided for neutral processes. When followed consistently, this relatively simple rule may require the modification of dosing points, and of piping and storage chests.

In the production of coated papers, the broke should be chemically treated to avoid the agglomeration of synthetic binding agents (formation of so-called 'white pitches'). Good results have been achieved here with short molecular chain, cationically charged polymers. A suitable monitoring parameter is the turbidity of the reject filtrate.

New parameters have also arisen for gradient monitoring. Thus the streaming current detector (SCD) can be used to measure the charge state of process water, and a quick test for the determination of chemical oxygen demand (COD) enables the contamination with organic material to be determined. On-line measurements are, however, still at an experimental stage.

However, absolute values resulting from SCD charge measurement, and also from COD determination, while suitable for assessing the equilibrium state, only provide superficial information on the potential for problems or runnability of a neutral system. Large amounts of cationic agents may certainly be used without perturbation, provided sufficient cationic polymers are employed for retention purposes. Not until an actual displacement of the equilibrium point takes place can a statement be made on the probable extent of perturbation.

Additionally, perturbations arising from wire and felt wear, following a change from acid to neutral papermaking processes, were intensively investigated. Thus further discoveries were made relating to the abrasion mechanisms along the route from the wet stage to the drying cylinders of a paper machine in the presence of calcium carbonate.

While still on the forming table, the fibre-filler web first passes through a hydro-dynamic, and then a vacuum-supported dewatering zone. Owing to the vacuum, not only water

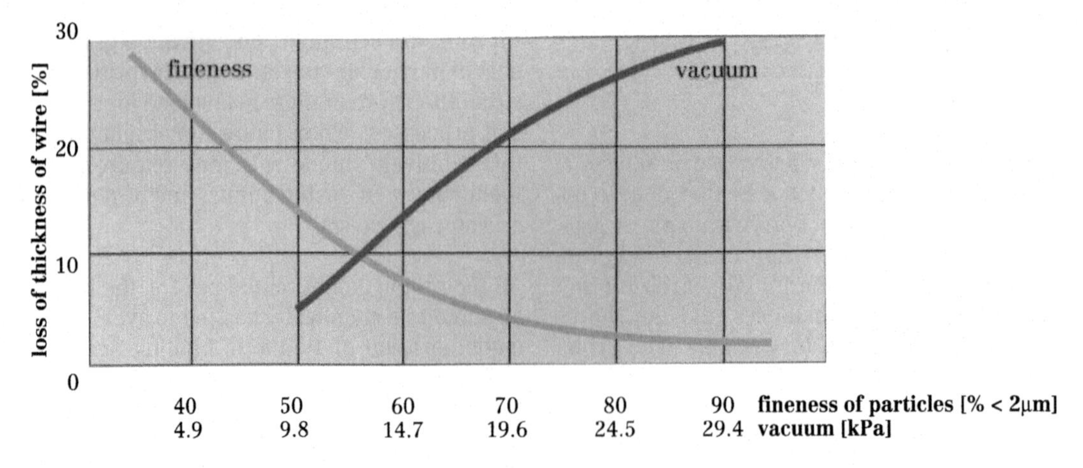

Loss of wire thickness through plane suction vacuum and higher granulation.

but also part of the filler is mobilised and drawn through the wire, whereby the loss of thickness rises with the degree of vacuum applied (see figure).

Under dynamic conditions, because of its particulate form (form factor ≈1), and surface charge, calcium carbonate dewaters more rapidly than the 'platy' kaolin, so that dewatering of a web containing calcium carbonate filler requires substantially less suction capacity.

Furthermore, vacuum reduction is necessary for reasons of process technology. Were the same suction capacity to be applied as for platy fillers, the loss of thickness using

Whiteness and opacity as a function of calcium carbonate content for a woodfree paper model.

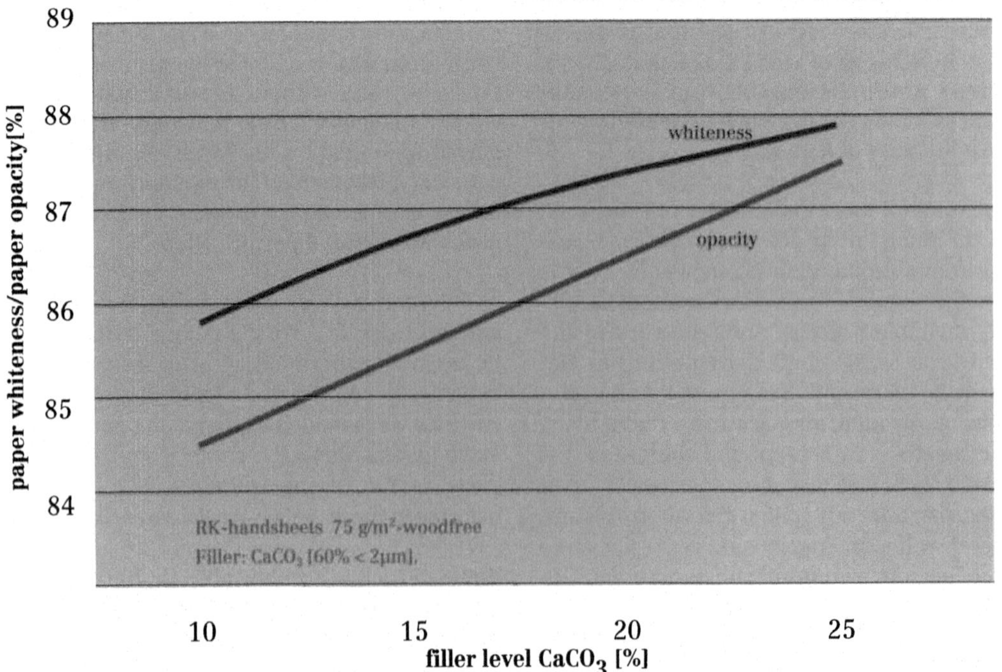

region	paper weight [g/m²]	ISO-whitenes [%]	opacity [%]	paper thickness [µm]
Europe	78-82	82-90	88-95	95-108
North America	74-78	80-86	88-93	95-102
South America	74-76	78-80	92	95-98
South Africa	80-82	80-86	92-93	102-107
South East Asia	64-82	78-85	84-96	87-112

Opacity and whiteness in copy and office papers worldwide — market perspectives in 1996 (without optical fluorescence activation).

calcium carbonate with its approximately rhombohedral structure would be considerably higher than for kaolin with its platy structure. Concomitant with the higher losses, wire abrasion would also increase markedly.

Thus in changing from kaolin to calcium carbonate as filler – assuming the same production speed and specific weight – the vacuum must be reduced and the bearing capacity of the wire increased. This usually has a favourable effect on total energy consumption in dewatering.

Optical properties with calcium carbonate

The optical properties obtainable with calcium carbonate, such as brightness, opacity and gloss for a uncoated paper, may be adjusted. Here, the specific physical properties of the filler and the filler level play a decisive role. If the filler level of a paper is increased, for example, by 5 per cent, brightness and opacity increase by 0.5-1 per cent (see figure).

Using PCC as a calcium carbonate source, a high paper thickness per unit area (bulk) may be obtained for non-calandered, woodfree papers. This effect is advantageous particularly for North American material models based on pine and spruce. These cellulose mixtures, as opposed to European fibre models and Asian hardwood material models, develop less thickness at the same filler level. The thickness of copy papers was raised by the Xerox company in the 1990s to a quality criterion. As a result, the demand for PCC in the production of copy papers also increased in Europe in some paper markets.

An international study of copy and office papers carried out in 1996 showed that the brightness niveau of uncoated, non-brightened, papers filled with calcium carbonate differs worldwide. Opacity, however, lies at the same level in all countries (see figure).

The reason for the higher degree of brightness detected in Europe is the higher specific weight (80 g/m²), which permits a higher filler level than in North America (75 g/m²). The reason that the opacity of European papers is nevertheless not higher than for North American papers is that a higher filler level leads to a denser paper structure. European papers have less 'bulk'. Conversely, North American papers with less filler have more 'bulk'. For the latter, light is also dispersed at the air-fibre surface. This tends to narrow the differences in opacity between these and European paper qualities.

The level of paper brightness depends of course on the brightness of the filler itself (see figure, next page). This phenomenon is particularly impressive in the case of wood containing papers. Thus, the paper brightness may vary by up to 7 per cent, depending on whether kaolin or calcium carbonate is chosen as filler. Calendering that is usual for SC papers reduces paper brightness in the same way for all fillers by approximately 2 per cent.

Unlike paper brightness, paper opacity does not increase beyond a certain point in a linear direction with calcium carbonate filler level. The opacity level reaches a maximum, and from then on any further increases in filler level no longer have any effect. When a filler with stronger light absorption, such as kaolin, is chosen, the opacity may be increased, but this is achieved at the expense of a considerable loss in brightness.

Calcium carbonate causes a change in pH value, whereby loss of whiteness may occur, particularly in wood containing systems. This is known as 'brightness reversion'. Depending on pH value, the loss of brightness amounts to 1 to 3 per cent . Celluloses and sufficiently bleached wood pulp usually have a reduced tendency to brightness reversion.

The net brightness of a wood containing paper containing calcium carbonate therefore depends on the total increase in brightness due to the filler minus brightness reversion. A practical guide for paper circulation systems with high wood content is as follows: the total tends to zero if the filler level for natural calcium carbonate is below 5 per cent.

Using kaolin in super-calendering SC papers, gloss is more readily developed and reaches higher values than with calcium carbonate. Hot-soft-nip calendering, which is still under development, indicates opportunities for the deployment of calcium carbonate as filler in SC papers without the drawback of reduced gloss.

Whiteness level for different fillers; wood containing material model for SC-papers.

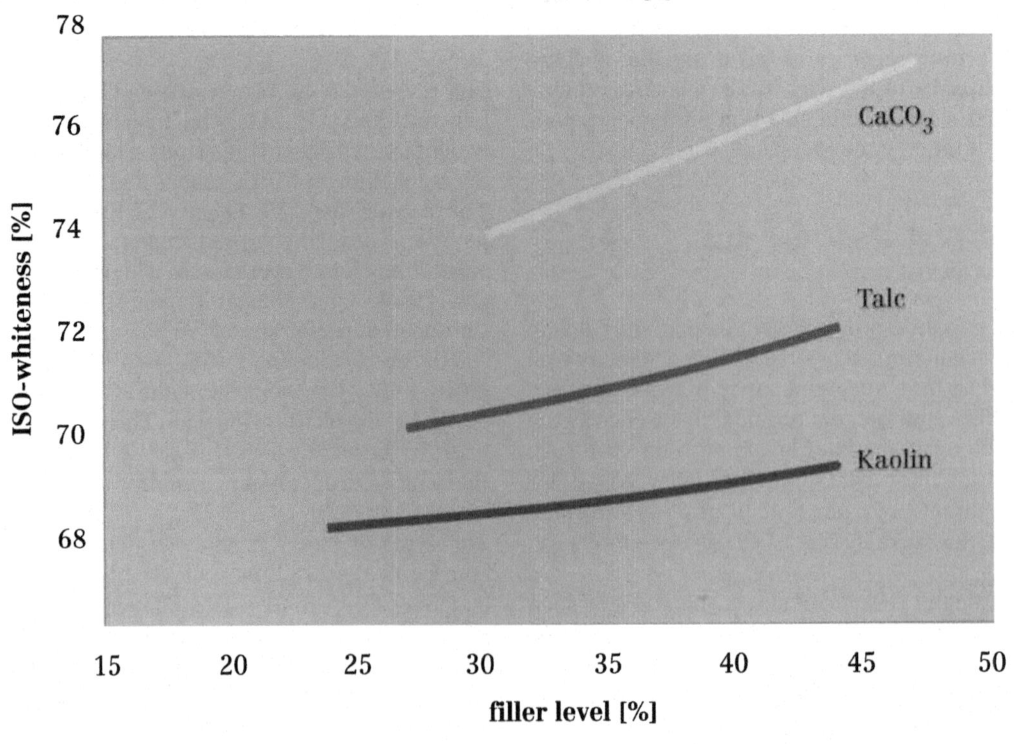

Paper properties using calcium carbonate

In addition to optical paper properties, such as brightness, opacity and gloss, other paper characteristics also alter when calcium carbonate is used in place of kaolin or talc.

The strength of a paper sheet is fundamentally reduced when minerals are used in place of fibres. The filler is deposited between the fibres, and thereby severs the fibre-to-fibre bond. The degree to which this occurs differs for the various minerals, and depends, among other things, on the fineness of the filler. Calcium carbonate is no exception here. However, calcium carbonate plays a supporting role in the development of strength during grinding of the fibre material. Owing to the higher pH value and the lower aluminium ion content, the grinding resistance of the fibre substance is reduced, and the strength of most cellulose materials is more rapidly developed. This outcome may be achieved either in order to reduce the quantity of energy required for grinding, or to obtain a higher filler level.

Papers containing calcium carbonate demonstrate an inherent resistance to acid decomposition. This resistance to ageing results from the buffer effect of calcium carbonate on mineral acids, such as sulphuric acid that is formed over the course of time from the aluminium sulphate used, and destroys the paper fibres through hydrolysis:

$$Al_2(SO_4)_3 + 6\ H_2O \rightleftharpoons 3\ H_2SO_4 + 2\ Al\ (OH)_3$$
(sulphuric acid formation)

$$H_2SO_4 + CaCO_3 \rightleftharpoons CaSO_4 + CO_2 + H_2O$$
(buffering)

The yellowing of the paper and loss of strength occur in parallel may be slowed down by 90 per cent, so that complete decomposition of the paper is avoided.

These properties of calcium carbonate are important in connection with archives and libraries. Many older papers are adversely affected by acid decomposition, as, particularly since the beginning of industrial paper manufacturing at the beginning of the 19th century, alum (aluminium sulphate) was added in large quantities to paper. By spraying these books with a calcium carbonate solution, their life expectancy may be markedly increased, and they can remain accessible for normal use.

Under modern industrial conditions, the use of calcium carbonate is almost indispensable in the production of durable paper.

1.2 Calcium carbonate as coating pigment

Coated papers are favoured whenever the label of a bottle, coloured advertising brochure, travel catalogue or illustrated business report for a large bank is printed, or when paper is required to convey an advertising message. Coated papers combine aesthetic appeal with an appearance of distinction.

1.2.1 Upgrading of paper and cardboard

In upgrading paper and cardboard, surface structuring, principally by coating with mineral preparations, is paramount. Coated surfaces are optically and mechanically more homogeneous, and are smoother and more readily printable than untreated papers (see figure). All this results in enhanced image projection for these papers. Even when unprinted, the appearance and texture of these papers are more appealing.

For coloured printing, coated papers are particularly important, as here, high contrast is a pre-requisite. In particular, white, neutral coloured or weakly tinted blue paper surfaces are favoured, and among the mineral substances employed, calcium carbonate best meets the high demands made on coating pigments.

However, minerals are not only indispensable in magazines and high-gloss brochures in the advertising sector. Also for packaging papers and cardboard, there is a trend

Coloured high-gloss paper products.

Uncoated (top) and coated (bottom) paper surfaces; view through the scanning electron microscope.

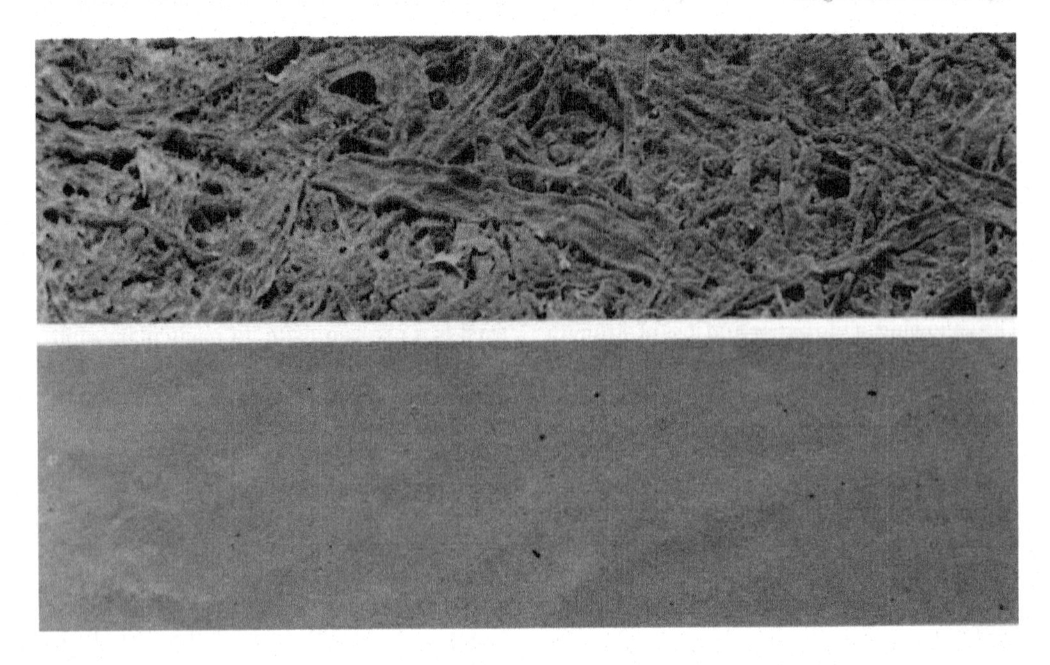

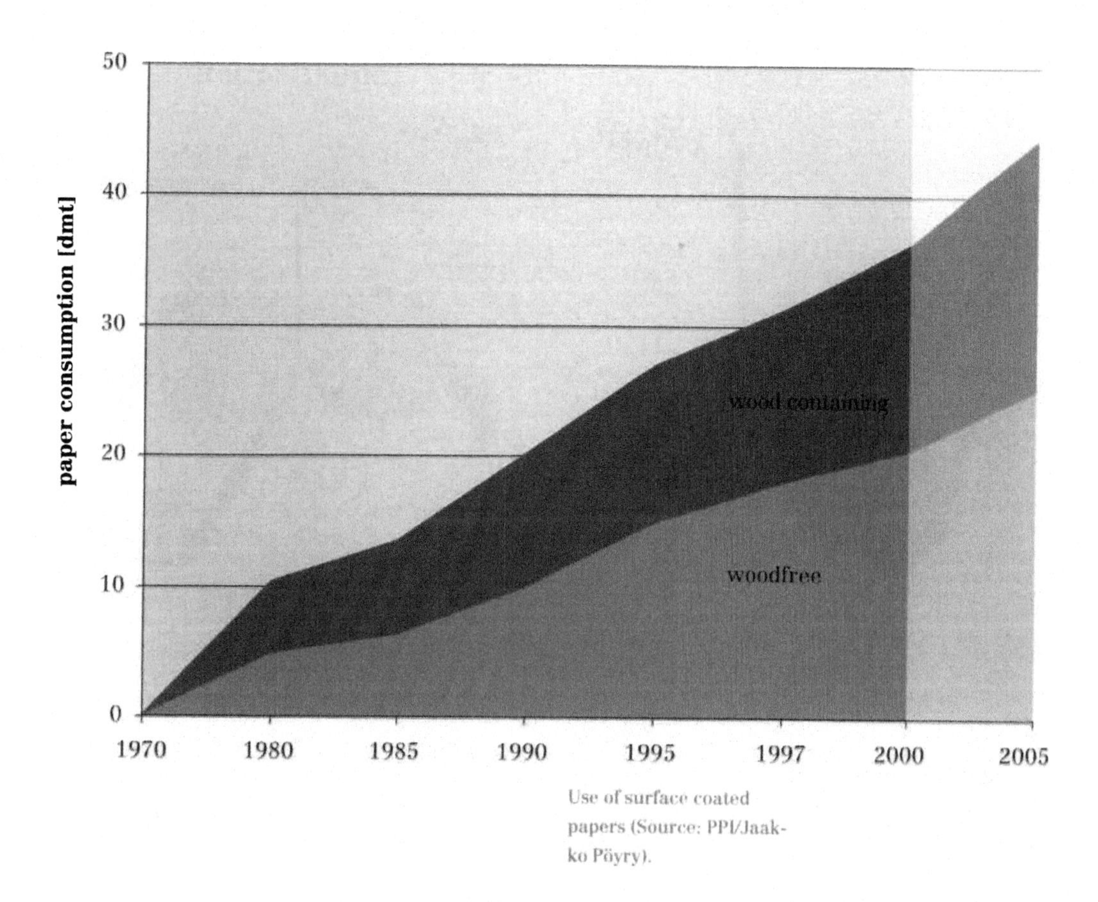

Use of surface coated papers (Source: PPI/Jaakko Pöyry).

towards using packaging surfaces as printed advertising medium – with corresponding demands on paper quality. The forerunners of these particularly distinctive packaging products are to be found in the cosmetics, foodstuffs and cigarette industries. Other industries also closed the gap a long while ago, so that over the last 30 years the demand for surface coated papers has profited from high annual growth rates (see figure).

As a result, the use of white calcium carbonate as an upgrading pigment has soared. Thus the proportion of mineral coating material in Western Europe has reached a high 37 per cent of total weight. Together with the minerals contained in the form of fillers, total mineral fractions of up to 50 per cent of total paper weight result. Chemical analysis of these papers shows that calcium carbonate has assumed a leading role, and cannot at present be replaced by any other natural mineral.

Upgrading processes and calcium carbonate

Paper upgrading processes are based on three process stages: application of coating, equalisation of the wet coating, and final drying. Only a few commercial coating processes dispense with equalisation, for example the metered size press coating and the curtain coating. While curtain coating is seldom used for paper upgrading with quantity papers, metered size press coating for the upgrading of LWC papers is gaining an increasing share of the market.

The following equipment is used today. Thus it is possible for different manufacturers to employ vastly different engineering principles and designs, and combinations thereof (see figure).

Equalization with a blade

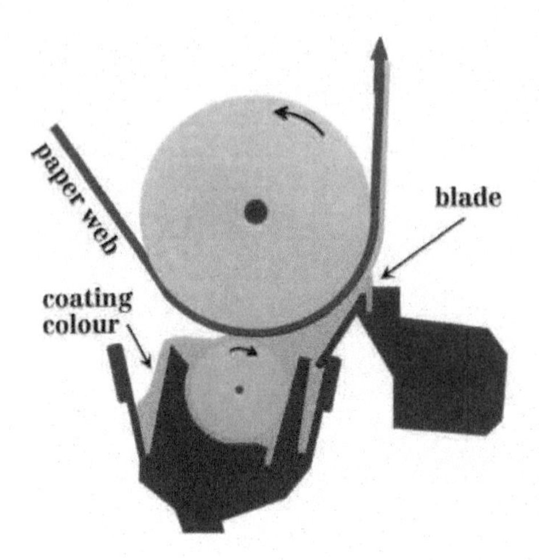

roller coating without equalization

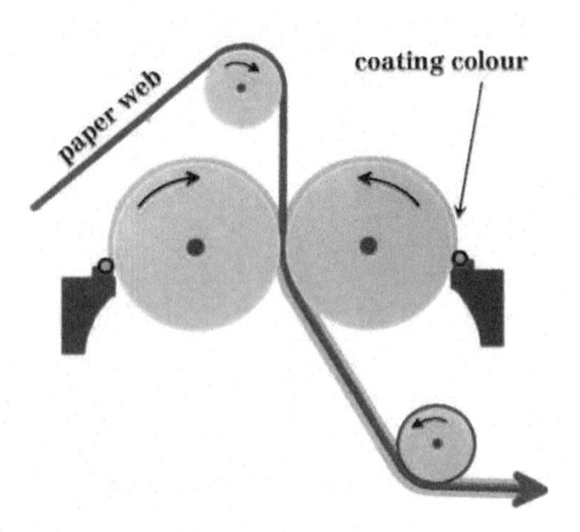

Paper upgrading processes.

Coating application with
- Roller
- Bar
- Free jet

Coating equalisation by
- Blade
- Air jet

Calcium carbonate may be used in all of the surface upgrading processes mentioned. It demonstrates unusual versatility for a coating pigment. This is because some of the physical properties of a mineral, such as its characteristic particle geometry, may hamper the application of a process, or, indeed, make it completely unworkable. A well-known example of this is the use of kaolin and talc for the formation of multiple coatings. If these platy minerals are used as coating pigments in the formation of complete multiple coatings, it can lead to such smooth coating surfaces following precoating that blade streaks appear in the top coating on equalisation with the blade. These problems can be avoided by using calcium carbonate.

As an alternative, from 1960 onwards, the wet grinding technique was modified to enable the necessary mineral fineness to be produced to suit all coatings using calcium carbonate. While coarser pigment fineness is used for precoatings, the finest pigment possible is used for top coatings.

1.2.2 Coated paper qualities

The variety of coated paper qualities worldwide is immense, and comprises several thousand different brands and specific weights. As far as the use of calcium carbonate is concerned, these qualities, for the sake of simplicity, may be classified into two groups:

- single coated papers
- multiple coated papers

This classification makes sense since the methods for processing minerals when making single and multiple coated papers are fundamentally different. Likewise, the requirements for the calcium carbonate used differ in each case.

It makes a considerable difference in determining the choice of a suitable type of calcium carbonate and the remaining constituents of the coating whether the coating is applied to an uncoated base paper or – as with multiple coatings – to a paper that has already been pre-coated.

Paper properties such as brightness, opacity, gloss, print gloss, porosity and smoothness are mainly the result of the optical and mechanical covering of the fibres at the paper surface. If a pre-coated (covered) paper is to be treated, these characteristics can best be achieved by applying a 100 per cent ultrafine calcium carbonate quality or blends of kaolin/calcium carbonate.

If the same preparation were to be used for the lightweight single coating of a coated base paper (< 7 gram per square meter), the coating penetration of the ultrafine particles in the fibrous surface would cause the paper characteristics to be considerably less well developed (see figure). It would make no particular difference with which of the possible process variants the coating was applied and equalised.

For single coatings, the state of the art today is to use calcium carbonate qualities with reduced fine particle content (selected particle qualities) to minimise penetration of the paper surface. These calcium carbonates also increase the optical covering, since they lead to better light scattering at interstices in the coating matrix (void volumes). Brightness, gloss, opacity and contrast are increased accordingly.

Coating preparations comprising several mineral types are also employed in single coatings in order to avoid the problem of excessively fast penetration (see figure, next page). Thus the properties of kaolin or talc may be used to produce gloss, opacity, colour, dull finish and printability effects in the single coating containing calcium carbonate. While calcium carbonate had not been used regularly in LWC paper coatings until the early ‘80s, the proportion in LWC at the end of the millennium in Europe reached between 50 and 70 per cent.

Paper gloss and print gloss as a function of calcium carbonate content.

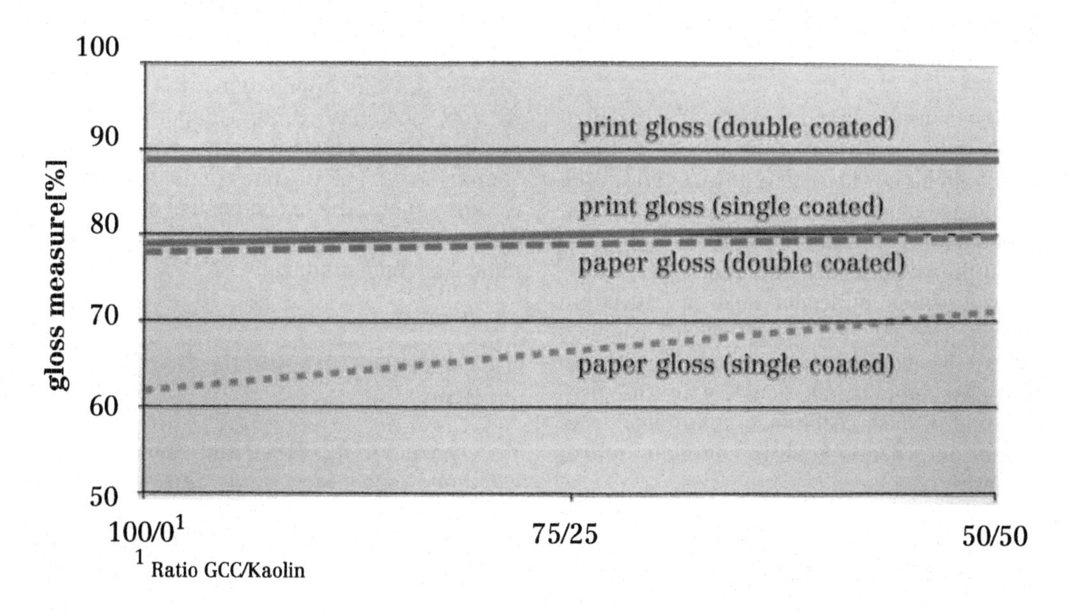

Preparation for single coating	Preparation for double coating	
	Pre-coating	Top-coating
40-80 parts $CaCO_3$ 20-60 parts Kaolin	100 parts $CaCO_3$ (60-75% < 2μm)	80 parts $CaCO_3$ (90-96% < 2μm) 20 parts high glossing clay
10-12% synthetic binder (latex) 0.7% carboxymethylcellulose (CMC) 0.5% thickener (e.g. polyacrylate)	8% natural binder (e.g.starch) 8% synthetic binder	11% synthetic binder (latex 0.4% carboxymehtylcellulose 0.5% optical brightener 0.4% polyvinylalcohol (PVOH)
Amount of coating 8-14 g/m² per side 58-64% solids content	Amount of coating 10 g/m² per side solids content 65%	Amount of coating 12 g/m² per side solids content 64-68%

Typical coating preparation for single and double coating.

For multiple coated papers, the coating function determines the selection of calcium carbonate quality. Therefore, it is important to distinguish between precoating and top coating. Whereas for top coating, the fine qualities are used, the coarser calcium carbonate qualities are suited to the formation of precoating. These allow better mechanical anchoring of the subsequent topcoat, and ensure adequate cleaning of the blade edge during equalisation of the top coating. They also have less demanding rheological properties – three advantages that considerably widen the margin for process parameters, such as coating speed, coating quantity and coating and equalisation processes.

Owing to the fact that the rough milled calcium carbonate qualities are also cheap to produce, it should be no surprise that throughout the world pre-coatings for paper and cardboard are currently based on almost 100 per cent calcium carbonate formulations. For cardboard, other minerals such as titanium dioxide and calcined kaolins are occasionally found. These must be used as additives to increase opacity, particularly in the pre-coating of base cardboard containing used paper, in order to achieve complete optical covering.

Irrespective of whether a single or multiple coated paper or cardboard is concerned, calcium carbonate may be used in all present-day printing processes to achieve high-quality, or at least adequate, results. This versatility has fuelled demand for calcium carbonate as a coating pigment, particularly as it is obtainable worldwide at an acceptable price and may be used in all coating processes. For specific applications, the extent of calcium carbonate content depends on the required paper characteristics. For multiple coated (i.e. up to four times per sheet), high-weight papers, the exclusive use of calcium carbonate in all European paper mills has today become standard.

1.2.3 Pigment properties for paper coatings

The characteristic form and chemistry of calcium carbonate, and the resulting mechanical properties, have changed paper coating technology since its introduction in 1973.

The pigment surface as paper constituent – mechanics and chemistry interact

The mechanical properties of the pigment surface play an important role in the upgrading and printing of paper. The form and size of the mineral particles influence the flow properties and also the ability to cover the paper fibres. They are also responsible for the flux of the applied particles in subsequent paper calendering, in which paper gloss and paper smoothness are produced.

The chemical character of the pigment determines the interactions with other coating pigments and the organic content of the coating. In addition to charge, ion exchange and complexing processes, genuine chemical bonds can occur. Here, calcium carbonate shows a much more pronounced reaction than the two other important coating pigments, kaolin and talc. Its solubility in water and the resulting alkaline pH value mean that each calcium carbonate particle is surrounded by a sphere of calcium and hydrogen carbonate ions that may interact with their surroundings at all times.

Thus calcium carbonate behaves as a buffer to acids, thereby raising the ageing resistance of papers (see section 'neutral paper manufacturing with calcium carbonate'). This buffer effect of calcium carbonate at the paper surface has also required a modification to the composition of the fountain solution in offset printing. The original pH of the fountain solution of 3-4 was raised to 6 to avoid decomposition of the mineral.

The use of an excess of calcium carbonate can retard yellowing of the paper induced by UV, thermal or heavy metal catalysis.

The sum of all these effects determines the final printing quality. Since the surfaces of

SEM cross-section through a woodfree double-coated paper (120g/m2).

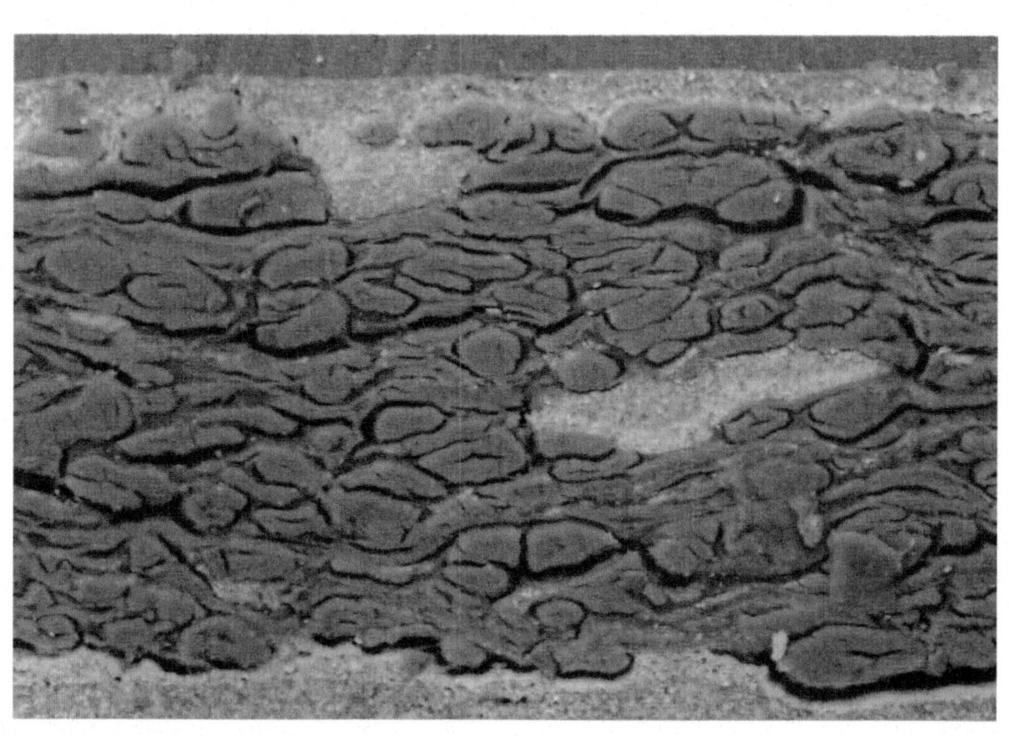

modern papers today consist of up to 95 per cent mineral (see figure), the printing ink hardly comes into contact with the paper fibres in those papers. The interaction of ink and mineral determines the final quality of the printed product.

Surface charge and coating structures

The objective of paper upgrading is to achieve a uniform coated surface in which, ideally, the pigment particles are statistically distributed at the paper surface. Owing to the specific properties of each mineral, this is only approximately possible and the differences between the different pigments are considerable.

Concerning the principal differences in the behaviour of kaolin and calcium carbonate as common constituents of a coating, Dahlvik of the Stockholm Institute for Surface Chemistry writes:

"At lower pH strong structure formation takes place if clay is present. This is partly due to the special chemical characteristic of the clay particle surfaces, but also to the interaction between adsorbed dispersing agent and calcium ion. Calcium carbonate did not show a pH-dependency to the same extent as clay."
TAPPI Coating Fundamentals, 1995, Dallas (TX)

Freshly ground, non-dispersed, calcium carbonate has a weakly positive charged surface and a pH value of 8.6. These two factors lead to comparatively weak chemical and physical interactions and low structuring. For kaolin, on the other hand, the edges of the pigment have a positive, the surfaces a negative particle charge, so that for kaolin coatings, macrostructures are readily formed. Thus kaolin coatings appear duller than calcium carbonate coatings prior to calendering.

For calendering, calcium carbonate and kaolin behave differently. For kaolin coatings, calendering produces a faster seal on the paper surface. The platy kaolin particles projecting from the coating surface are flattened into the x-y-plane by virtue of the external pressure of the calender, and the paper surface is thereby closed.

Calcium carbonate coatings, on the other hand, form more or less spherical packs owing to the rhombohedral or spherical particle form. Thus the particles cannot significantly re-orient along the x-y-plane. Rather, compression takes place in the z-plane, i.e. normal to the paper surface, and following calendering, the surface remains more open in conformity with the packing density of dense sphere packs.

Compact coatings consisting of 100 per cent calcium carbonate are therefore characterised by a fine-pore capillary system that often extends down into the base paper, permitting unhindered transport of water vapour in all directions.

These open coating structures of calcium carbonate were responsible for the development of multiple coatings in the 1970s. This was because, even for several superimposed coating layers, water vapour could still escape unhindered during drying. Unhindered escape is not possible for dense kaolin coatings. Particularly with rapid, i.e. very hot, drying processes following printing, blisters may appear at the paper surface (blistering).

In the wake of historic advances in coating speeds, the introduction of calcium carbonate with its characteristic pigment surface had extensive technological ramifications. After it had been recognised that calcium carbonate coating colours ease the control of coat weight better then kaolin when increasing machine speed, the final barrier to very high coating speeds was removed.

It was possible to raise coating machine speeds from 500 meter per minute (1970) to 1500 meter per minute (1998) without impairing pigment rheology. The explanation for this may be found in particle geometry. Thus calcium carbonate when flowing does not have a preferred direction, whereas kaolin alters its film structure continuously with increasing speed. Computer simulations have shown that different cluster types are formed in the film interstices in kaolin and calcium carbonate, depending on speed.

Particle size and Particle size distribution – more than just statistical material parameters

Observed microscopically, an individual calcium carbonate particle has a rhombohedral basic structure. Even with very fine grinding, rupture occurs in the grating plane or axis of symmetry of the natural crystal structure. Particles that appear to be spherical when viewed macroscopically prove to be similarly formed rhombohedral microstructures when viewed microscopically.

The functional differences in ground calcium carbonate products used in paper manufacturing arise from the agglomeration of these 'micro rhombohedrons' with different particle diameters to form large agglomerations. The particles in the agglomeration may be oriented statistically or selectively, and may be further modified by chemical surface treatment (dispersion, flaking, etching, recrystallisation).

The term 'fineness' in papermaker's language

Whether or not a calcium carbonate product is designated as coarse or fine depends primarily on the respective use. A 'coarse' grade of calcium carbonate in paper manufacturing would be regarded in the paint, varnish and plastics industries as having a high degree of fineness, and this situation has existed now for decades. Even so, these classifications are indispensable in the day-to-day parlance of papermakers for simplification purposes, and may also be understood without further definition.

Should, however, a more precise classification of calcium carbonate products in regard to fineness be required, two main methods are used worldwide in the paper industry, i.e. sedigraphic x-ray diffraction and laser diffraction.

Both these methods provide a Gauss frequency distribution of particle diameters (see annex "Definitions and Measurement Methods"). Today, 'coarse' calcium carbonates are in general understood to mean those products for which a maximum of 80 per cent of all particles are smaller than 2 micrometers, whereas 'fine' calcium carbonate products contain more than 80 per cent fine particles. This distinction proves very helpful in understanding the complex implications for paper coating and finishing.

Particle size distribution and paper characteristics

The fineness of the calcium carbonate used influences numerous paper characteristics, and also certain of the important process parameters in paper upgrading.

Modern paper research is based on the concept that the coating structure is formed early in the wet upgrading process and that the inner base structure 'survives' calendering. It is primarily the surface layers that are affected by calendering.

When it is considered that fine calcium carbonate qualities lead to a denser inner coating structure, it is clear that fine calcium carbonate produces a higher paper gloss at the same calendering pressure. To calender a coarse calcium carbonate to the same paper gloss, a much higher calendering pressure must be applied. In this case, the fibre constituents lying below the coating are necessarily affected. This can lead to a loss of brightness and opacity, i.e. to so-called calender blackening. To avoid this, paper gloss is limited for coarse calcium carbonate (see figure, next page). Therefore coarse qualities are particularly suited to dull finished surfaces and to use in precoating.

The mechanical characteristics of the paper surface, such as smoothness, roughness, porosity and vapour permeability, also change with the particle size distribution of calcium carbonate. With increasing fineness, paper smoothness increases and roughness decreases. A certain microporosity remains, however, even for ultra-finest calcium carbonate types (colloidal limit). This also applies following calendering. This porosity keeps the coating layer open and facilitates the passage of water vapour, even with high-gloss papers, which have heavily compacted surfaces.

At increasing drying temperatures in heat-set offset printing, the layer structure typical of kaolin is a barrier to the escape of water vapour from the fibre layer. As the synthetic latex bond holds the particle layers together, a blister is formed in the paper surface.

The significance of this characteristic of calcium carbonate at the speeds typical of heat-set offset printing presses was recognised as early as the 1980s. At that time, blistering occurred more frequently during printing if the paper had been coated at high kaolin content and the print had been dried at high temperature.

In Europe, this problem gradually diminished with the increasing substitution of calcium carbonate for kaolin in paper surfaces. At the same time, the fineness of the calcium carbonate had to be increased to maintain the niveau of paper gloss and to avoid calender blackening. As a result, the level of brightness in the paper in European paper markets in the 1980s and 1990s rose markedly. Since the mid 1990s, the fineness of ground calcium carbonate products has peaked. In the high-quality high-gloss carbonates, more than 90 per cent of all particles are less than one micrometer in size. Incremental improvements to the paper gloss values achieved in this way may only be attained by expensive, synthetically produced minerals, such as satin white or special crystallised calcium carbonate.

Particle size distribution and printability

During printing, the printing ink interacts both mechanically and chemically with the paper surface. The particle size distribution of the calcium carbonate used in the coating plays an important role in determining the extent of both these interactions.

Gloss production with coarse and fine calcium carbonate products.

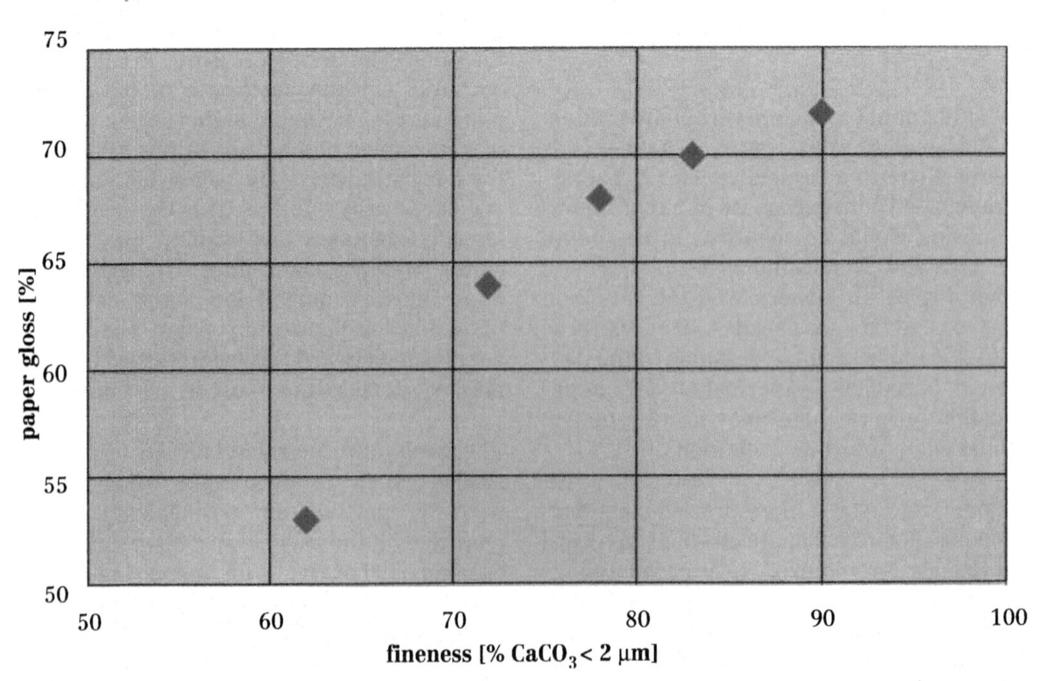

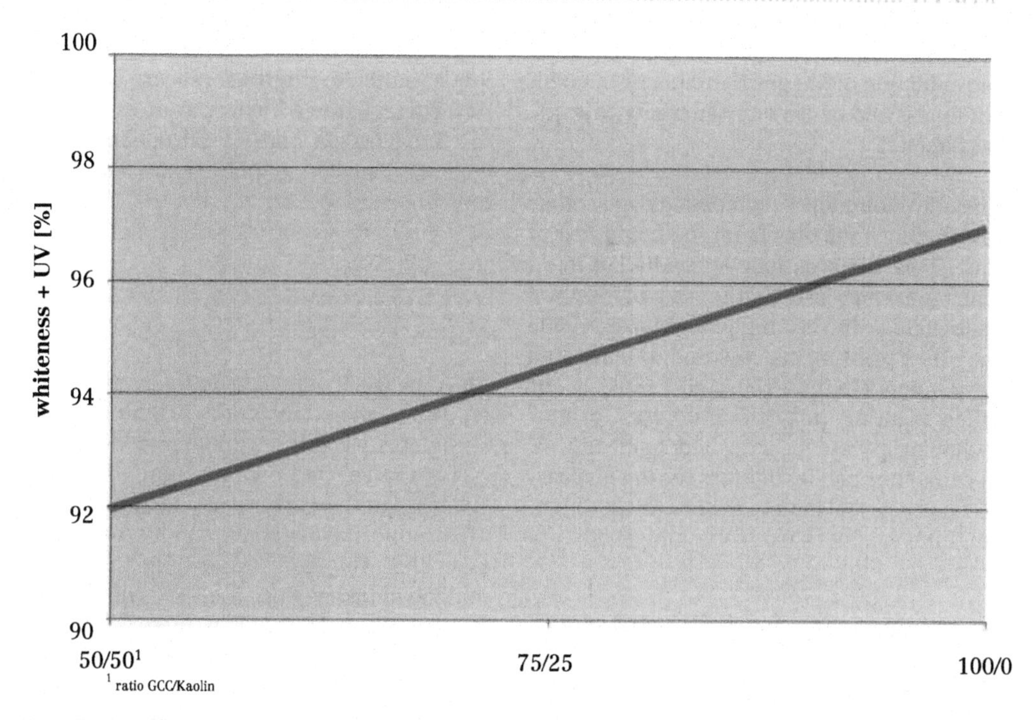

Higher whiteness from increased calcium carbonate content (single coated, woodfree 100 g/m2, coat weight 12 g/m2 per side).

Fine calcium carbonate qualities lead to an enhanced microporous paper surface and contain larger quantities of dispersion agents than coarse calcium carbonate qualities. Correspondingly, differences may occur between both qualities in respect of printing ink quantity, print gloss, half-tone dot expansion, printing uniformity, drying behaviour and colour scuff resistance.

Differences of this nature occur particularly with lightweight coated rotogravure and with four-colour cold-set offset newspaper printing. In both these printing processes, a colour pigment is used that during printing forms a film on the paper surface following absorption of the organic solvent. In both cases, ultrafine calcium carbonate fractions may hamper the rapid absorption of the solvent, i.e. toluene or mineral oil. Significantly better printing results may be achieved by using coarse or particle-selected qualities in which the ultrafine fractions have been reduced.

Brightness and chromaticity

With the availability of fine and ultrafine calcium carbonate as coating pigment, the brightness demands on coated papers changed rapidly in the industry in Europe. As calcium carbonates from chalk and limestone became available at the same or higher whiteness, and were 25 per cent cheaper than the kaolin commonly used up to then, a higher degree of paper brightness could be offered to the market at no additional cost.

This led to an alteration in the chromaticity of paper. While paper surfaces coated with kaolin display a marked yellow chromaticity, using calcium carbonate, a more neutral location may be achieved. In addition, with the introduction later of products based on marble, it was possible to dispense with a proportion of the so-called shading substances, the use of which had always led to the sacrifice of brightness (see figure).

In the wake of these developments, significant advantages arose for colour printing with offset papers, which had become increasingly popular from the 1970s onwards. Greater opportunities are open to the designer and printer for precise detailed reproduction of images by virtue of the high whiteness and neutral chromaticity with offset paper.

The development of inexpensive colour printing technologies from the beginning of the 1970s onwards has generally led to an enormous increase in the use of calcium carbonate with very high brightness. While the offset printing process underwent rapid development in the 1970s and 1980s, colour office printing (ink jet and laser printer technologies) and the so-called 'home offices' became the chief instigators for the increased use of calcium carbonate in high-brightness papers from the mid 1990s onwards.

Stimulated by new opportunities for producing whiter papers with calcium carbonate, optical brighteners experienced a boom in Europe. Whiter and whiter papers were demanded, and where coating pigments no longer sufficed, brighteners were called for. At the turn of the millennium, an average 10 to 15 brightness units in European, double coated, woodfree, paper are attributable to optical brighteners.

The spirit of the age transforms resource deployment

The new spirit of the age in the paper industry in connection with brightness and chromaticity was not without its consequences for the coating pigment manufacturers. Whereas in the early years calcium carbonate produced from chalk and limestone had been the preferred products in the paper-making industry, since the 1990s marble has become the principal raw material. Today, the proportion of marble in paper-making is about 80 per cent (see figure). The increasing proportion of PCC is also attributable to this fact.

The result was a shift not only in brightness and chromaticity but also in the opacity of

Chromaticity with kaolin, chalc and marble.

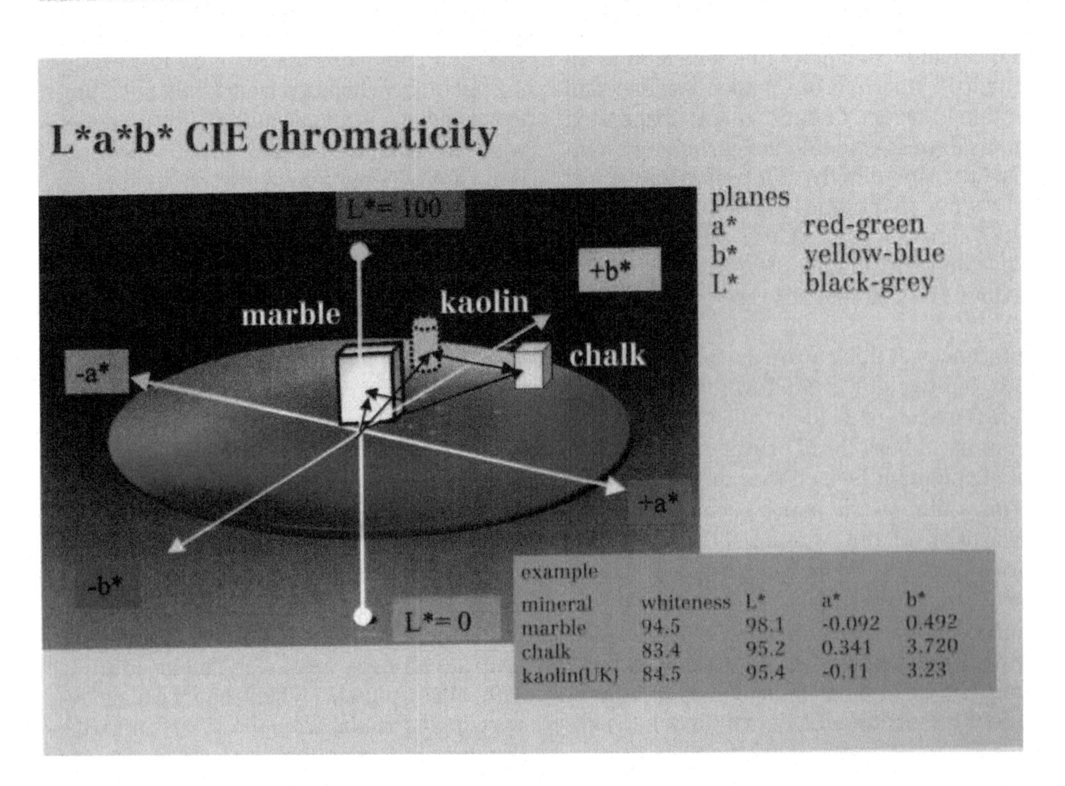

mineral	whiteness	L*	a*	b*
marble	94.5	98.1	-0.092	0.492
chalk	83.4	95.2	0.341	3.720
kaolin(UK)	84.5	95.4	-0.11	3.23

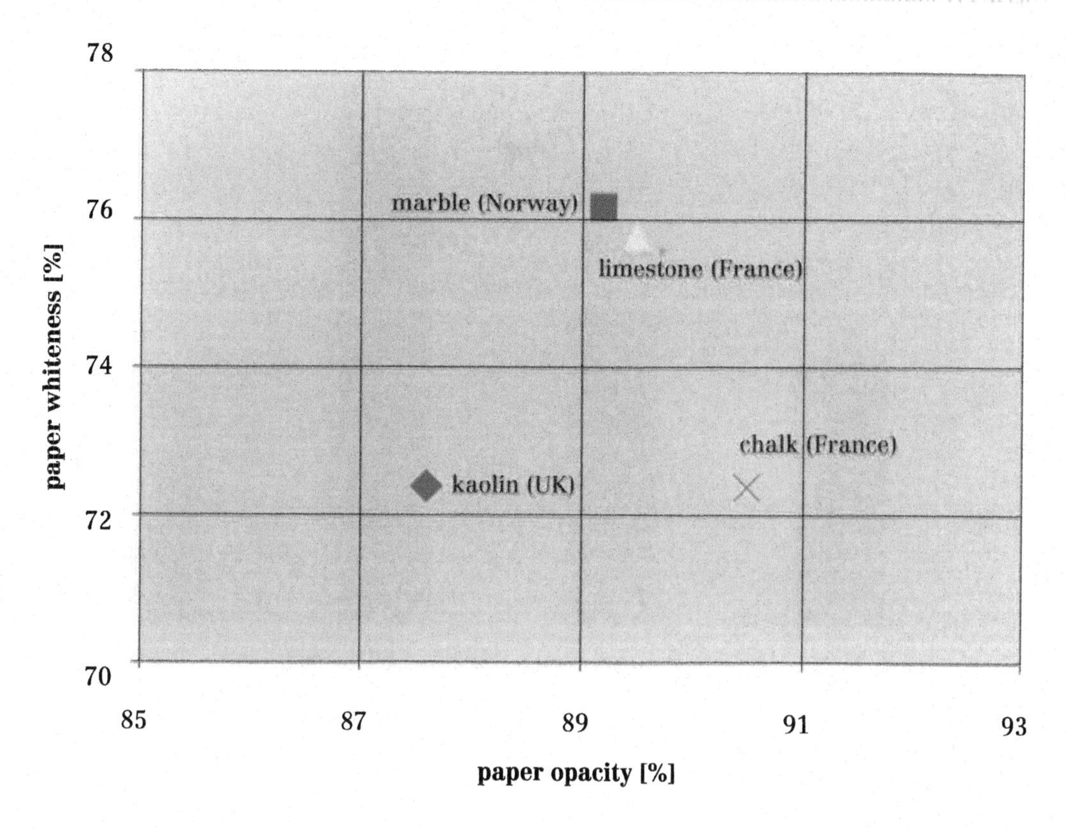

Influence of stone raw material on whiteness and opacity.

papers. By opacity, papermakers understand the ability of a paper to hold back light at the surface of the paper so that the text on the reverse does not shine through. For a satisfactory paper, as little light as possible should pass through to the reverse side.

By the addition of minerals such as titanium dioxide and dye, high opacity of the coating may be achieved. While the former increases light scattering, the latter absorbs light. In the case of calcium carbonate, the opacity of the coating may be influenced by the choice of stone raw material (see figure).

Coating opacity with calcium carbonate

For the same coating thickness, coating pigments based on chalk display 1 to 3 units higher opacity than those based on marble. The reason for the greater opacity is the higher light absorption of chalk caused by impurities such as iron oxide, manganese oxide, kaolin and organic constituents. On the other hand, the paper brightness imparted by chalk is up to 6 units lower.

To avoid this deterioration in paper brightness, the path to higher opacity must be sought via improved light scattering of the coating minerals. By means of particle-selective grinding or by crystallisation, calcium carbonate products offer considerably improved light scattering. They may now be manufactured, while still maintaining the brightness characteristic of products resulting from conventional ball-mill grinding (see figure, next page).

The increased light scattering coefficient of calcium carbonate produces a paper coating with improved optical covering. The greater light scattering arises here from the concentration of calcium carbonate particles, offering diameters optimally suited to the optical requirements. Furthermore, the

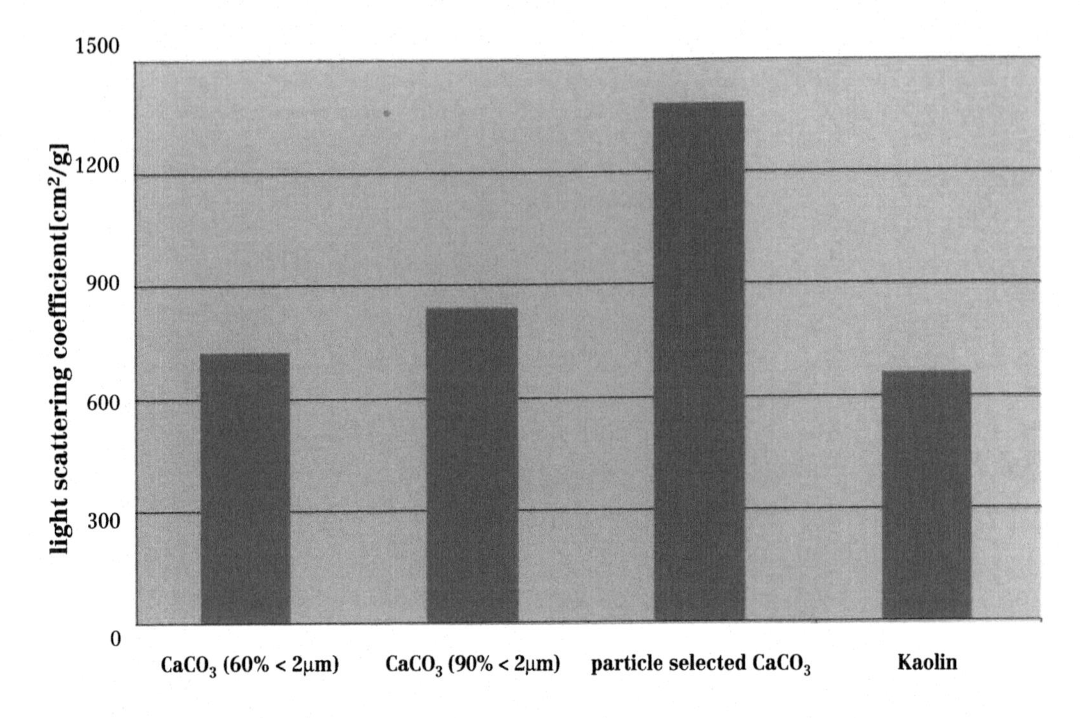

Light scattering coefficient for conventional and particle-selective ground coating pigments.

internal coating matrix contains cavities in which incident light is also scattered.

This effect may be explained by taking the optical properties of a head of beer as an example. The foam on beer consists of yellow beer which is transparent. When mixed with air, a foam is produced with opacity of 100 per cent, with a whiteness even higher than that of beer itself. Through the addition of air, spheroid cavities are produced whose diameter leads to optimum scattering, as shown by the theory according to Mie, Holst and Weber (see figure). Light waves with wavelengths in the visual range cannot traverse this foam layer, but are instead dispersed at the surface. Thus the optical impression of an opaque and white beer layer arises.

This principle was investigated in the laboratories of the Omya, and led in 1992 to the development of a new family of calcium carbonate products that were introduced commercially in paper applications in 1994. In the meantime, it has become clear that no other calcium carbonate product has ever achieved such rapid introduction to paper manufacturing on a worldwide scale. At the turn of the millennium, around one million tonnes of this product with enhanced light scattering was being produced and utilised worldwide.

Abrasion phenomena with calcium carbonate

Abrasion arises at almost all points in paper upgrading, whether this be in the mixers, tanks or piping used for processing and storage of calcium carbonate, or on coating equipment, the rollers of coating plant, drying equipment, covering material or cutting tools. The extent of abrasion is determined principally by two factors: the material properties of the abrading opponents and the total system energy of the various materials at the time of interaction.

Using Mohs' hardness as abrasion criterion, pure calcium carbonate having a Mohs' hardness of between 3 and 4 does not differ significantly from kaolin or talc. Nevertheless, considerable differences can, for example, arise in the running times at the equalising blades during coating.

Long-term experience in paper coating with calcium carbonate has shown that ultrafine calcium carbonate products considerably lengthen blade lifetimes compared with coarser carbonates or coarser coatable kaolin that has also been delaminated. The reason may be traced to the principal lack of specific oversize particles with diameters above 10 micrometers. At high coating speeds, the physical momentum of such particles increases quadratically, and this is apparently sufficient to damage the finely ground blade edge in a relatively short time.

This led to the introduction of the so-called 'topcut' as a measure for the proportion of oversize particles in high-quality calcium carbonate products. As a result, the lifetime of blades in coating technology was extended, and thus also the total number of unperturbed paper production cycles. Using coating pigments consisting of 100 per cent calcium carbonate, blade lifetimes in excess of 10 hours at coating speeds of 1400 meters per minute are common.

Slurry versus powder

In paper upgrading with calcium carbonate, the use of dispersed pigment slurries now prevails. The advantages of slurry in respect of transport, storage, handling and quality at the requisite quantity flows of present-day paper upgrading installations are entirely convincing, not least because their use has permitted many logistic processes to be largely automated.

At the turn of the millennium, calcium carbonate is supplied in the form of dry powder and dispersed on site for paper manufacturing in only a few countries, such as Venezuela, Columbia, China, Russia and Japan. Powders of this type do not, however, display the same quality features as a wet-ground slurry in terms of fineness, rheology and top cut.

Using dry grinding, it is not possible to obtain the ultrafine granulation quality

Light scattering and particle diameter according to Mie, Holst and Weber.

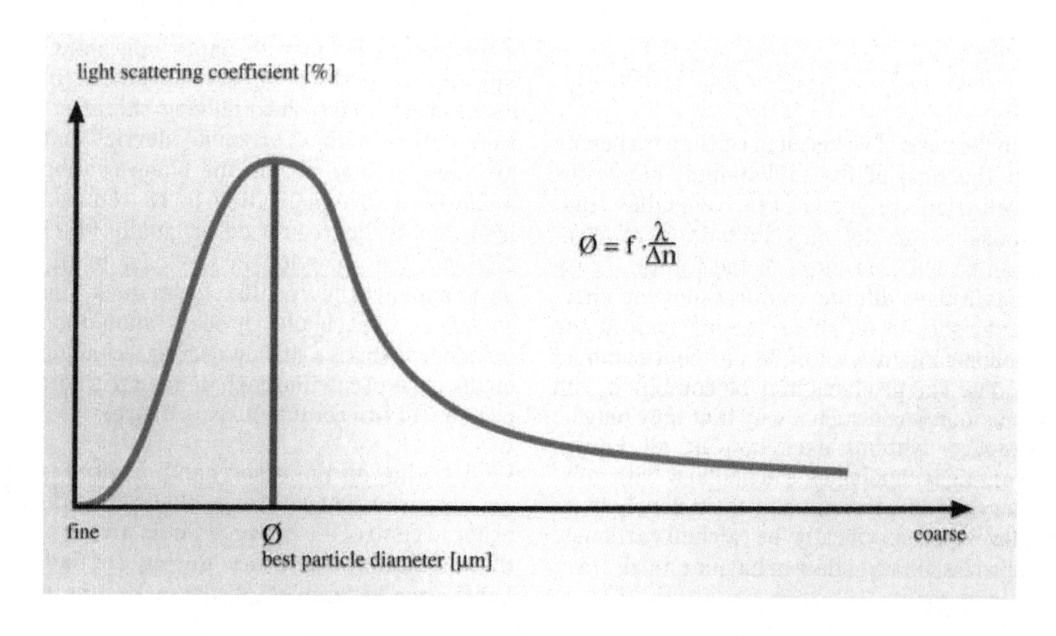

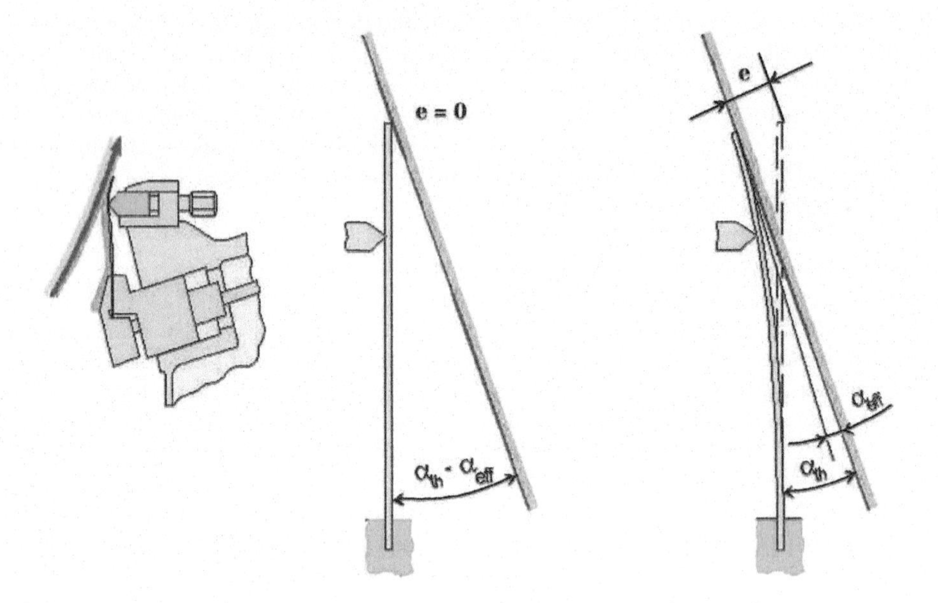

Stiff blade and bent blade adjustment. The greater the bend the smaller the coating angle (left: stiff blade; right: bent blade).

necessary for achieving the gloss and surface qualities of modern papers. On the other hand, it seldom makes sense from a technical, economical and ecological point of view to go as far as to: wet grind a coating pigment to the necessary degree of fineness; then dry it under high energy expenditure; and finally to mix the powder obtained in this way with water to obtain the dispersion required for coating.

Solids content as pacesetter in new coating technologies

On the basis of wet ground calcium carbonate at the turn of the millennium, numerous well-tried products are available that measure up not only to today's requirements, but also those of the future. These retain their diluting and free-flowing characteristics in all shear ranges even at the solids content level of 78 per cent common today. The products may be combined with one another in such a way that they may be applied without exception in all known coating technologies. Particularly with polyacrylate dispersions and their copolymers, developed specifically for calcium carbonate slurries, dilatant flow behaviour hardly ever occurs now.

From this, several consequences immediately arise in connection with paper upgrading. For example, the high solids content in connection with the 'shear diluting' characteristics of the calcium carbonate slurries that are now possible permit the blade coating angle used in equalisation to be reduced from the 45 degrees used originally to 25 degrees. This had led in the past to the development of new coating techniques. The so-called 'bent blade' process should be mentioned. This is a shallow coating technique on the basis of calcium carbonate that is now used on all five continents (see figure).

With kaolin, on the other hand, coating is mainly performed with large blade angles in order to control the hydrodynamic pressure that constantly increases during coating. Large-area (delaminated) coating kaolins in

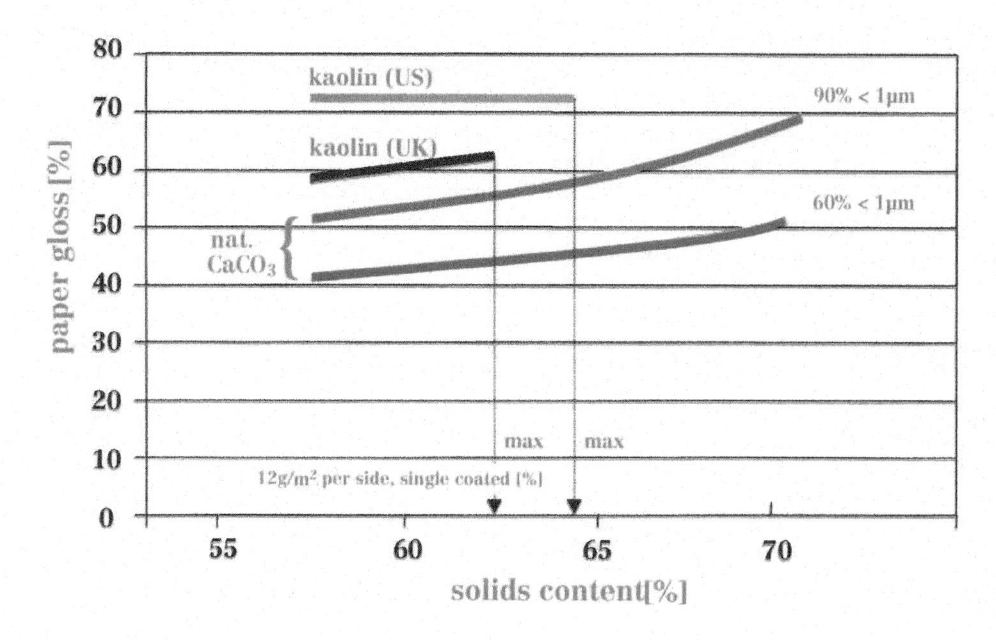

Production of paper gloss by increasing the solids content in a calcium carbonate coating.

particular create very much higher hydrodynamic forces on the coating blade than calcium carbonate, even at low solids content and low speeds. To prevent an undesired increase in the coating quantity applied, in the case of pure kaolin coatings, the blade is often bolstered with a reinforcing blade. However, this may sometimes lead to considerable variations in coating homogeneity at the paper surface. This particularly explains the phenomena of perturbations due to mottling and brightness that arise with pulsating blade pressure variations.

At times, water was added to the coating when the hydrodynamic pressure on the equalising blade rose due to an unexpected change in the flow behaviour of the coating. Although it enabled coating quality to be maintained, this unregulated introduction of water was in some cases detrimental to coating quality. Owing to the fact that the relationship between solids content and coating angle was no longer optimal, heterogeneous particle distribution could result, resulting in an open and rough paper surface.

Seen in this light, it becomes evident how advantageous it is to use a dispersed calcium carbonate slurry with its rheological shear-diluting properties. Even at high solids content, low blade angles may be used, and these may be maintained during the entire period between blade changes, resulting in a marked reduction of the forces impinging at right angles to the paper web. This leads especially at high web speeds to fewer ruptures and a smooth paper surface with predictable qualities. Owing to the low water content of calcium carbonate coatings, expensive drying energy may be conserved.

However, even at high coating angles of 40-45 degrees, a high solids content has its advantages. Thus, the high packing density of calcium carbonate particles that is already present in the wet coating, favours rapid coating immobilisation, and has a positive effect on the emergence of paper gloss in subsequent calendering (see figure).

The compact calcium carbonate film also permits the use of less synthetic binding agents, reducing costs compared to pure kaolin film, since there is less open volume in the film structure that must be filled with binder.

As calcium carbonate permits stable quantity coating application even at higher solids contents, more compact films result. This is an advantage in the manufacturing of high-gloss papers. If the papermaker exploits the whole range of solids content in the coating for gloss development, he may partly or wholly dispense with the introduction of special synthetic pigments (plastic pigments).

The production of multiple films with up to four coatings per paper side would be unthinkable without the employment of calcium carbonate grades of different granulation at high solids contents. Only by their use is proper film quantity management possible. The shear diluting flow properties of calcium carbonate exert a self-cleaning effect on the blade edge. By this means, contamination that may occasionally occur with fibres and agglomerates on particularly smooth, precoated, paper may be avoided before the production of blade streaks. Contrary to this, many coating kaolins display a tendency towards densification with increased speed, and this hampers blade cleaning.

Interactions with calcium carbonate

Already Paracelsus knew that all effects have their side effects. Calcium carbonate is no exception to this rule, whereby even here, its chemical character dictates the limits of exploitation with regard to desired and undesired interactions. In this, it differs significantly from the chemically less sensitive kaolin.

In aqueous solution, calcium carbonate has a pH value of 8.6. If the pH value of a coating falls below this value, calcium carbonate decomposes, giving off carbon dioxide. Down to a pH value of 6.8, the decomposition process proceeds very slowly, so that in normal day-to-day coating practice with processing times of only a few hours, it is hardly noticeable. However, shock introduction of acid substances leads to the direct release of gaseous carbon dioxide, and of the corresponding calcium salts at the point of injection. The concentrated addition of acid substances to calcium carbonate slurries should therefore be avoided.

The acid sensitivity mentioned above is one of the reasons why cationically dispersed calcium carbonates have not met with success for coating purposes. In cationic coating, acid binding agents alone are available that must be introduced at a pH value of 3-4, causing unpredictable effects.

When dispersing with polyacrylic acids, the surface of calcium carbonate is to some extent protected against spontaneously occurring processes. Heavy dilution of the dispersed calcium carbonate slurry can, however, diminish this protection. By virtue of the desorption of dispersing agent that occurs, the equilibrium between the dispersed particles is disturbed, so that sedimentation or flaking phenomena may arise, with concomitant disadvantages for paper quality. The most noticeable results of destabilised slurries of this kind are: paper dusts, film picking and piling, calender deposits and printing colour mottling as a result of the heterogeneous paper surface.

Modern calcium carbonate slurries have been tested for compatibility with most of the known auxiliary coating agents, and found to be applicable without restriction within the usual production range. When mixed with kaolin, they generally reduce the viscosity of the kaolin coating. For this reason, viscosity-regulated coating processes can make it necessary to introduce thickening additives, or to adjust the solids content to suit the prevailing conditions.

In offset printing, decomposition products of calcium carbonate may be deposited on the rubber blanket of the printing press in cases where extremely acid fountain solution additives (pH 4) are used. This problem may be avoided by use of complexing substances and neutral or alkaline fountain solution.

Owing to their more open surface structure, coated papers coated with calcium carbonate accept printing colours more readily than papers coated with kaolin. Correspondingly, using a high calcium carbonate fraction in the film, the drying time in the printing process may be shortened.

The scuffing phenomenon with dull offset printing papers known as 'chalking' has, however, nothing to do with the use of calcium carbonate, and also occurs with completely calcium carbonate free, coated, papers. Chalking results from insufficient bonding between the mineral particles within in the film or paper surface, and is influenced both by technological processes and mechanical parameters.

Dynamic dewatering and sedimentation

When calcium carbonate powder is dispersed in water, rapid sedimentation takes place when left standing, while at the same time dewatering of the sediment takes place. This process may only be retarded by the use of dispersion agents. The objective in dispersing calcium carbonate slurries is to prevent sedimentation and topostatic dewatering for approximately 6 weeks to permit the slurries to be transported and stored until they are used.

Differences in dewatering rates arise from the prevailing solids content, fineness and type of stone of the calcium carbonate. Among these, the solids content of the coating has the greatest influence on water retention capacity (see figure).

The dewatering speed decelerates with increasing fineness. Thus, ultrafine calcium carbonate qualities sediment more slowly than coarse qualities, whereas calcium carbonate particles in colloidal solution do not sediment at all.

In paper upgrading, dynamic dewatering of the entire coating takes place beneath the equalisation blade. During application and subsequent film equalisation, hydrodynamic forces impinge upon the dispersed particles, and almost completely compensate for the slight differences in water retention capacity caused by dispersion and stone type. With increasing web speed, pressure dewatering and flow behaviour in the inner capillary film structure become dominant.

Under dynamic conditions, calcium carbonate as a rule dewaters more readily than kaolin. This may be explained, on one hand, by the rhombohedral base form of the calcium carbonate particles that results in increased mobility and more ready paper penetration, making the particles amenable to direct contact dewatering at the fibre surface. On the other hand, the surface of calcium carbonate is less hydrophilic than that of kaolin.

raw material	drying time [sec]	
	fineness ($d_{60} < 2$ µm)	fineness ($d_{90} < 2$ µm)
chalk	10	20
limestone	17	23
marble	9	14

Water retention capacity (method SD-Warren; 67% solid content + 1% CMC).

In summary, coatings based on calcium carbonate are immobilised more rapidly at the paper surface for the same solids content than pure kaolin coatings. Since with calcium carbonate, higher solids contents are generally possible, the immobilisation point can be chosen within a larger range, so that the extent of film penetration at high web speeds may be influenced.

Although rapid film immobilisation is advantageous when coating with calcium carbonate, the water retention capacity should be adjusted by means of thickener, since uncontrolled dewatering can lead to a heterogeneous paper surface at the next coating equalisation stage. This applies particularly to processes with low coating speeds of less than 600 meter per minute that are usual with cardboard coatings and precoatings with coarse calcium carbonate.

Thickeners currently in use are natural starches, semi-synthetic carboxymethylcellulose (CMC) and synthetic substances based on polyacrylate.

Cationic coatings with calcium carbonate

Never before has a topic of research led to so much discussion and intensive research both within and outside the European mineral industry than the development of cationic coatings.

Cationic coatings are defined as coatings with an excess of positive charge. This is usually measured with a zeta potential measurement set-up. With natural, dry ground, and also with freshly precipitated, calcium carbonate, a weakly positive, i.e. cationic supercharge may be detected at the particle surface. During dispersion, this is, however, equalised by the anionic polyacrylates, so that for commercial calcium carbonate slurries, an anionic total charge results. For kaolins, a strongly negative excess charge of this nature at the surface is already found in the natural, non-dispersed, state (see figure).

Since the paper surface displays a negative excess charge owing to the anionic fibre surfaces, a cationically charged film should theoretically be immobilised at the fibre boundary prior to film equalisation, thereby significantly reducing film penetration. SEM shots of different films have confirmed this conjecture (see figure).

Initial coating tests with cationic coaters were based on kaolins and anionically dispersed calcium carbonates that were 'charge-shifted' by the addition of large quantities of cationic polymers. The coating pastes that resulted, however, were so heavily flaked that the laboratory results could not be implemented in day-to-day coating practice, since the processing times for cationic coatings of this type were simply too short.

It was not until 1989 that Omya AG developed and patented a production process for homogenous, rheologically stable, cationic calcium carbonate slurries. Thanks to this process, practicable, reproducible, coating tests could be carried out in the early 1990s that made possible the handling, transport and storage of cationic coatings under practical conditions. Anticipating new sales

Zeta potential of selected minerals.

	zeta potential without dispersing aid [mV]	zeta potential with dispersing aid [mV]
Kaolin	-25 bis -45	-25 bis -45
$CaCO_3$	0 bis +6	-25 bis -55
PCC	0 bis +8	-25 bis -55
$CaCO_3$ (cationically dispersed)	0 bis +6	+33

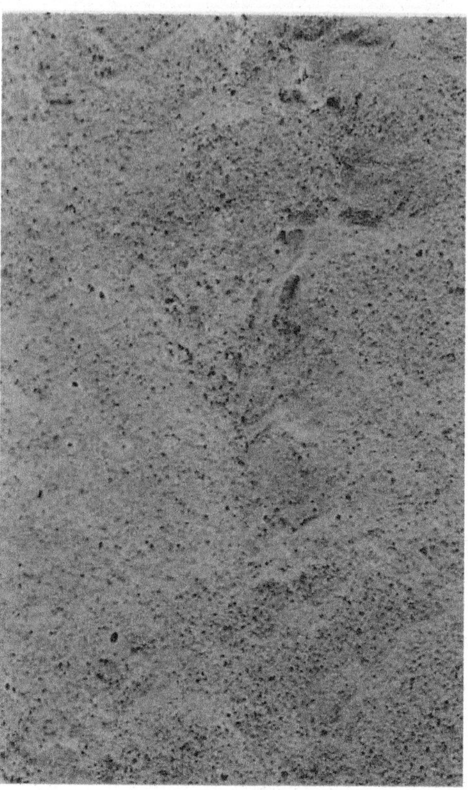

SEM comparison of anionic (top) and cationic precoating (bottom) on wood containing paper (10 g/m2 x Seite).

markets, binder and additive manufacturers developed numerous new additives, such as latex binders, thickeners and optical brighteners for these cationic calcium carbonate coatings. However, the results were disappointing.

The introduction of 100 per cent cationic calcium carbonate in precoatings in double-coated papers indeed displayed the expected rapid and satisfactory immobilisation of the coating. However, after applying the topcoat, the paper characteristics showed no significant advantages over conventional anionic coating practice. The significantly higher costs resulting for cationic additives using this strategy could therefore not be justified.

The situation is somewhat different for the manufacturing of lightweight, single-coated, papers. For this, cationic calcium carbonate must be combined with charge-shifted kaolin in order to achieve the intended high gloss. Although this results in an enormous cost increase for the preparation, it yields significant advantages in regard to fibre cover at film weights of less than five grams per square meter. It is therefore conceivable that the introduction of new coating technologies, such as the film press in combination with on-line calendering, could lead to the resurgence of cationic coatings.

At the turn of the millennium, however, no papermaker known to us has adopted a cationic coating based on calcium carbonate.

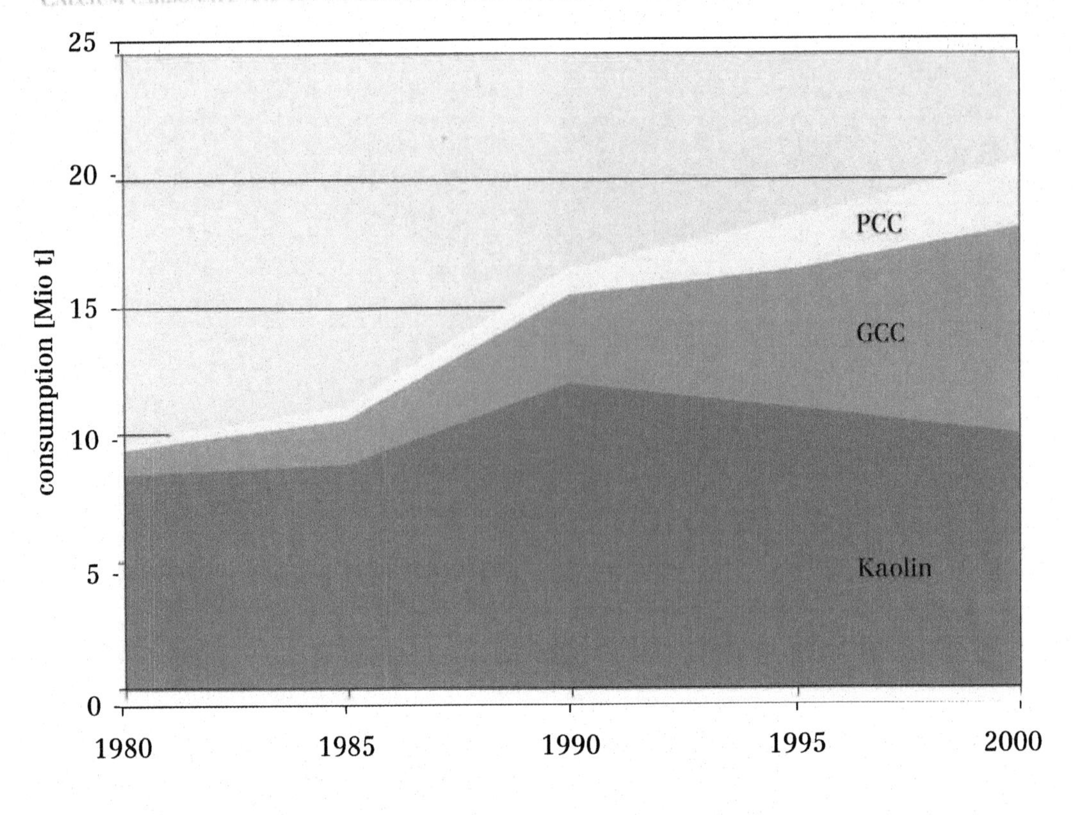

Worldwide mineral consumption in the paper industry from 1980-2000.

11.3 Industrial use of calcium carbonate in the paper industry

Within the last 20 years, calcium carbonate has achieved the same worldwide production volume in the paper industry as kaolin. Today, over 10 million tonnes of calcium carbonate products are used annually in paper manufacturing. This represents approximately 44 per cent of the total consumption of minerals in the paper industry (see figure). Contained in this figure are approximately 2.5 million tonnes of calcium carbonate based on PCC and 8 million tonnes of natural, ground, calcium carbonate.

In the mid 1980s in North America, PCC played a significant role as filler (see figure), albeit not for coating applications. As limestone is common in the whole region, the on-site production of PCC in the vicinity of paper mills proved the most cost-effective option. In this, PCC occurs at a solids content of 25 per cent, and can be introduced directly to production as a filler. A total of 48 PCC plants were built on the American continent within approximately 15 years.

Meanwhile, the concept of 'economy of scale' has led to mature and convincing strategies for the manufacturing of ground calcium carbonate, ensuring adequate supplies to most paper mills. Today, the largest production units manufacture 2.5 million tonnes of ground calcium carbonate per year for the paper industry at a single site.

The growth prospects for calcium carbonate in the paper industry are buoyant. Although coated papers account for the greater part of material consumption, initial steps have now been taken to introduce high-whiteness minerals into newsprint and the remaining

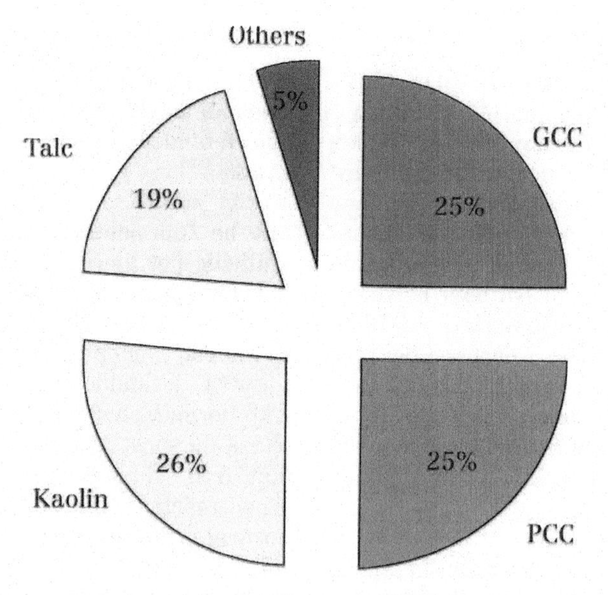

Worldwide filler (top) and coating pigment consumption (bottom) for paper manufacturing in 1997.

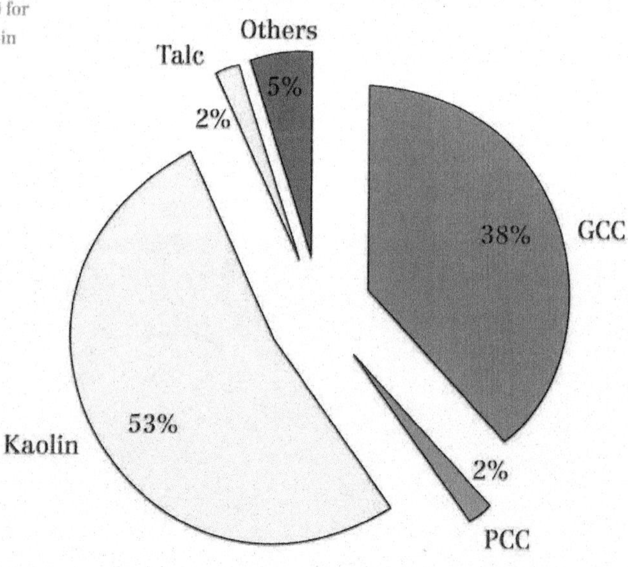

wood containing natural papers. For newsprint in particular, the mid 1990s saw a new trend. The introduction of up to 15 per cent by weight of minerals in certain high-quality products is already quite common. Owing to its natural whiteness, ready availability and highly developed application technology, calcium carbonate will continue to play an important role in the future.

2. Plastics

Rubber and celluloid were, at least in chemical terms, the first plastics to be manufactured industrially. Both have the same characteristics as all synthetic plastics. They are high-molecular polymers consisting of many regularly linked building blocks, i.e., monomers. Nevertheless, rubber and celluloid are strictly speaking not synthetic polymers but converted natural materials. In 1839, the American chemist Charles Goodyear vulcanized the dried sap of the Hevea tree to obtain a useful rubber, and the Hyatt brothers (USA) manufactured their celluloid in 1869 from collodion, which they obtained by nitration of cotton.

These "converted raw materials" had their heyday in the 19th century. But that does not mean that they disappeared from the market in the 20th century. Films of "Zellglas" (a cellulose hydrate), developed in 1910, were a popular packaging material for a long time and were ideal for use as synthetic sausage cases.

But the 20th century belonged to the fully synthetic polymers. As early as 1907, the Belgium chemist Leo Baekeland obtained a patent on his bakelite: thermosets produced by the polycondensation of phenol and cresol with formaldehyde. Polycondensation with formaldehyde subsequently became one of the most important reactions for the industrial manufacture of thermosets. When reacted with urea, it produced urea formaldehyde resins (1910), also known as "organic glass"; flexible polyvinyl acetate could be "hardened" with formaldehyde (1926) and the extremely versatile melamine resins could be obtained by polycondensation with melamine (1935).

But there was longer to wait before the synthesis of the first thermoplastics. Hermann

Tapping a rubber plant.

Staudinger first had to create the scientific foundations for "macromolecular chemistry" with his work from 1922 to 1926, before the countless new compounds with a huge variety of properties could be synthesized on an industrial scale.

By 1928, the first polyacrylic polymers had been introduced, and between 1930 and 1935, polyvinyl chloride (PVC), polystyrene (PS) and polyvinyl acetate (PVA) were produced on an industrial scale. At about the same time, the industrial-scale production of synthetic elastomers from butadiene and other monomers started. After thermosets and thermoplastics, they were the third important plastics group (see box).

Developments followed inexorably: from 1936, ICI (Imperial Chemical Industries) produced polyethylene, now the most important polymer in terms of volume. By 1938, the polyamides Nylon and Perlon had appeared on the market, and during the Second World War, other important polymers, such as polyurethane (PUR), epoxy resins, fluorinated polymers and silicones, made the leap from the laboratory to industrial mass production. Finally, in the fifties, they were supplemented by the polycarbonates (PC) and polyesters.

In the years since then, there has not been a significant increase in the number of entirely novel polymers. Development work has rather concentrated on improvements in properties and more efficient production and processing techniques. In particular, the development of low-pressure polymerisation by Karl Ziegler and Giulio Natta at the beginning of the fifties considerably extended the possibilities of the plastics industry. The technically elaborate high-pressure processes disappeared and plastics production became simpler and cheaper. Production volumes increased and plastics were able to enter new fields of application, such as medical technology, automotive engineering, building construction and furniture production.

However, it also became more important to tailor each polymer to its particular application, while at the same time keeping costs low. These challenges could only be met by the widespread use of fillers and reinforcing agents.

Fillers and reinforcing agents were incorporated into polymers from the start. For example, already in the 19th century, rubber floor coverings were filled with cork flour or wood flour, and at the beginning of the 20th century, Leo Baekeland had even patented the use of fillers and reinforcing agents. The addition of wood or asbestos fibres, textile or paper scraps, allowed him to reduce the brittleness of his bakelite and other phenolic resins: wire mesh increased the strength and toughness of polymers. By the begin-

Thermoplastics:

Polymers that are hard or soft at their service temperature but pass through a glass transition range above the service temperature, in which they repeatably soften without undergoing any chemical change. In the softened state they can be processed by pressing, extrusion, injection moulding or other shaping processes to produce parts. Thermoplastics are the main plastics category.

Thermosets:

Generic term for all plastics that are produced from curable resins and crosslink irreversibly, by polymerization, polycondensation or polyaddition when heated or on the addition of catalysts, and are thereby converted to an insoluble state.

Elastomers:

Natural or synthetic polymers with elastomeric characteristics that can be elongated to twice their length and immediately return to their starting length when the force necessary for elongation is removed.

Adhesives and sealants:

Organic and inorganic, viscous compounds that are capable of joining two substrates together. Adhesives form a durable surface bond between two materials. The purpose of sealants is to prevent the penetration of gaseous or liquid substances between two substrates.

Organic fillers		Inorganic fillers	
Synthetic	**Natural**	**Synthetic**	**Natural**
Carbon black Carbon fibres Carbon spheres Coke dust Polymer fibre	Wood flour Cellulose fibres Starch Rice husks Cork flour	Silica gels Precipitated calcium carbonates and barium sulphates Aluminium hydroxide Glass spheres, glass fibres Metal oxides Slag wools Aluminium and calcium silicates Barium ferrite Whiskers	Calcium carbonates Talc Kaolin Asbestos Silicates Quartz flour Baryta Mica Nepheline, syenite Wollastonite Metal powder

Typical fillers in the plastics industry.

ning of the forties it had become standard practice to use mineral fillers.

Plastics producers often spoke of "dividend powder" to describe fillers, since the most important, often the only, argument in favour of their use was to cheapen the formulation. A correspondingly wide variety of materials was used (see figure).

Nowadays, by contrast, fillers fulfil a variety of functions, and mineral fillers in particular have become indispensable. In our daily lives we continually encounter filled and reinforced polymers: for example, coatings on carpet backings and other floor coverings contain up to 70 per cent fillers; electric cable sheathing is filled with large quantities of fine mineral filler. And in all these applications, calcium carbonate fillers are the most important.

2.1 The Plastics Market

Plastics include a large variety of different products that play an important role in technology and everyday life (see figure). As materials, they are often superior to traditional products such as wood and metal. Plastics have a low density and mechanical properties that can be easily controlled. They are corrosion resistant and their mouldability allows them to be economically processed into commodity products of a variety of forms and colours.

There are a number of approaches to a systematic classification of plastics. They can be classified according to their chemical structure and the fundamental reactions forming them, or according to physical characteristics and the associated aspects of processing and applications.

Chemically speaking, plastics are classified according to the type of reaction that underlies the formation of the polymers from its monomers. A distinction is made between:

- Polycondensates (phenolics, amino resins, polyamides, polyesters, silicones, etc.)
- Polymerisates (PE, PP, PVC, polystyrene, acrylic compounds, etc.)
- Polyadducts (PUR, epoxy resins, etc.)

In addition, there are semi-synthetic polymers, such as cellulose derivatives, which are obtained not from starting products such as coal, petroleum or natural gas, but by the chemical conversion of high-molecular natural materials to obtain the macromolecular structure.

Plastics products.

However, in practice, plastics are usually classified into three groups according to their mechanical-thermal behaviour: thermoplastics, thermosets and elastomers, but these groups do not cover the entire range of polymeric materials since, in chemical terms, polymers also comprise sealants, adhesives, chemical fibres and synthetic rubber, which are therefore also included. Even synthetic resins (paints and coatings, dispersions and additives) cannot be clearly distinguished from plastics on chemical grounds. It is only the applications that make a distinction possible.

However, many publications discriminate between the plastics market, covering the two groups comprising thermoplastics and thermosets: the elastomers market, and adhesives and sealants are usually dealt with separately. For reasons of clarity, these distinctions are also maintained below.

Market Survey

If we measure the economic importance of a polymer solely by the volumes produced annually, it is obvious that thermoplastics and thermosets hold an outstanding position (see figure). Of the 183.7 million tonnes of plastics produced worldwide in 1998, 158 million tonnes could be attributed to these two plastics groups. Natural and synthetic rubber products (elastomers) saw total production of just 17.7 million tonnes, about 40 per cent of which was natural rubber. Adhesives and sealants were processed in volumes of a good 8 million tonnes globally in 1998, however this market shows particularly high growth rates. In the last three decades alone, their production volume has increased approximately fivefold.

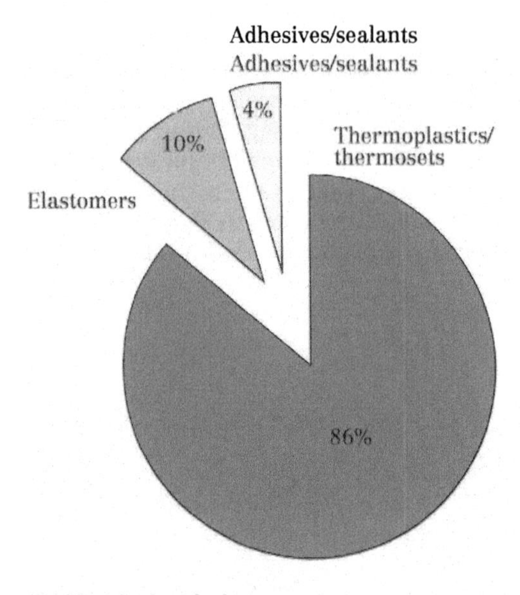

World production of polymers in 1998.

All plastics, sealants and rubber products have one thing in common: their main sales areas are the highly industrialized regions. According to sector, 65 to 80 per cent of the respective market volume goes to the states of the European Union, the USA and Japan.

2.2 Fillers and Reinforcing Agents

Fillers and reinforcing agents generally comprise additives of fixed shape that are distinct from the polymer matrix in both composition and structure. The air bubbles in foamed plastics can be considered as a special case of a filler.

It is difficult to draw a clear distinction between fillers and reinforcing agents. Classifying them according to their chemical composition or differentiating between natural and synthetic materials only results in confusion. A much clearer picture is obtained by classifying fillers and reinforcing agents according to their function.

In general, the term 'reinforcing agents' is used to describe all additives that improve the ultimate tensile strength. Fillers, on the other hand, decrease these properties and are added for other purposes. For example, fillers (with the exception of air) lead to the following property changes in thermoplastics: they increase the density and stiffness, reduce thermal expansion and increase thermal conductivity and thereby improve the stability of the final parts.

The reason why a particular filler changes the properties of a plastic in the way it does can be best understood if we consider the following physical aspects:

- Particle shape (aspect ratio)
- Mean particle diameter
- Gradient of the particle size distribution curve
- Coarsest particle fraction (top cut)
- Specific surface area
- Surface energy/surface tension
- Surface coating
- Packing density (at high filling ratios)

These criteria are adequate for describing the mode of operation of fillers.

Particle Shape

The so-called form factor, or aspect ratio, describes the ratio of a particle's length to its thickness. For fillers used in non-polar or semi-polar polymers, such as polyethylene, polypropylene and PVC, it is the most important physical parameter and determines whether an additive acts as a reinforcing agent or filler. A distinction is made between five basic particle shapes, which can in turn be subdivided into two groups (see figure).

Spheres, cubes or cuboids act as fillers and, apart from increasing stiffness, usually do not improve the mechanical properties of plastics. If there are very strong adhesive forces between the filler surface and polymer chains, however, an additive of small aspect ratio (spheres or cubes) can have a reinforcing effect, such as increasing tensile strength. A typical example is the reinforcing effect of uncoated calcium carbonates in polyamides.

Platelet structures and fibres with high aspect ratio, on the other hand, produce a reinforcing effect in all plastics. For example, talc, with a platelet shape, is used as a reinforcing agent in bumpers for automobiles, which must withstand extreme mechanical loads. In modern composites, the polymer matrix itself is even of subsidiary importance. The mechanical loads in carbon-fibre-reinforced epoxy resins are borne almost entirely by the carbon fibres. The polymers serve mainly for separating the individual fibres from one another and for transferring shear forces to the fibre surface.

The effect of the aspect ratio is shown by the stiffness of filled polypropylenes. Glass fibres significantly increase the stiffness, expressed by the modulus of elasticity (E-modulus). In the case of talc, with its platelet structure, the effect is less pronounced, and is least in the case of cubic calcium carbonate. At higher filler contents, the differences are even greater: for example, at a 10 per cent filler content, the reinforcing effect of the glass fibres is a factor 1.5 greater than that of calcium carbonate (see figure, next page). At a filler content of 30 per cent, it is a factor of 2.5 greater. However, this increase in stiffness is usually obtained at the cost of embrittlement: the impact strength or notched impact strength decreases.

Particle-Size Distribution and the Top Cut

Most fillers are produced and sold with precisely defined particle size distribution curves (see appendix: "definitions and test methods"). The gradient of the particle size distribution curve allows a precise prediction of the effect of the filler in the polymer matrix. The curve gradient not only indicates how big are the average deviations from the mean: it also determines the amount of coarse particles and the proportion of fines. And these two parameters have a significant effect on the mechanical and optical properties of the filled part.

Particle shapes (aspect ratio) of fillers and reinforcing agents.

Shape	Sphere	Cube	Cuboid	Platelet	Fibre
Aspect Ratio	1	~1	1.4-4	5-100	>10
Examples	Glass spheres silicate spheres	Calcium carbonate	Silica Barium sulfate	Mica Talc Kaolin Graphite Aluminium trihydrate	Glass fibres Asbestos Wollastonite Cellulose fibres C fibres Whiskers

In the case of fillers with a too high proportion of fines, higher shear forces occur in the extruder, which may lead to thermal damage of the polymer (and also affect its appearance). If, on the other hand, the fines are removed from the filler, there is no deterioration of the plastic, even at very high filler contents.

In addition, modern classifier technology now allows filler producers to maintain a precise top cut in the particle size distribution curve. The top cut, too, has a significant effect on the properties of filled polymers, since the coarsest foreign particles act as points of greatest stress concentration, where the crack or fracture occurs under load. Reducing the top cut improves the following properties of thermoplastics:

- Impact strength and notched impact strength
- Ultimate tensile strength of sheet and films
- Gloss of the part surface
- Abrasion at screws and cylinders of the processing machinery

The degree of improvement in mechanical stability is shown by the following example: a typical cable duct of uPVC is filled with - 20 per cent of a coated calcium carbonate. Reducing the top cut of this filler from 25 to 5 µm leads to an increase in notched impact strength of 125 per cent.

Specific Surface Area – A Measure of the Amount of Fines

The specific surface area – expressed in square meters per gram [m^2/g] – is an extremely important physical parameter, whose

Effect of aspect ratio on the stiffness of polypropylene.

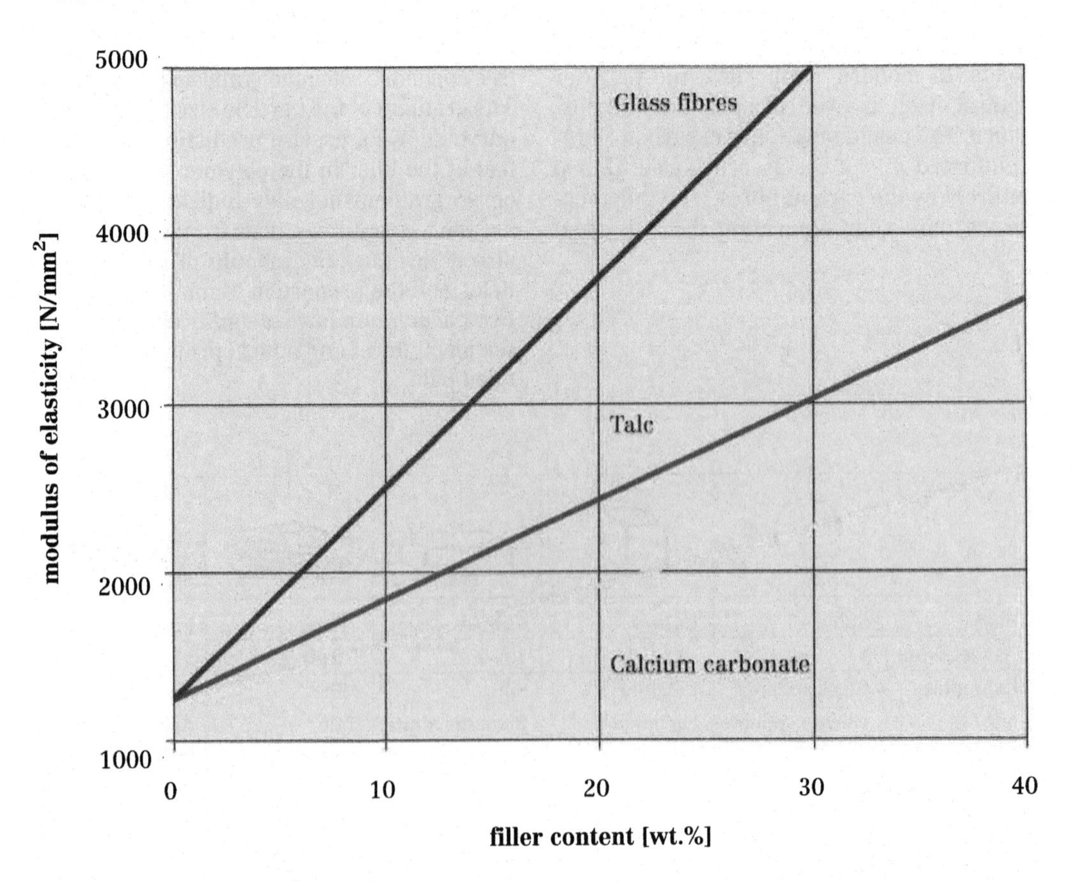

significance for the successful use of the filler is often underestimated. It is a measure of how many points of bonding are theoretically possible between the polymer chains and the additive material. A large filler surface area allows a large number of bonding points, and therefore better mechanical properties, than a small surface area. In general, an increase in the specific surface area of fillers or reinforcing agents, improves the following properties:

- Higher stiffness (higher modulus of elasticity)
- Better tensile and ultimate tensile strength
- Better impact and notched impact strength
- Higher surface gloss of the plastic

However, there are limits on how much the specific surface area can be increased, since a larger surface area also means a greater proportion of fines, and as filler particles become smaller they show an increased tendency to agglomerate.

This behaviour is the result of physics, since a solid particle is subject to two opposing forces. Gravity separates the particles from one another, while the so-called "van der Waals forces" cause attraction. The smaller the particles, the greater is the effect of the attractive van der Waals forces. The effect of gravity is reduced. Particles with a mean statistical particle diameter of 1 micron are held together by van der Waals forces ten million times more strongly than they are separated by gravitational forces. Once filler aggregates have formed during processing, they can be difficult to disperse, and can lead to corresponding defects in the finished part.

The specific surface area of the fillers and reinforcing agents must therefore be matched to the possible processing conditions (shear forces). The surface area must not be too large. If, however, undispersed filler agglomerates result, they act as a very coarse top cut. The notched impact strength in particular decreases greatly.

Surface tension of some materials.

Material	Surface tension [mN/m]
Diamond	10 000
Mical	2 400
Glass	1 200
Titanium dioxide	650
Kaolin	500-600
Calcium carbonate	200
Talc	120
Urea formaldehyde resin	60
PVC	40
PP	32
PE	30

Surface Energy and Surface Tension

The surface energy of fillers – expressed in millijoules per square metre (mJ/m^2) – corresponds to the amount of energy that must be applied to form a square metre of filler surface. The harder the material, therefore, the greater is the amount of energy that must be applied (see figure).

Although the surface energy cannot be measured directly on solids, in purely mathematical terms it corresponds to the surface tension in millinewtons per meter [mN/m], which, in the case of mineral powders, can be measured by means of gas chromatography processes. For general purposes, therefore, the surface energy is regarded as equal to the surface tension.

The surface energy – or tension – determines the magnitude of the mutual forces between the individual substances, and is thus the critical factor in the filled or reinforced plastics part. If the difference between the surface energies of filler and polymer is too high, the forces of adhesion prevent good mixing of the different materials. To facilitate incorporation into the polymer melt, fillers are therefore preferred whose surface tension does not depart too much from that of the plastic. It can be adjusted by surface coating of the filler particles.

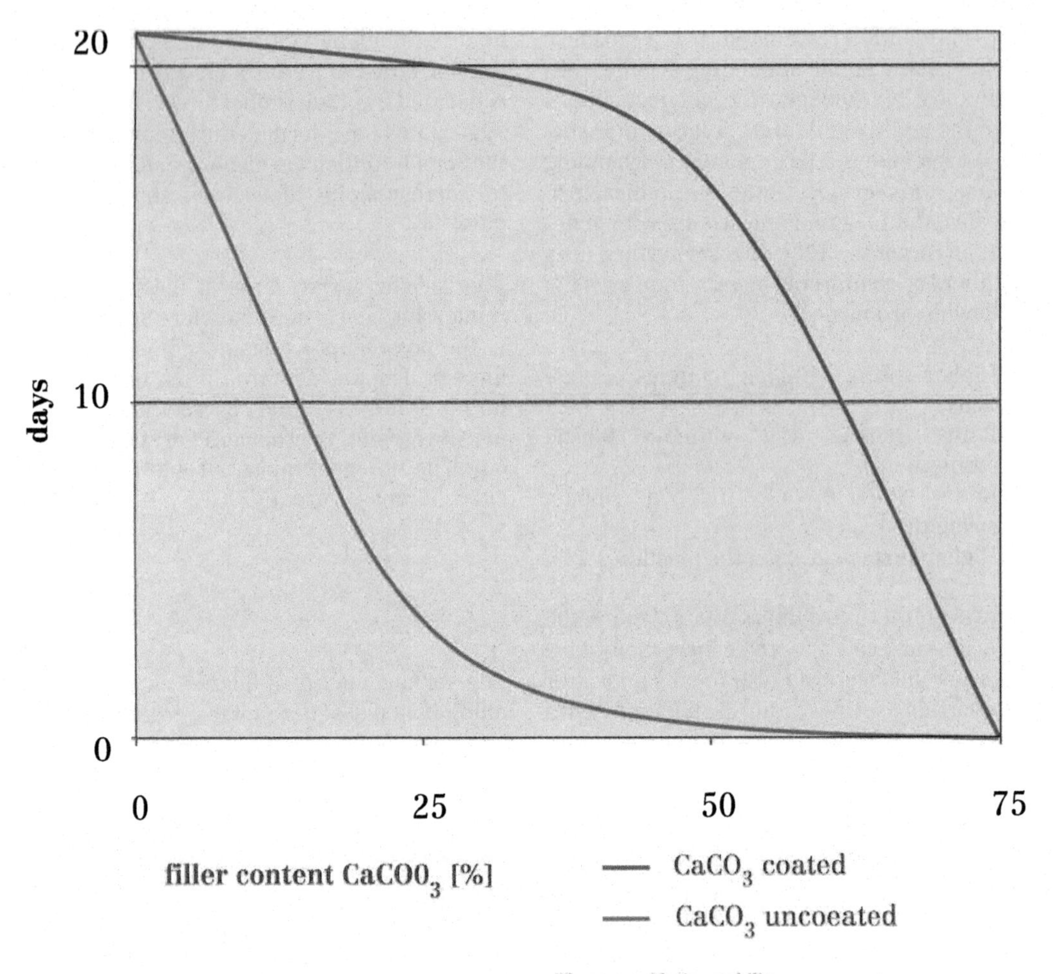

Thermo-oxidative stability of filled polypropylenes at a temperature of 150 °C.

Surface Coating

The technical literature contains many references to surface coating agents and coupling agents for fillers. Similarly, there are many substances available on the market for these purposes:

- Fatty acids and fatty acid esters (for carbonates and oxides)
- Silanes (for silicates and hydroxides)
- Titanates (for silicates and hydroxides)
- Zirconates (for silicates and hydroxides)
- Lubricants such as esters and waxes (for carbonates and silicates)

So far, there are only a few compounds that are able to form a genuine chemical bond between the filler and polymer matrix. Usually they only form a weak bond between the filler and polymer that does not extend much beyond electrostatic attraction (van der Waals forces). Among the few exceptions are silanes with sulphonyl azide groups, which make possible a covalent bond between the polypropylene and OH-Si groups on the surface of silicate fillers, such as mica or glass fibres.

But even with a rather weak coupling, surface-coated fillers and reinforcing agents provide some important advantages. They impart the following properties:

- Water repellency
- Reduction of the forces of attraction between the filler particles
- Decrease of the surface tension
- Decrease of the melt viscosity
- Improvement of dispersion
- Reduction of the amount of stabilizers and lubricants required
- Improved surface quality of the finished product

Most coating agents have the function of reducing surface tension. A milled calcium carbonate filler has a surface tension of 200 millinewtons per meter, if one takes into account both the polar and dispersive components of the surface tension. However, stearic acid coating can reduce this tension to 40 millinewton per meter, corresponding to the surface tension of PVC. This allows excellent dispersion of the surface-coated calcium carbonate in the PVC matrix.

In addition, surface coating reduces the absorption of stabilizers and lubricants at the filler surface. This contributes to the much better heat stability of coated calcium carbonate compared to its uncoated counterpart (see figure).

Packing Density

In many liquid polymer applications, such as latices, unsaturated polyester resins (UP), PVC plastisols, epoxides, thiokols and silicones with high filler contents, the packing density of the fillers plays an important part. Since fillers increase the viscosity of the liquid polymer systems, they are often only used in limited proportions. This limitation can be circumvented by optimizing the packing density of the particles. Optimum space utilization by the individual filler particles liberates liquid polymer matrix, which otherwise would be used for filling the interstices. The result is a lower viscosity for the same proportion of filler, or a higher filler content for the same viscosity.

Cooling time as a function of filler content in the injection moulding of $PP/CaCO_3$ compounds.

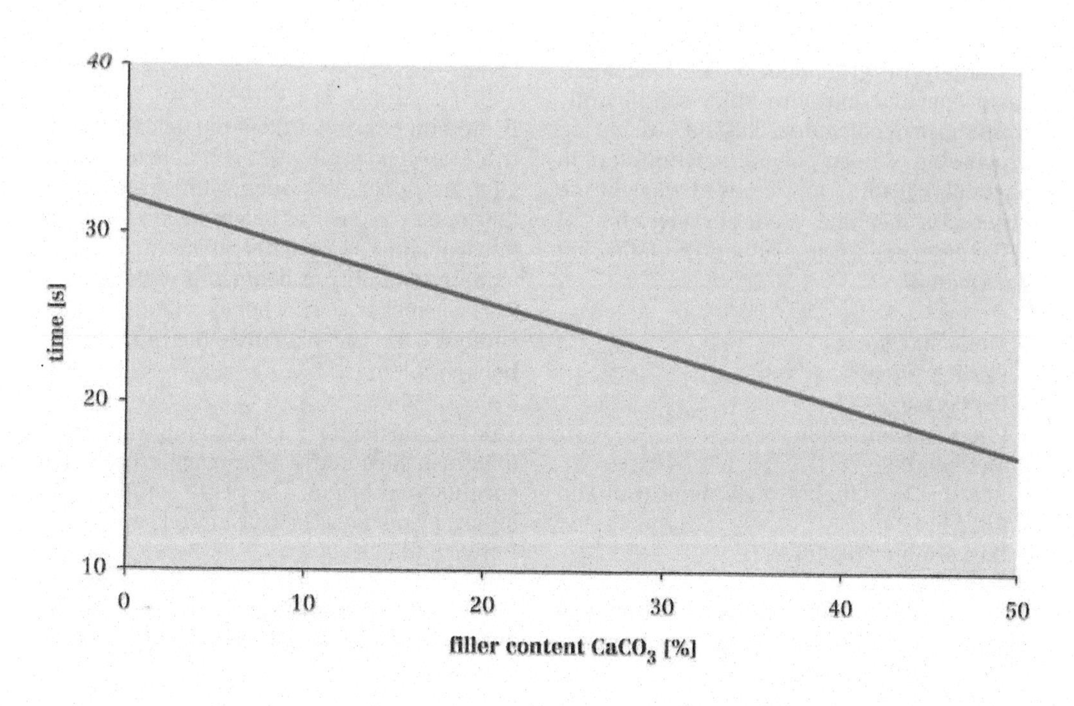

Specific Heat and Thermal Conductivity

Mineral fillers have entirely different energetic properties from traditional polymers. This is particularly true of the thermal conductivity and the specific heat. In general terms, it can be stated that the thermal conductivity of mineral fillers is approximately ten times greater than that of commodity polymers. The specific heat of the minerals, on the other hand, is only half as much as that of the polymers. Therefore, compared with commodity polymers, only half the amount of energy must be supplied to heat a kilogram of filler to a certain temperature (or removed to subsequently cool it again).

The addition of mineral fillers improves thermal conductivity of the system significantly, with a linear dependence between filler content and cooling time (see figure). This results in certain advantages for the manufacturing process. In particular, it is possible to significantly speed up processes involving heating or cooling operations. These include injection moulding, blowing of hollow articles, thermoforming or welding. For extrusion, the cooling zone can be kept shorter.

Abrasion in Processing Machinery

Mineral fillers are still crystalline, even in extremely fine ground form. Their mechanical properties therefore differ significantly from those of polymers. This also affects the processing of filled polymers, with fillers in particular leading to increased wear of the processing machinery. The abrasive effect of a filler depends primarily on the following parameters:

- Mohs' hardness
- Particle shape
- Particle size
- Filler concentration
- Melt viscosity
- Relative velocity between metal wall and filler
- Corrosion by the filler

The crucial role in the abrasion is certainly played by the Mohs' hardness. However, it is not the only deciding factor. Just as important are the particle shape and size of the filler particles.

In general, an appreciable abrasive effect is only shown by particles with a diameter greater than 5 microns. Alpha quartz shows by far the greatest abrasivity, but barytes ($BaSO_4$), wollastonite and titanium dioxide also have very high abrasion values. These values can be determined for the individual fillers by various test methods. However, the abrasion values determined in aqueous suspension have only a limited applicability to mineral-filled polymer melts.

In reality, none of these methods is suitable for the quantitative determination of the abrasivity of fillers in practice, since too many parameters play a role. The best method is still to measure the screw and cylinder dimensions after a specified service time of the extruder.

Chemical Aspects for Processors

Chemical average analyses are not necessarily relevant for characterizing fillers. Methods such as ESCA (electron spectroscopy for chemical analysis) or secondary ion spectroscopy are often more valuable since they give information about the chemical composition of the outer crystal layer.

If the outer layer of filler particles contain traces of heavy metals, such as iron, manganese, copper, nickel or cerium, in sensitive thermoplastics such as polypropylene or polyethylene, they can greatly affect the thermo-oxidative stability or behaviour with respect to UV exposure. The heavy metals act as catalysts for the oxidative degradation of polymers.

Talc is a particularly good example of how a filler's surface activity can vary greatly according to its origin. The result is that stability problems must be considered separately for each filler deposit.

Water-soluble compounds such as sodium and potassium salts should also not be present in fillers and reinforcing agents. This is particularly true when they are used in cable compounds, where traces of sodium or potassium salts measurably depress insulation properties even under minimum humidity conditions.

2.3 Calcium Carbonate as Plastics Fillers

A wide selection of natural ground calcium carbonates, in both uncoated and surface-coated forms, is now available (see figure); in terms of volume, they are the most important fillers in the plastics industry. In thermoplastics and thermosets, more than 3.5 million tonnes are used annually. For elastomers, consumption is almost 2.5 million tonnes and even adhesives and sealants consume about 900 000 tonnes of calcium carbonate as filler.

Calcium carbonate fillers are distinguished particularly by the following applications:

Mean particle diameter for calcium carbonate in plastics. The bandwidth of the calcium carbonates used – coated or uncoated – is as broad as the range of applications for plastics, adhesives and sealants.

Application	Mean particle diameter [µm]
PUR	2-5
PE	1-5
PVC	1-5
Technical rubbers	1-10
Cable insulations	1-3
Auto underseal	2-160
Latex coatings	10-160
Calendered films	1-5
Coatings	2-15
Floor coverings	5-160
Adhesives	2.5-90
Jointing compounds	5-30
Sealants	2.5-50

- High chemical purity, which eliminates a negative catalytic effect on the ageing of polymers
- High whiteness and low refractive index, so that calcium carbonates help to reduce consumption of expensive abrasive pigments such as titanium dioxide, but on the other hand are also well suited to the manufacture of coloured final products.
- Low abrasivity, which contributes to low wear of machine parts such as extruder screws and cylinders

Furthermore, surface-coated grades are particularly easy to disperse, and therefore, even at very high filler contents, the polymers retain excellent processing properties. In thermoplastics, finely ground calcium carbonates increase stiffness without seriously affecting impact strength. In the case of thin-walled manufactured parts, they help to avoid shrinkage and distortion.

Not least, natural calcium carbonate is non-toxic, odourless and tasteless, as well as conforming to food legislation in plastics packaging. It is available globally in ample quantities and is one of the most economical fillers for plastics applications.

Precipitated calcium carbonate (PCC) for plastics applications should have the following properties:

- Average particle size 0.07 to 2.0 µm
- Whiteness 95 to 96 %
- Oil absorption value 35 to 40 g/ 100 g powder
- Surface area 10 to 25 m^2/g

The fine PCC grades with a d50 value of 0.07 microns are usually used with a stearic-acid-coated surface.

The main applications for PCC fillers are PVC plastisols, such as those used for automotive underseals. Here, PCC mainly affects the rheological properties of the plastisol, preventing dripping of the plastic. Further applications include sealants and rubbers.

Until a few years ago, PCCs were often encountered as fillers in PVC rigid profiles (windows, cable channels). Here, they have

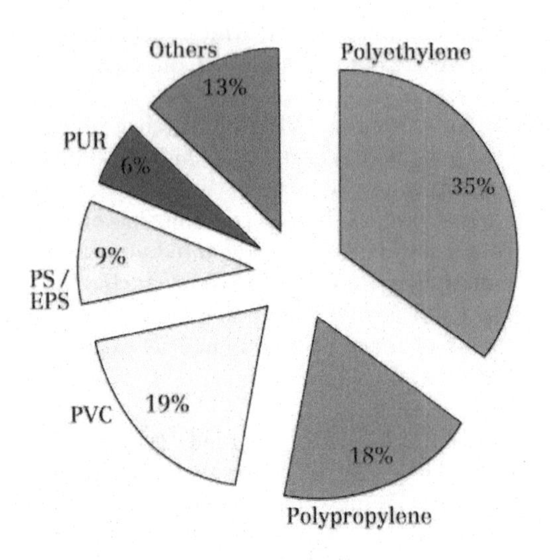

Market shares of thermoplastics and thermosets in 1998 (total volume 158m tonnes).

been almost entirely replaced by natural calcium carbonates, since the latter allows a considerable reduction in formulation costs. This is because of the lower purchasing prices for natural calcium carbonates. Furthermore, filler contents in the polymer matrix can be greatly increased because of the easier dispersion of natural calcium carbonates compared with PCC. Even two or three times the normal amount is not rare.

The energy crisis of the eighties resulted in a rethinking in the plastics industry. Until then, mineral fillers had mainly been used in PVC, thermosets and elastomers, but developments with fillers have since been made for other thermoplastics. The most economically interesting polymers for filler manufacturers are polyvinyl chloride (PVC), polypropylene (PP) and polyethylene (PE), as well as the engineering materials, predominantly polyamides (see figure).

2.3.1 Calcium Carbonate in Thermoplastics

Although, worldwide, almost twice as much polyethylene as PVC is produced and consumed, for manufacturers of mineral fillers, PVC is the most important thermoplastic because of the higher filler contents (see figure).

There are several reasons for the great success of calcium carbonate fillers: PVC is the only bulk polymer that is sold and processed as a powder. It is therefore technically simple to add powder-form calcium carbonate fillers during dry blend compounding, and there are no extra overheads.

Typical filler contents in thermoplastics.

Application	Filler content
• Plasticized PVC (cable, films, sealing profiles)	10-50 %
• PVC pipe	3-30 %
• PVC window profile	5-15 %
• PVC profile for indoor applications (cable channels, skirting board and curtain rails, furniture profiles)	20-30 %
• PVC calendered film and sheet	5-20 %
• PVC floor covering	45-80 %
• PVC plastisols (leathercloths, wallpaper, underseals)	15-50 %
• PP garden furniture	20-40 %
• PE film and sheet (packaging film and construction sheeting)	2-30 %
• Polyamide parts	20-40 %

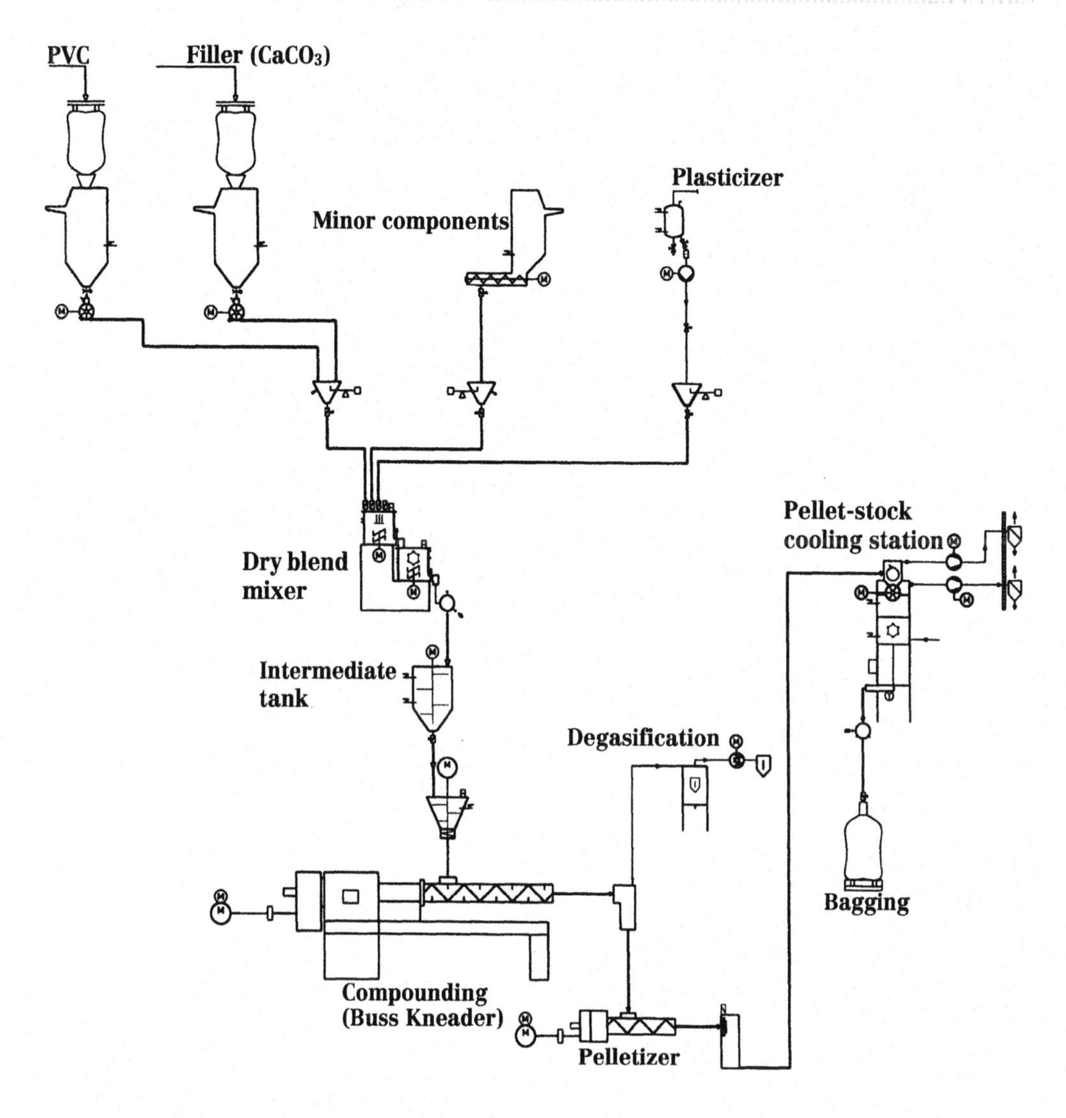

Schematic diagram of a compounding plant.

In addition, PVC, with a density of 1.4 grams per cubic centimetre [g/cm^3] is a comparatively heavy polymer: polyolefins, for example, only have a density of 0.9 g/cm^3. If in a plastic, part of the polymer is replaced with calcium carbonate (density 2.7 g/cm^3), the relative weight increase is significantly lower in the case of PVC than in the case of a polyolefin. Therefore, the service properties of a PVC product will change less than those of a more lightweight polymer.

Not least, the different densities also have financial consequences. Since the density of calcium carbonate is only twice as high as that of PVC, filling is worthwhile if the price of PVC is only twice that of calcium carbonate. In the case of polyolefins, on the other hand, a saving is only obtained when the polyolefin price is three times as high as the price of the filler.

But it is not only the low densities of polymers such as polypropylene, polyethylene or engineering materials such as polyamide that militate against the use of mineral fillers. Since all these polymers are processed as pellet stock, the mineral fillers, too, must be converted to pellet form by a technically complicated compounding step in order to prevent separation during processing (see figure). In addition, the filler-containing pellets, so-called masterbatches and compounds, must be readily dispersible in the polymer matrix for further processing. Such high demands lead to considerable overheads in comparison to PVC processing.

If we consider the low density of polyolefins compared with mineral fillers, it is clear that fillers such as calcium carbonate are not used as cheapening additives. Fillers are only appropriate here if there are significant technical advantages in using them.

To permit the widespread use of calcium carbonate in the field of polyolefins, more economical processes are currently being developed.

PVC window frame.

Unplasticized PVC

The most important uPVC products are pipes, window and roller-blind profiles, cable channels, calendered film and panels. In all these applications, only very fine-particulate surface-treated natural calcium carbonates (d50 < 1.5 µm, top cut < 7 µm) should be used, which allow high filler contents and a significant cheapening of the formulations. Since fillers increase stiffness (Young's modulus), for example in the case of PVC roller-blind profiles, it is also possible to reduce wall thickness, and thus save material and costs.

Fillers that are too fine, such as precipitated calcium carbonate (PCC) result in high energy input during shearing in the extruder, which cannot be absorbed. This often results in thermal damage to the sensitive PVC system with possible discoloration. For this reason, the maximum permissible filler content for fine-particle PCC grades must be limited to about 7 per cent.

Plasticized PVC

In plasticized PVC products, either uncoated or surface-coated calcium carbonate fillers come into consideration, depending on the application, (cable, floor coverings, profiles, film) and the type of plasticizer.

For example, in the case of cable production, there are three fundamentally different applications: insulation, sheathing and filling compounds. Practical experience has shown that optimum mechanical and electrical properties can only be achieved with calcium carbonate fillers. Their mean statistical particle diameters should be about 2 microns. The top cut of the particle size distribution curve should not significantly exceed 10 microns.

The filler content for individual cable grades depends on their use. Low-voltage cables (up to 1000 V) are nowadays usually insulated with PVC. The lifetime of an insulation compound should be at least 30 years under electrical, mechanical and thermal loading. To ensure this, in the case of low-voltage cables, filler contents must not exceed 40 per cent at service temperatures below 105 degrees Celsius, or 25 per cent calcium carbonate at higher temperatures. In the case of medium voltage cables (1000 to 10 000 V), up to a maximum of 10 per cent calcium carbonate filler may be employed. This presupposes that calcined kaolin is also added as ion scavenger, to adsorb cations contained in the PVC compound and render them harmless. This improves the electrical insulation properties of the formulation. Insulation compounds for high-voltage cables are not made of PVC.

PVC cable sheathing.

Normal cable sheathing compounds contain between 40 and 50 per cent calcium carbonate. In this case, the filler primarily improves the flow behaviour, and therefore the extrusion properties, particularly the take-off rate. In addition, the final product has significantly improved thermal conductivity in comparison to unfilled plastic, and heat generated during electrical power transmission can be better dissipated to the surroundings.

The interstices between the insulated electrical conductors are filled with a filler compound. These compounds do not have to meet any mechanical requirements, calcium carbonate filler contents can therefore easily be as high as 90 per cent, which is also desirable for reasons of cost. Particularly desirable cable filler compounds include PVC wastes or highly filled rubbers, such as EPDM (ethylene propylene diene monomer).

Apart from in the cable industry, flexible PVC also has a wide technical importance for floor coverings and profile extrusion. Profiles of flexible PVC are used widely in the automotive industry as edge protection strips or decorative trim, and in the construction industry as jointing tape and skirting board. But door seals for refrigerators or flexible tubing are nowadays also made of PVC. Filler contents in the individual products are between 15 and 40 percent.

PVC Plastisols

Dispersions of PVC in plasticizers such as phthalates are called PVC pastes or PVC plastisols. The major advantage of these plastisols lies in their ease of processing. For example, fabrics and similar backing materials can be coated with PVC plastisols and fused at temperatures between 130 and 180 ° Celsius to convert them to flexible, resistant products. Leathercloth, wallpapers, tarpaulins, floor coverings and automotive underseal systems are just a few examples of the applications of plastisols.

The most important mineral fillers for plastisols are uncoated calcium carbonates, whose average particle diameters lie between 1.5 and 40 microns. In addition, other fillers are also used. Surface-coated PCCs and fine-particulate silicates (colloidal silicas) facilitate the adjustment of the rheological properties. Barytes ($BaSO_4$) is particularly used for anti-resonance compounds, and, lastly, aluminium trihydrate (ATH) serves as a mineral flame retardant.

Polyolefins

Calcium carbonate fillers are used in polyolefins as so-called compounds or masterbatches. Compounds are produced from the respective polymer and up to 40 per cent fillers, with stabilizers and pigments usually also being added when appropriate. Compounds are used directly for the production of plastic parts. In contrast to compounds, filler masterbatches must be diluted with pure polymer pellet stock; their filler content is 70 to 88 weight per cent calcium carbonate.

Polyethylene

In the polyethylene sector, calcium carbonate fillers now play a role preferably in films and sheets. Low-density polyethylenes (LDPE and LLDPE) are usually filled with very pure calcium carbonate grades. The optimum average particle diameter of 2 to 3 microns allows them to be used in films up to a thickness of about 20 microns without mechanical problems occurring or the transparency being affected. The addition of 2 to 3 per cent calcium carbonate is enough to improve the antiblocking and slip behaviour of the films. The term "blocking" describes the sticking together of finished films and sheet on the rolls, making it impossible to unroll the films. The slip properties characterize the phenomenon of films sliding over one another on stacking.

Pigment masterbatches are usually produced with fine-particulate, surface-coated calcium carbonates. In most cases, the consumption of titanium dioxide or coloured pigments can be reduced, since the filler particles act as milling balls, disrupting the pigment agglomerates.

From practical experience to date, it has been found that surface-coated carbonate used in amounts of 4 to 10 per cent also leads to the following processing advantages:

- Increasing output by 5 to 15 per cent
- Self-cleaning of the tools, and therefore shorter interruption times on pigment exchange
- Improved printing properties of the films
- Easier further processing of the polymer thanks to increased stiffness
- Reduction of corona treatment by 80 per cent (with a calcium carbonate content of 10 per cent)

In the case of high-density polyethylenes (HDPE), the aforementioned property improvements apply in the same way. For applications of films and sheets of HDPE, the following points have proved advantageous:

- Matt surfaces that can readily be written on or printed (e.g. for packaging materials)
- Opaque, white colour with reduced consumption of expensive white pigments
- Handle resembling tissue paper and relatively high stiffness
- Better folding and bending properties (e.g. for margarine and fat packaging).

High stiffness, low shrinkage, relaxation of internal stresses and shorter cycle times are the main advantages gained by using calcium carbonates in injection moulding and blow moulding of HDPE. The upper limit of filler content is about 40 per cent calcium carbonate for injection moulding, but only

20 to 30 per cent for blow moulding. They are enough, however, to significantly improve thermal conductivity, and thus increase the productivity of blow moulding by a third.

Computer simulation of the heat distribution during injection moulding. Red colours indicate the highest temperature zones.

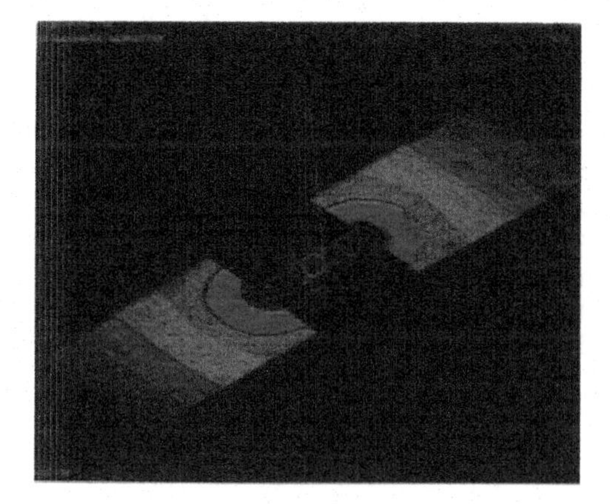

Polypropylene

Polypropylenes (PP) filled with calcium carbonate or talc are among the more recent developments on the polyolefin market, and in the last decade have shown impressive growth rates. As in the case of polyethylene, fillers are used wherever physical properties such as stiffness, density, shrinkage or creep behaviour cannot be met with unfilled polypropylenes.

Natural calcium carbonate-filled polypropylenes are mainly used for garden furniture, domestic appliances, such as coffee makers and refrigerators, as well as automotive interior trim. Woven tape and cord contain large amounts of calcium carbonate. Filler contents vary between 5 and 60 per cent, depending on the application, with surface-coated grades showing high whiteness and an average particle diameter of between 1.5 and 3.5 microns.

Polyamides

Polyamides are engineering materials. Since they are used as parts for many different kinds of machinery, they must not only be capable of being turned on lathes, milled or sawn, but subsequently withstand high service loads. Therefore, only reinforcing agents or fillers with a reinforcing effect come into consideration. The substances most used have a high aspect ratio, such as glass fibres, kaolin, mica and wollastonites. Nevertheless, uncoated calcium carbonates can be used to obtain a significant reinforcing effect: the amide groups of the polymer, as a dipole, interact with the calcium carbonate surface, which is also polar. This increases the tensile strength and ultimate tensile strength of the polyamide. If, on the other hand, the calcium carbonate surface is covered with stearic acid or other conventional coating agents, the polarity is largely lost, and with it the reinforcing effect.

2.3.2 Calcium Carbonate in Thermosets

Thermosets usually have very high filler contents (see figure), with calcium carbonates being principally used as fillers in unsaturated polyester resins and polyurethanes: the following discussions apply in principal for all other crosslinking polymer systems.

Typical filler contents in thermosets.

Application	Filler content
• PUR	10-50 %
• Unsaturated polyester resins (automotive and equipment parts of synthetic marble)	55-70 %
• Polymer concrete (only calcium carbonate)	10-20 %
• SMC/BMC	10-30 %
• Phenolic resins	approx. 10 %
• Melamine resins	approx. 20 %

Unsaturated Polyester Resins

Unsaturated polyester resins (UP) generally consist of two components: an unsaturated polyester and styrene monomer, which is also unsaturated and at the same time acts as a solvent.

As supplied, the two chemicals are pale, viscous liquids. If stored correctly, they are only converted to polyester resin after the addition of appropriate reagents: polyester and styrene react exothermically to form a three-dimensionally crosslinked molecule. In the process, they undergo volume shrinkage, which can be explained entirely by the improved packing density of the reaction counterparts in the new product. No liquid or gaseous products are released during the reaction.

Most polyester resin applications are currently reinforced with glass fibres. The resin fills the interstices between the glass fibres, resulting in a part with a homogeneous surface. The use of filler reduces the shrinkage of the polyester resin during curing and ensures improved heat dissipation – despite the fact that temperatures reach up to 200 Celsius. This prevents cracking during curing.

In general, glass fibre-reinforced polyester resins are characterized by high strengths, impact resistance and dimensional stability. Since they are also insensitive to temperature shocks and chemical attack and have a relatively low weight, they are used as substitutes for metals in a large number of applications. To obtain optimum properties, the filler content in the product must be about 55 to 70 per cent. The particle size distribution must be chosen so that a good packing density is achieved without disruption of the bond between the resin and glass fibres.

Calcium carbonates are used in significant amounts in SMC/BMC products (sheet moulding compounds/bulk moulding compounds), artificial marble and polymer concrete. Chalk products, on the other hand, are not very suitable for use in unsaturated polyester resins because of their high oil absorption value, since viscosities become high at only relatively low filler contents.

In recent years, a market has opened up, particularly in the automotive industry, for products produced by the so-called SMC/BMC process: engine cowls, rear ends or headlamp housings are currently often made of compression-moulded or injection-moulded thermosets.

For SMC/BMC products, calcium carbonates with a mean particle diameter of 1.5 to 10 microns are used. Thanks to the combination of fine-particulate grades ($d50 = 1.5$ µm) with relatively coarse grains ($d50 = 10$ µm), the packing density can be optimally adapted; the viscosity is correspondingly low.

Artificial marble and polymer concrete are also highly filled products. For example, polymer concrete comprises only 9 to 12 per cent polyester resin. The rest comprises quartz sand, quartz flour and calcium carbonates, blended to obtain optimum packing density. The advantages of polymer concrete compared with traditional cementitious concrete are its faster hardening, better mechanical properties, better chemical resistance, and the lower water absorption.

With highly white calcium carbonate fillers, even bathroom fittings, such as washbasins, bathtubs and other parts can be made of plastic.

Polyester slabs containing marble chips are marketed as synthetic marble. The resin content of synthetic marble is about 12 to 15 per cent. The fillers mainly used are calcium carbonate, as well as dolomite and aluminium trihydrate (ATH). The range of particle sizes in calcium carbonate fillers is huge: the coarsest fractions of marble grains are several millimetres in size, while the lower limit is about 10 microns.

Polyurethanes

Polyurethanes are produced by the polyaddition of di- or polyisocyanates with polyhydric alcohols (polyols). Crosslinked polyurethanes are insoluble, unmeltable polymers that are suitable for the production of cold-curing two-pack lacquers for surface protection, as adhesives for sand-

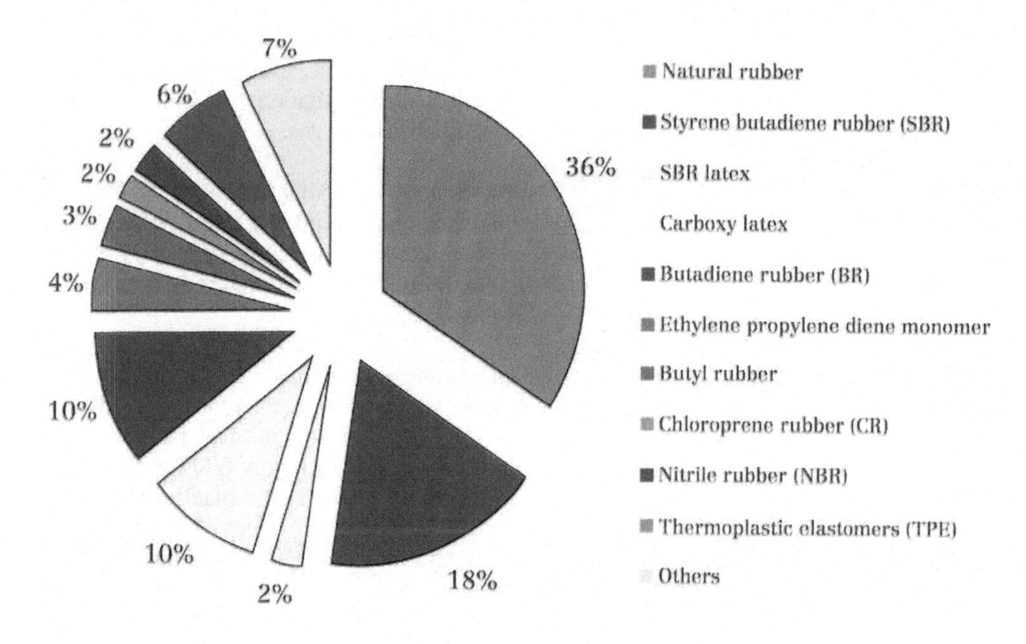

Market shares of individual elastomers in 1998 (total volume 17.8 million tonnes).

wich structures and for automotive engineering, and in particular for producing foams. This is where the highest volumes of calcium carbonate fillers are used. Filler contents of 20 per cent are by no means rare in foams.

In foam production, the calcium carbonate filler is first predispersed in the polyol and then metered directly into the mixing head or the polyol feedstock. Products based on chalk can also be used, since because of their low abrasivity, they do not damage the mixing head of the production plant.

2.3.3 Calcium Carbonate in Elastomers

The number of elastomers is large (see figure) and almost all technical rubber products contain natural calcium carbonates as fillers – for different purposes in each case.

For example, calcium carbonate fillers with d50 values of between 5 and 15 microns [µm], a top cut of 40 to 80 µm and a specific surface area of at most 1 square metre per gram of filler [m^2/g] are mainly used as extenders and cheapening agents. Important applications include cable filling compounds and carpet backings of SBR or NBR latex. Filler contents vary between 20 and 80 per cent, depending on application (see figure).

On the other hand, calcium carbonates with d50 values of 4 to 8 µm, a top cut of 16 to 40 µm and a specific surface area of 1 to 2 m^2/g are first and foremost reinforcing agents. Their prime function is to improve the dimensional stability of calendered and compounded unvulcanized goods or webs, as well as of vulcanized, thick-walled mass-produced products. As much is added as permitted by the processing conditions and mechanical and physical properties of the vulcanizates. Amounts of 40 to 60 per cent

Application	Filler content
• Tyres (inliners)	50-60 %
• Conveyor belts	10-40 %
• Floor covering	40-60 %
• Roofing sheets	35-55 %
• Tubing	30-40 %
• Cable	20-80 %
• Moulded articles	10-20 %
• Shoe soles	5-15 %
• Carpet backing	60-80 %

Typical filler contents in elastomers.

are easily feasible, even higher filler contents are occasionally used.

Natural calcium carbonates with d50 values of 1 to 3 µm, a top cut of 7 to 15 µm and a specific surface area of 4 to 10 m^2/g act as dispersion and processing aids. Since 15 to 50 per cent of the filler particles have particle sizes of below 1 micron, these filler grades already influence the colloidal range of rubber compounds. They provide improved homogenization of the aggregates during compounding and assist the calendering or extrusion properties of the rubber compound. They are mainly used in combination with other fillers, such as kaolin, silicas or carbon black, in which case they are added in amounts of between 15 and 30 per cent.

Very fine calcium carbonate grades with d50 values of below 1 µm, a top cut of below 6 µm and a specific surface area of 7 to 10 m^2/g have the character of additives. In rubber compounds, their effects mainly develop in the colloidal size range. Calcium carbonate particles of the order of 1 to 600 nanometres [nm] are comparable in effect to secondary silica aggregates and structural carbon blacks: they are primarily added to improve dispersion of active fillers. Amounts of 10 to 15 per cent calcium carbonate are sufficient to achieve such effects.

Calcium carbonate particles also have a partial reinforcing effect in the colloidal range as well as, secondarily, improving the homogeneous distributions of other components, such as vulcanization accelerators, resins, zinc oxide, peroxides, antiageing agents, antiozonant waxes, pigments or sulphur. This has other advantages:

- Reduction of mixing times
- Lowering of the mixing temperature
- Constant vulcanization quality and reproducibility of the mixing parameters
- Saving of energy costs

The application of very fine calcium carbonates is particularly recommended for EPDM, NBR, CR and natural rubber compounds, which are highly filled with silicas and carbon black. These plastics are used for tubing, roller coatings, V-belts, tyres, seals, shoe soles, cables and foam rubber.

Of secondary importance is the use of calcium carbonate as dusting and release agents. Relatively small amounts of calcium carbonate are applied to the sheeted-out rubber hide when talc or release agent mixtures are too expensive for the particular application.

2.3.4 Calcium Carbonate in Adhesives and Sealants

Adhesives and sealants have changed their face in the last thirty years. Traditional putty and universal adhesives have been replaced with a diverse array of products for a growing number of applications in virtually all branches of industry. Not all these adhesives and sealants necessarily include fillers – as technical requirements become more stringent, the proportion of filler is tending to decrease. But wherever a balanced ratio between cost and performance is required, fillers are indispensable. Filler contents of 50 per cent and more are by no means rare (see figure). Fillers may include not only calcium carbonate, but also kaolin and quartz flour: parquet-flooring adhesives mainly contain calcium sulphate (gypsum).

The use of calcium carbonates mainly improves processing properties, such as pot life and sag resistance, and increases both the durability and strength. Fillers can also be used to adjust the rheology of a sealant

Application	Filler content
• Tile adhesives	
- dispersion bound	60-80 %
- cementitious	40-60 %
• Gypsum adhesives	0-20 %
• Packaging adhesives	0-50 %
• Carpet adhesives	30-50 %
• Dispersion-based textile adhesives	50-70 %
• Dispersion-based parquet adhesives	40-55 %
• Joint fillers	
- dispersion bound	65-85 %
- cementitious	45-60 %
• Automotive window adhesives	10-30 %
• Sealants	
- acrylic sealants	40-60 %
- silicone sealants	30-60 %

Typical filler contents in adhesives and sealants.

and minimize shrinkage after curing. This prevents cracking. Further properties that can be affected by fillers include: adhesion, compressive strength, electrical conductivity, heat resistance and, not least, water absorption.

2.4 Recent Developments

Modern catalyst technology with metallocenes allows the production of polyolefin copolymers with a narrow molecular weight distribution and a constant co-monomer proportion. This permits their absorption capacity for calcium carbonate fillers to be increased with no loss of mechanical properties. Alternatively, improved catalyst technology, as in the case of polypropylenes, also allows the production of polymers with significantly improved crystallinity and stiffness.

The use of calcium carbonate or other mineral fillers in the field of polyolefins still requires an expensive compounding step. Only the direct use of fillers in powder form, without precompounding, would permit significant cost reductions and open up new, attractive applications for calcium carbonate in the polyolefin sector.

In the field of extrusion (pipes, thermoformed sheets, cable), the direct processing of mineral fillers with so-called compounding extruders will become established, since this process also allows the processing of plastics recyclates without additional overheads (see figure). However, direct extrusion will only be worthwhile when production has been unified, and throughput rates have become greater. This is necessary to make the required investments attractive.

In the field of injection moulding and in extrusion applications with low output rate, inexpensive, newly developed masterbatches containing about 88 per cent of calcium carbonate will make it economically attractive to fill a larger number of final products with calcium carbonate.

Co-rotating twin-screw extruder for direct extrusion.

3. Surface Coatings

The term "coating material" as defined in DIN EN 971-1 includes paints and other types of coating materials such as primers, surfacers, stoppers and plasters. To enable them to be described more precisely, coatings are often subdivided on the basis of the type of binder used (acrylic paints), the solvent (waterborne paints), their constitution (powder coatings) or one of many other criteria, such as film-forming, gloss or usage.

The areas in which surface coatings are used are no less diverse than the names by which they are described (see figure). Coatings are applied in metal working, in plant and equipment construction and in the electrical industry. Motor vehicles of all types, as well as ships and aircraft, all have to be painted. Coatings are as essential in structural and civil engineering, steel structures, concrete buildings and in wood working as they are in modern packaging products, whether these are made from paper, plastics or sheet metal. Even leather goods require surface finishing with varnishes or lacquers in many cases. All these and countless other products could not be used without being coated.

Surface coatings perform two key functions: they protect the coated substrate (wood, metal or other mineral substrate) from destructive external influences, and they impart colour and gloss to the objects. Of primary importance is protection against weathering influences such as sun and rain, cold and heat, but the aggressive pollutants in the atmosphere also make coatings essential. In addition, the surface protection must be able to withstand severe mechanical stresses such as chipping by flying stones and brush strokes from cleaning devices; a coating film must even be resistant to attack from micro-organisms and biochemical influences.

The principal function of colour and gloss is to improve appearances and hence to en-

Areas of application for surface coatings.

House and DIY paints	Industrial coatings
• Interior emulsion paints	• Metal coatings for domestic appliances, prefabricated claddings and coil coatin
• Exterior emulsion paints	• Automotive paints and refinishing paints
• Alkyd resin paints for interior and exterior use	• Timber and furniture coatings
• Synthetic resin bound plasters for interior and exterior use	• Anti-corrosive and marking paints concrete coatings and renovation
• Primers and stoppers	• Marine paints and coatings for drilling platforms

Automotive paints.

hance our surroundings. By drawing on physiological and psychological knowledge in choosing the colours for tools and equipment, for offices and living spaces, we can make a significant contribution to easing the stress of daily work and improving our quality of life.

At the same time, however, coatings often perform a safety and guidance function. Standardised colours can be used to indicate the content of containers or of pipelines. Luminous yellow stands out well from the background and has long been used in many countries to identify telephone booths and letter boxes from a distance. Bright red stands for danger and helps to clear the road

Colours stand out.

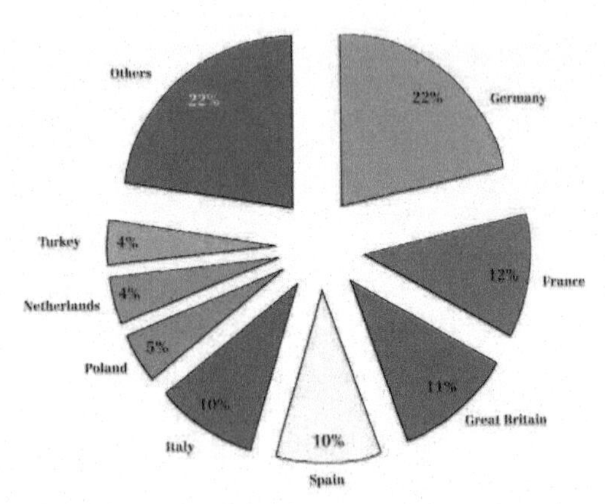

Production figures for surface coatings in Europe (excluding CIS) in 1998.

for fire engines. Conversely, an unobtrusive colour or a colour that matches its surroundings can shield an object from curious eyes, effectively camouflaging military objects in particular.

These examples highlight the economic importance of the coatings industry, whose annual production in 1998 in Europe alone (excluding CIS) was 7.8 million tonnes (see figure).

3.1 Building blocks for surface coatings

Huge demands are made today on the properties of coating systems; particularly important criteria influencing the quality of a paint include gloss, hue, adhesion, hiding power and wet scrub resistance. Binding standards exist for the testing of these and

Building blocks for coating systems.

Solvents	Aromatic/aliphatic hydrocarbons/ Esters/ketones, alcohols/water
Binders	Polymer solutions: alkyd resins, reactive resins Dispersions: polyvinyl acetate (PVA), styrene acrylate
Additives	Driers, anti-settling agents, anti-skinning agents, wetting and dispersing agents, preservatives, matting agents
Extenders	Calcium carbonate, barium sulphate, dolomite, talc, kaolin, silicates
Pigments	Titanium dioxide (anatase/rutile), iron oxide red (natural/synthetic), organic and inorganic coloured pigments

Manufacture of paints and coatings

The objective of paint manufacture is to combine the vast number of components with different physical and chemical properties to form a durable and homogeneous blend. No coarse solids particles (flocculates) may appear in the paint, and it must not separate during storage or transport. These objectives can only be achieved by means of a complex, multi-stage process.

The **mixing** stage involves combining binders, pigments, extenders and solvents in a mixer to form a homogeneous paste. This paste is then transferred to the next and most important production stage, that of dispersion.

During **dispersion** any pigment or extender agglomerates are broken down into smaller particles and then wetted by the dispersion medium (binder and solvent). The resulting dispersion is stabilised against flocculation by means of dispersing agents.

The complexity of the processes that occur during dispersion means that even today experience and above all the dispersing equipment used play a critical role: the most important types of dispersing equipment are rotor-stator mills for primers and emulsion paints, agitators (mixing blade devices), which are likewise used for emulsion paints, attrition mills for high-grade coatings and a variety of heavy-duty mixers for highly viscous stoppers and synthetic resin bound plasters.

In the final stage, **let down**, additional binders and solvents are added to the dispersed paste along with additives. The coating material is now ready for use.

Inside an agitator.

many other properties. Increasingly complex formulations are required in order to meet the ever-increasing demands placed on the various types of coating systems. The manufacture of surface coatings can often involve more than ten individual components, each of which has a greater or lesser influence on the properties of the end product (see figure).

The various components used in the vast range of different coating systems can be assigned to five categories of building blocks:

- Binders
- Pigments
- Extenders
- Solvents
- Additives

Of these five building blocks, particular importance is attached to binders, pigments and extenders.

For example, binders determine the characteristics of the liquid coating material, the film-forming process and the properties of a cured paint film. They are responsible for the durability of the coating, its mechanical strength, hardness and flexibility and for its adhesion to the substrate.

Pigments have one principal task: as a coloured or achromatic colorant they determine the optical properties of a coating system, such as colour, gloss and opacity. Their excellent hiding power means that modern pigments such as the white titanium dioxide are used in only small concentrations in some coating systems – a thoroughly desirable objective not least in terms of cost.

Finally, extenders have a significant influence on both the liquid paint and the cured paint film. This ranges from the flow characteristics and shelf life of the paint through to the corrosion resistance and weather resistance of the coating. Extenders are the main component of many modern coating systems in terms of volume. Thus, for example, almost 2.8 million tonnes of extenders were used in the manufacture of 7.8 million tonnes of surface coatings in Europe in 1998. This shows the importance attached to extenders in the coatings industry. Nowadays it is impossible to imagine modern coating systems without them, and many coatings simply could not exist without extenders.

3.2 Extenders in surface coatings – functions and properties

According to DIN EN 971-1 and DIN EN ISO 3262-1, extenders are materials in granular or powder form, practically insoluble in the application medium and used as a constituent of paints to modify or influence certain physical properties. An enormous variety of extender grades are used in the coatings industry: calcium carbonate, dolomite, barium sulphate, kaolin, quartz, talc and mica. In addition to individual physical properties such as refractive index, density, hardness and particle shape (see Chapter III, Section 2 "Calcium carbonate – pigment and extender") the fineness of the extenders is one of the most important criteria for their use. For practitioners, the particle size distribution, the average particle diameter (d50) and the top cut (d98) are of particular interest in this respect.

It all began with chalk

Natural chalk – suspended in water with the addition of glue – was already used in simple wall paints right back in the time of Ancient Egypt. However, growing demands in regard to the optical properties of paints, in particular, have meant that since the 20th century, if not earlier, cheap chalk has primarily been used to dilute and to extend expensive white pigments. With its low refractive index (n = 1.59), chalk could certainly not compete with white pigments such as titanium dioxide (n = 2.75).

The same also applied to other minerals such as talc, barytes or kaolin, which were likewise used only as extenders. For a long time the refractive index was actually a key deciding factor in whether a mineral should be used as a pigment or an extender – provided that the inherent colour of the mineral corresponded to the desired hue or at least did not conflict with it.

Nevertheless, until after the Second World War the minerals used as extenders were rarely prepared especially for that purpose. Coatings manufacturers generally used mineral powders that were a by-product in the processing of grit or gravel for the building industry. Quarry owners saw a welcome source of additional income in the sale of these waste products, and coatings producers too profited from these "dividend powders".

Since surface coatings were sold by weight alone, manufacturers added as much of these extenders to their products as possible in order to economise on the use of binders. Barium sulphate was particularly popular, as the high density of this mineral meant that it increased the weight of the paint and at the same time cut raw material costs. The influence of an extender on the quality of a paint was of negligible importance in those days.

Things began to change in the 1950s, however. Rapid industrial growth and high demands on the properties of surface coatings awakened an interest in extenders on the part of coating chemists. The "dividend powders" from the gravel industry soon exhausted their usefulness, and in their place there grew up an extenders industry which still produces tailor-made mineral powders with specified properties to meet the various requirements of the coatings industry.

This has become especially evident in developments during recent decades. Whereas in the 1950s limestone powders with an average particle diameter of 5 micrometers were available on the market, today diameters of well below 1 micrometer can be achieved.

Although the definition based on refractive index has lessened in importance, extenders are still used, and their rather disparaging name makes it clear that being cheap is still an essential characteristic for an extender. Nevertheless, since the coatings industry

▪ Flow characteristics	▪ Hiding power
▪ Pigment suspension capacity	▪ Packing density
▪ Shelf life	▪ Filling capacity
▪ Density	▪ Sandability
▪ Gloss	▪ Wet scrub resistance
▪ Surface smoothness	▪ Anti-corrosive effect
▪ Whiteness	▪ Weather resistance

Properties that can be influenced by extenders.

not only wanted cheap extenders but also steadily raised its requirements in terms of the quality and properties of mineral powders, pricing pressure on extender manufacturers was and still is acute. Individual companies could only withstand this pressure by automating and rationalising their processes on an ongoing basis. Only in this way could they keep their prices relatively stable over the years whilst improving the performance of their products.

The function of extenders today

The list of properties required of surface coatings is very long, and a good many of them can now be influenced with extenders (see figure). The majority of the functions fulfilled by extenders in coatings relate to mechanical and physical properties.

As solids particles dispersed in binder, extenders manifest themselves spatially in three different ways:

- Within the coating film they contribute as framework substances towards improving the mechanical structure. They increase the density and hardness of the film and reduce permeability to gases and capillary water.
- They improve adhesion at the interface with the substrate.
- At the surface they influence sandability, improve abrasion resistance and regulate the desired degree of gloss.

In addition, extenders also enhance the chemical stability of the coating. As natural buffers or ion exchangers they protect against corrosion or acid attack. Finally, extenders also possess optical properties that have a positive influence on the quality of the paint: some minerals increase whiteness, for example, whilst others absorb damaging UV radiation.

Selection criteria

Since not every extender satisfies all these requirements, the coating manufacturer is compelled to make a choice. A wide range of characteristic parameters can serve as criteria for selecting the appropriate extender, and they can also be used to influence the ultimate properties of the coating:

- fineness
- particle shape
- Mohs' hardness
- oil absorption value and density
- whiteness
- water solubility and pH
- chemical purity and toxicology

Not all of these parameters are invariable, however. Although density and hardness like pH and toxicity are inherent properties of a mineral and cannot be altered, many parameters can certainly be varied between particular limits.

For example, chalk and limestone powders coated with stearic acid, which are characterised by a low oil absorption value and improved dispersibility in organic solvents and binders, have been available since the 1950s. What is more, modern grinding and classifying methods allow defined fine grinds to be produced for individual applications. This is the principal factor behind the great success of extenders.

Pigment volume concentration and CPVC

Many properties of coating materials can be regulated by binders, extenders and pigments and by the choice of pigment volume concentration (PVC). The PVC describes the volume ratio of extenders and pigments (V_{E+P}) to the total volume of all non-volatile components (V_T) of a coating system.

$$PVC = \frac{V_{E+P}}{V_T} * 100$$

Hiding power as a function of pigment volume concentration.

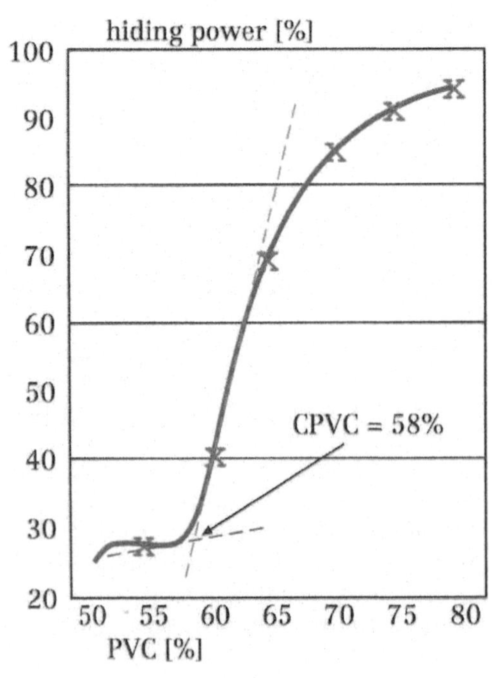

The greater the volume content of extenders and pigments in the film volume of a dry coating film, the higher the PVC.

As the PVC increases, a threshold value is reached that is known as the critical pigment volume concentration (CPVC). This occurs when the volume of binder is no longer sufficient to completely fill the voids between the extender and pigment particles.

The CPVC is of especial importance for coatings, since the properties of a paint can alter dramatically when the CPVC is exceeded. It is known, for example, that the hiding power of emulsion paints increases steadily as the PVC rises and then jumps suddenly in the area of the CPVC (see figure).

For that reason the CPVC is traditionally also determined from the hiding power, but other methods, such as internal tension or porosity measurements and the Gilsonite test, can also be used.

For any given PVC the position of the CPVC can be influenced by the choice of raw materials and by means of the formulation to obtain various PVC-CPVC intervals. This is important for the formulation of coatings, since many film properties are also dependent on this interval.

Packing density

In a number of paint systems, the packing density needs to be as high as possible in order to obtain good hardness and low porosity in the film. High-build systems must ideally also demonstrate no mud cracking.

If the wrong extender/pigment packing is selected in synthetic resin bound plasters, this can lead to cracking or to holes in the surface. It is therefore extremely important to choose the correct extenders or combination of extenders.

In an ideal packing state, the voids of coarser extenders are filled up by smaller and smaller extender and pigment particles until only a minimal volume content remains for the binder. At the same time, however, the

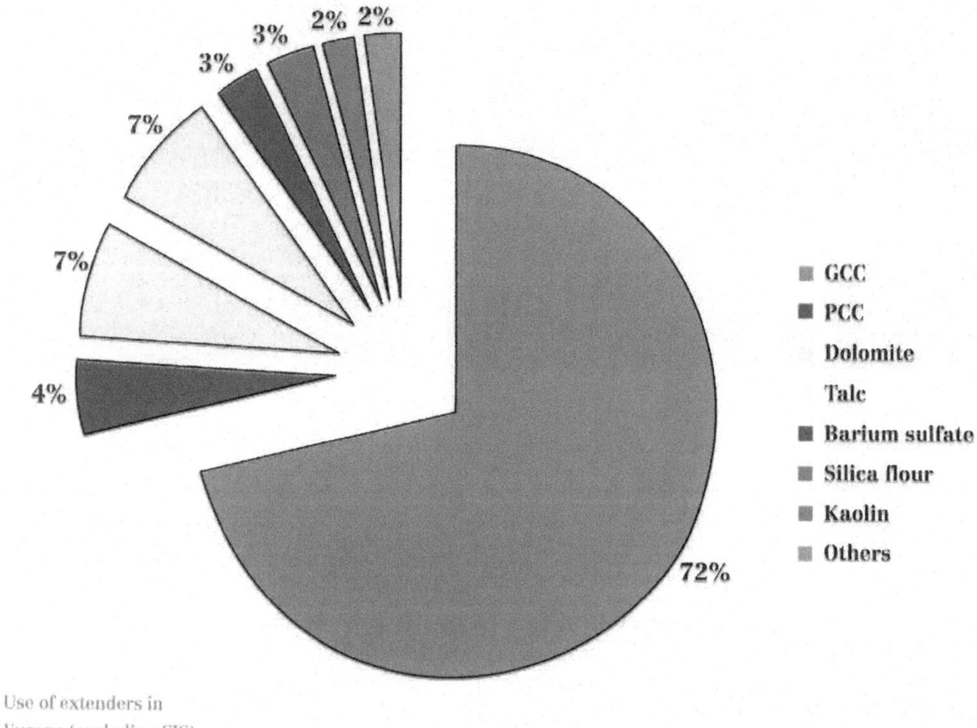

Use of extenders in Europe (excluding CIS) in 1998.

extenders and pigments must be sufficiently wetted by binder to ensure good workability. Detailed knowledge of the fineness and particle size distribution of the extenders is vital if this optimum packing state is to be obtained.

Calcium carbonate extenders

Calcium carbonate-based extenders are defined in EN ISO 3262 as follows:

- Whiting, a natural calcium carbonate derived from chalk, a sedimentary rock of soft texture. It is characterised by microcrystalline calcitic crystals (up to 1 µm across). Chalk is formed mainly from shells and skeletons of small maritime organisms.
- Natural crystalline calcium carbonate: a calcium carbonate derived from limestone and marble. The trigonal rhombic crystals tend to be rather larger than those of chalk.

Taking the two grades together, calcium carbonate is the most important extender for surface coatings, since unlike any other extender it meets almost all the requirements of the coatings industry. In 1998 it took 72 percent of the market share in Europe (76 percent if we include the synthetically precipitated PCC). A long way behind it come dolomite and talc, with 7 percent each (see figure).

Calcium carbonate is non-toxic and weather-resistant; it demonstrates good whiteness and low density, low interaction with pigments, binders and other components of surface coatings, and low electrolyte content and oil absorption. It has an anti-corrosive effect due to its alkaline pH and its low

Calcium carbonate – from fine grind to granules.

abrasivity prevents excessive machine wear, but the most striking property of the mineral is that with conventional processing methods calcium carbonate extenders are available in almost any desired particle size distribution and fineness (see figure). This is especially useful for regulating physical properties such as dis- persibility, gloss, gloss retention and hiding power.

The only noteworthy disadvantage to set against these applicational benefits is that of acid sensitivity. Otherwise calcium carbonate would be something of a "universal extender" for surface coatings, particularly as its cost-performance ratio is virtually unsurpassed. The list of typical applications for this product is therefore correspondingly long (see figure).

3.3 The use of calcium carbonates in selected coating systems

The number of different types of surface coatings is immense, and the variety of calcium carbonate extenders that are used in the individual applications is almost as large. The principal difference between the various grades of extenders lies in the particle size and particle size distribution (see

Typical applications for calcium carbonates in the coatings industry.

Coating system	Calcium carbonate content [%]
Emulsion paints	
– interior	50-70
– exterior	40-60
Synthetic resin bound plasters	70-80
Stoppers	70-80
Anti-corrosive primers	10-30
Road-marking paints	30-40
Silk-finish paint systems	20-30
Powder coatings	10-20

Application	Average particle diameter [µm]
Emulsion paints	0.9-70
Primers	0.9-5*
Stoppers	2.5-90*
Trade paints	0.9-5*
Corrosion protection	1.5-5*
Industrial paints	0.9-2.5*
Textured paints	30-160
Powder coatings	0.9-20
Road-marking paints	0.9-160*
Silicone resin paints	2.5-10
Printing inks	0.9
Plasters	
– brush and spray plasters	500-1500
– rubbing and grooved plasters	1000-3000
– plasters with open grain	1000-3500
– roller plasters	500-2000
– decorative plasters	1500-2500

*coated and uncoated

Average particle diameter for calcium carbonates in surface coatings. The range of calcium carbonates used – coated or otherwise – is no less diverse than the applications for surface coatings.

figure), but quite apart from this each application requires its very own calcium carbonate.

Emulsion paints

Emulsion paint is the lay term used for polymer dispersion paints in which the organic binder is dispersed – or very finely divided – in water. In addition to the polymer dispersion, an emulsion paint primarily contains extenders and pigments.

Emulsion paints are one of the most important coating systems because they demonstrate a whole range of positive properties which are of particular benefit in the area of building conservation.

Emulsion paints

- are environmentally friendly because they are water-based and contain little or no organic solvent,
- are easy to use,
- are quick-drying because of the generally high pigment volume concentration, and are inexpensive.

A distinction is generally made between interior and exterior emulsion paints. Given the number of different formulation options for the various areas of application, however, further subdivisions are unavoidable.

In the case of interior emulsion paints, for example, scrub and wash resistance are common criteria, but they are also typically classed by degree of gloss (high gloss to matt), brightness or contrast. The individual properties of an emulsion paint can be regulated via the binder, extenders and pigments and the PVC.

Applying an emulsion paint.

Matt paint surfaces with the maximum possible hiding power often require the addition of matting agents, although the matting agents that have been used for many years – such as kieselguhr (diatomaceous earth) – have recently come under increasing criticism because of their content of crystalline silicon dioxide, which with particle diameters below 5 micrometers can cause silicosis.

A non-toxic matting agent based on a natural calcium carbonate has been commercially available since 1998, however. It can be used to formulate matt interior emulsion paints that fulfil the requirements of DIN 53 778.

Exterior emulsion or masonry paints are used on surfaces made from concrete and concrete blocks, exterior mineral plasters, cellular concrete, limestone, brickwork and wood. As with interior emulsion paints, a grading system has been developed that classifies paints by gloss, flexibility or physical properties such as water vapour permeability.

Since the development of waterborne styrene-acrylate dispersions in the mid-1960s, calcium carbonates made from marble, which exhibit a particularly high degree of whiteness, are the most widely used extenders for both interior and exterior emulsion paints. Styrene-acrylate dispersions are characterised by good pigment compatibility and high saponification resistance, which means that high extender contents of up to 70 percent calcium carbonate are easily obtainable. What is more, with calcium carbonate extenders the hiding power, gloss and wet scrub resistance of interior emulsion paints can be adjusted by the judicious selection of particle size.

For matt interior emulsion paints, the most important paint type in terms of volume, the preferred calcium carbonates have an average statistical particle diameter of 2-5 micrometers and are characterised by good dispersibility. Up to 5 percent talc is commonly added to improve their workability. Additions of chemically precipitated calcium carbonates are occasionally also used to further enhance the whiteness of interior emulsion paints.

The appearance of masonry coatings alters over time as a result of weathering. Whereas external influences principally cause the film surface to become dirty, phenomena such as chalking and change in hue can be indicators of premature film degradation, which is attributable largely to the composition of the formulation but also to the extenders used.

The principal factors influencing the weather resistance of an extender are its chemical nature and its particle size distribution. Significantly better results can thus be obtained with natural crystalline calcium carbonates and dolomites than with silicate extenders (see figure). Since the particle size distribution in calcium carbonates is also very readily adjustable, white calcium carbonates made from very pure marble deposits are currently the most popular type of extender. Calcium carbonates with an average particle diameter of between 2 and 7 micrometers are predominantly used for matt white masonry paints, whilst products with an average particle diameter of between 10 and 20 micrometers are preferred for paints in deep shades.

Anti-corrosive paints

The function of an anti-corrosive paint is to provide corrosion protection for valuable metal surfaces that are highly sensitive to aggressive components in the atmosphere. Corrosion refers in very general terms to the destruction of metals and alloys with formation of metal oxide compounds, during which process the metallic material is converted to a thermodynamically more stable state.

The anti-corrosion problem is particularly severe in highly industrialised regions such as the USA, Japan and western Europe. In the Federal Republic of Germany alone the cost of damage due to corrosion in 1996 ran to DM 70 billion. Given these figures it is not surprising that increasing attention is being paid to the general issues of corrosion protection, and the demands on anti-corrosive paints are rising all the time.

Dark masonry paints (PVC 41%) after one year's weathering.

Paints do not form completely impermeable films. Instead they have capillaries running through them which provide access for air, water vapour, volatile acids and aqueous solutions. Not only does this provide a favourable environment for corrosion on the surface of the metal, it actually accelerates the process.

If calcium carbonate is added to a coating (anti-corrosive primer), it gives the film a deposit of a mildly alkaline substance that can counteract both H_2 and O_2 corrosion; in some cases a partial substitution of active pigments with calcium carbonates may even be recommended. For example, replacing 50 percent of iron oxide red with a surface-treated calcium carbonate grade in shop primers produced paint films with significantly improved corrosion resistance after 2 years' weathering (see figure, next page).

Combinations of extenders, where the addition of another mineral augments the basic potential of calcium carbonates by introducing other important physical properties, can also be helpful. For example, a mixture of calcium carbonate and platelet-like talc increases the packing density and thus prevents corrosive media from penetrating into the paint film.

In addition to their corrosion-inhibiting effect, calcium carbonates also improve the film properties of anti-corrosive paints. They reduce blistering, generally improve adhesion and, in comparison to barium sulphates, calcium carbonate has a positive influence on the weldability of the primer, usually enabling smooth pore-free weld seams to be obtained.

Gloss paint systems

Many coating systems, especially those for topcoats, offer gloss grades, which can range from high gloss to silk matt. Since gloss comprises physiological as well as purely physical components, it is not always possible to evaluate it precisely. Nevertheless, the gloss of coatings can be measured to within certain limits by the proportion of incident light reflected by the surface.

The degree of reflection of a paint system can be adjusted by a number of parameters, among which, in addition to the pigments used, the choice of a suitable extender is also extremely important. The particle size distribution and dispersibility of an extender are particularly important in determining gloss, as a result of which calcium carbonates are increasingly being used now in gloss paint systems. In white industrial gloss paints fine-particle calcium carbonates with a particle diameter of 0.9 micrometers can even replace a part of the opaque pigment.

These fine-particle calcium carbonates are also the preferred extender in high-gloss powder coatings, but any other desired degree of gloss, from high gloss to silk matt, can be obtained in powder coatings using natural crystalline calcium carbonates of different particle sizes (see figure).

Synthetic resin bound plasters

Synthetic resin bound plasters or synthetic resin modified plasters essentially consist of

Anti-corrosive primer after 2 years' weathering. Left: an iron oxide red with no extenders. Right: a combination of iron oxide red and calcium carbonate (50:50).

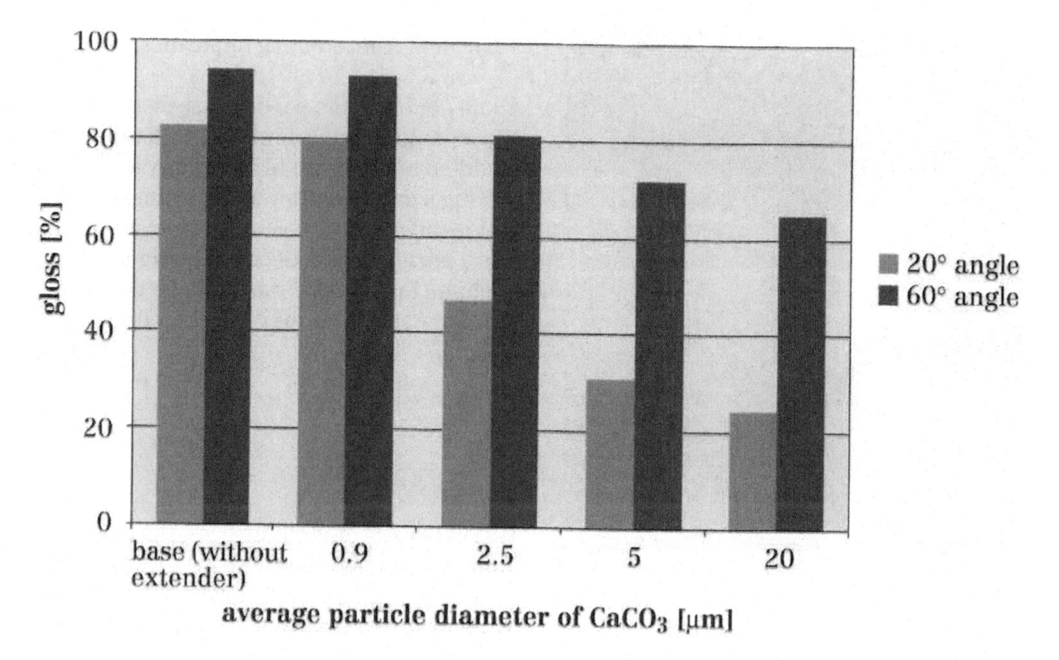

Gloss of powder coatings.

an organic binder, varying amounts of finer and coarser extenders and small quantities of pigments. Some of the most important properties of synthetic resin modified plasters are listed below:

- short setting time
- good opacity, which makes an additional finishing coat unnecessary
- good build, avoiding the need for separate treatment of minor defects in the substrate
- very good adhesion to virtually all substrates
- excellent flexibility
- very good weather resistance

As a result of these properties, synthetic resin modified plasters are an important material today for wall, ceiling and masonry coatings.

An important factor in the rapid growth of these plaster systems was the availability of new extender grades, since the silica sands that were formerly used were marred by a number of disadvantages that prevented the widespread use of plasters: they were dark in appearance, tended to grey, were ex-

Applying a masonry base plaster.

tPowder coatings are becoming increasingly important.

tremely abrasive and presented an elevated risk of injury. Since they contained extremely fine fractions, they were also suspected of causing silicosis.

These disadvantages were eliminated by a wide choice of calcium carbonate granules made from very white and pure marble and with accurately determined particle fractions of between 1 and 7 millimetres. These enabled very white plasters to be manufactured for a vast range of uses and application methods. Depending on the size of the granules used and the method of application, synthetic resin modified plasters with very diverse surface textures can now be produced.

The quality requirements for the calcium carbonate granules are the same for all plasters, however:

- high whiteness
- narrow particle size distribution
- preferably no oversize particles
- round particle shape
- freedom from staining impurities

A knowledge of the particle size distribution is particularly important, as synthetic resin modified plasters are high-build systems requiring a mixture of several calcium carbonates of different particle sizes. In order to obtain a perfect finish, the combination of fine, medium and coarse particles in the composition must be graded very carefully.

3.4 Trends

Just like many other industries, the coatings industry is today shaped by the three big Es: Economy, Energy and Ecology. Of these three factors, the most important is the ecological perspective for the future: environmentally friendly paint systems are increasingly becoming the measure of things. This issue primarily concerns solvents and binders, of course, since they represent the greatest hazard to health and to the environment. That is why water-thinnable coatings and powder coatings will continue to grow in importance in the future.

However, the development of improved extenders for these coating systems represents a particular challenge to the extender manufacturer too. Extenders that are not produced by environmentally friendly technologies or that are toxic will in the future increasingly be replaced by minerals that are harmless in all respects. The leading candidate for this is calcium carbonate.

Another option for the future is to enhance the optical properties of calcium carbonates. The objective here will be to use modern processing methods to develop new grades of extender that have a pronounced influence on the optical properties of the paint systems and thus enable the content of expensive white pigments to be further reduced. Calcium carbonates would then not simply be extenders with important functional properties, they would once again be pigments, just like chalk once was two thousand years ago.

4. Calcium Carbonate - A Versatile Mineral

Fillers and pigments for paints are extremely valuable products derived from calcium carbonate, but there are also many other areas where this mineral enjoys immense economic importance. Nowadays, crushed stones, sand, granules, grains and fine flour from limestone have all become indispensable raw materials. In fact there is hardly an industry that does not depend to a greater or lesser extent on calcium carbonate.

A consumption rate in excess of 3 million tons makes the building industry the largest buyer of calcium carbonate in Germany. Devonian limestone is used to produce mortar and flooring plaster. This dark mineral is also extensively used in concrete blocks and for paving stones. Limestone is preferred for super-white marble products, among them stucco plaster and marking-out paints for sports grounds.

The degree of whiteness is less significant for glass and ceramic products, but rather the chemical purity of the calcium carbonate. Even small traces of metal ions, such as iron or manganese, excludes the use of these minerals on account of their colouring effect.

Calcium carbonate is extensively used in the chemical industry, particularly in two raw material synthesis processes that are based on calcium carbonate. Solvay's ammonia-soda process transforms calcium carbonate into sodium carbonate, i.e. soda, while calcium carbide is produced from calcium carbonate and carbon which are heated to temperatures of 2500°C.

In mining more than 50,000 tons of limestone flour are used for rock-dust barriers to increase safety in mines. Some 2000 tons are used each year in fire extinguishers which is nearly twice as much as used for explosives. Each year several hundred tons of calcium carbonate are used for abrasives, welding electrode coatings and flushing liquids for gas and oil drilling holes.

It has now become almost impossible to manufacture a product that did not, at some time, come into contact with calcium carbonate. Yet, in most cases, the presence of this mineral remains hidden, either as an auxiliary in a production process, as it is the case in rice hulling, or it is chemically changed during the course of production. Consequently, there are very few final products which still contain calcium carbonate.

Next to building materials, calcium carbonate is primarily used in fertilizers for agriculture and forestry, and in chemicals for the protection of the environment. There are also many household articles, medicines and foods that still contain calcium carbonate, but usually only in small quantities and very few of these products have a future.

4.1 The use of calcium and magnesium carbonate in agriculture

The first reports on the use of fertilizers in farming date back to antiquity when efforts Owere made to find substances that would help promote plant growth. The studies were neither systematic nor scientific, but rather haphazard and based on trial and error. During these observations different processes and substances were tested. The results were quite remarkable! 2200 years ago the Roman statesman and writer Cato The Elder wrote in his treatise on agriculture "De agri cultura" a principle that continues to apply to this very day:

"Good farming means good tilling, good cultivating, good manuring."

The "fertilizers" known at the time can be subdivided into two classes:

- **Organic substances** of human, animal and plant origin such as dung and liquid manure.
- **Inorganic substances** of mineral origin (rock meal) or combustion residues (ashes).

In the first century AD a growing number of documents referred to targeted lime treatment of the soil or the use of calcareous marl. For instance, Pliny The Elder stressed in his "Natural History" that there was more to successful farming than merely the right climatic conditions. The state of the soil was also important. Eventually, experience showed that the chalky soil of Alba Pompeia (nowadays Alba in northern Italy) should be preferred to any other type of soil for wine-growing. Pliny also stated that soil was generally more useful after a fallow period during which it was able to recover and "smell more earthy".

However, Pliny also went further than these general descriptions. He put forward a series of suggestions as to how the soil could be improved. He reported about the use of marl (*marga*) in Britannia and Gallia, referring to this marl as the "fat of the land" through which the peoples became rich. He also praised the Ubians of Trier and Jülich who had "created fertile soil for themselves through the use of marl."

Several types of marl were known at the time:

- **White marl** which includes the fine sandy and stoney marl that is mixed with stones, glitter marl (*gliso-marga*) and silver chalk (*creta argentaria*), which were mined in pits down to depths of 100 feet. The latter retained its fertilizing effect for 80 years, stoney marl for 50 years, while the application of glitter marl had to be renewed after 30 years.
- **Blue marl** that was disintegrated by the sun and frost.
- **Sandy marl** that lasted for less than 10 years which is why it was only used for fertilizing when nothing else was available.

The fertilizing effect is readily understandable when one considers that all marls contain loam and clay to a greater or lesser extent, and that the term chalk in antiquity denoted different types of clay, including a mixture of lime, clay and other constituents. This is because when clay and loam are mixed in the correct ratio with lime, they form a nutrient store in the soil (see 4.1.1 "The influence of lime treatment on the soil").

Even though the growth-promoting effect of chalk and marl was not solely attributable to the calcium carbonate content but also to the entire soil mixture, it was nevertheless Pliny who was the first to draw attention to the fact that exhausted soils could be improved for continued use if they were treated with fertilizers containing lime.

Pliny's advice concerning fertilizers applied to the cultivation of field crops and to animal fodder. The purpose of marl treatment was always to increase soil fertility. Moreover it should always be applied "after ploughing" in a "one foot thick scattered layer". The application of marl fertilizer in a "one foot" thick layer is an inconceivable quantity because this would equal approximately 1000 to 1500 tons of fertilizer per hectare.

Very little is known about fertilizers except of Pliny. A number of natural substances were tested with regard to their value as fertilizers, but regular application was rare.

Success remained marginal as long as the application of fertilizers was not widespread. Consequently, frequent minor and major famines in the Middle Ages and in the early Modern Times were often the result of crop failures attributable not only to bad weather, but primarily to inadequate plant growth conditions. It was only the experience of these famines that forced the people to concern themselves more intensively with the possibility of increasing and safeguarding harvests. Fertilizers then became the focal point of interest. Yet all efforts were still mainly concentrated on short-term success. There was no long-term planning in farming.

The proverb "The application of marl (lime) makes the fathers rich and the sons poor" has its origin in this period. The application of lime mobilises the soil to release the remaining constituents it contains. This meant that the "fathers" were still able to reap rich harvests, but no longer the sons. Consequently, German tenant contracts in the 16th and 17th century prohibited the use of lime fertilizers as it was feared that they would "suck all nutrients out of the soil". However, this prohibition only applied to certain areas in Silesia, as opposed to the kings of Hanover and Prussia who supported, and even promoted, the use of marl fertilizers. Nowadays, this proverb has lost its validity due to the targeted use of mineral fertilizers.

At the latest by the beginning of the 19th century fertilizers and their long-term effects were the object of intensive research. Scientists, such as the French chemist Jean Baptiste Boussingault or the German botanist Christian Konrad Sprengel, made very important advance contributions in this respect. However, it was the German chemist Justus von Liebig who was the first to formulate basic rules for the successful use of plant nutrients, and these rules have remained valid to this day. His book "Organic Chemistry and Its Application to Agriculture and Physiology", published in 1840, was the first and for a long time the most important standard manual on the use of fertilizers for plant nutrition.

Justus von Liebig's "minimum barrel" which he used to demonstrate in a very lucid manner how the nutrient that was least available in the soil was limiting the ultimate yield. Targeted fertilizing requires previous analysis of the soil so that the individual nutrients can be correctly dosed on the basis of the analysis results.

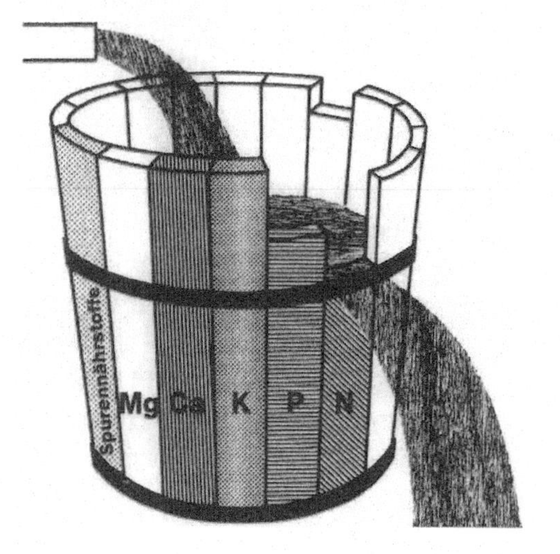

The soil as a nutrient store

Plants require numerous nutrients for their growth, and they extract most of them from the soil. The soil functions as a nutrient store, and its quality is primarily determined by its content of nitrogen (N), phosphorous (P), potassium (K), calcium (Ca) and magnesium (Mg), but also by a number of trace elements without which there would be no successful growing of crop plants.

It is not so much the actual quantity of each nutrient that is decisive, but rather the correct ratio between all the individual nutrients. Justus von Liebig realised this with his famous substitution theory in conjunction with the "minimum barrel" (see figure). Accordingly, the weakest link in the chain of growth factors is the element that is least present or available. And it is this element that determines growth.

However, Liebig related growth impediment primarily to nitrogen, potassium and phosphate as the main nutrients. It was only towards the end of the 19th century when it was realised that growth was also determined by a number of the so-called secondary nutrients, namely calcium, magnesium sulphur and trace elements. And this realisation has remained true until today except that the average yield as the basis for nutrient requirements has risen significantly. In 1840 it was in the order of 30 "double hundredweights" per hectare for wheat. This compares with 70-80 decitons per hectare (dt/ha)[1] in 1997.

This has an influence on the supply of nutrients in the soil. The higher the achieved yield, the more sensitive the plants react in yield and growth to an imbalance in the supply of nutrients.

[1] Right up until the 1970s harvest yields per hectare (ha) were specified in "double-hundredweights" (dz) which equalled the weight of a sack of grain. Since then the double-hundredweight has been replaced by the deciton (dt) so that the previous dz/ha unit has since been replaced with dt/ha.

Lime – One of the first natural mineral fertilizers

Calcium and magnesium carbonate in the form of marl and chalk are readily available from deposits that are close to the surface. It is therefore hardly surprising that special importance was attributed to the use of lime and marl when, towards the end of the 18th century, reports on mineral fertilizers became more frequent. For instance, in 1769, Andreä listed 300 different types of marl in his reports on "Lime and Marl Fertilizing" which he had been commissioned to draw up by the Hanover Chamber of Agriculture. However, the application of lime was also widespread in other areas of Germany and in Europe in general.

Liebig's minimum theory soon resulted in a decline in the application of lime. This is because the targeted application of the principal nutrients nitrogen, potassium and phosphate resulted in short-term success in terms of yield. Moreover, it had been assumed that the supply of calcium in the soil was plentiful everywhere.

The importance of balanced fertilizer application was only rediscovered at the beginning of the 20th century when many farmers were able to confirm in their own fields what the agricultural scientist Albert Orth had already predicted in 1896 in his paper "Lime and Marl Fertilizing" when he reported about the "experienced farmers" who, for years, "had applied fertilizers according to the [Liebig's] substitution theory and had near-

A combine harvester harvesting the wheat crop.

A trade brochure of Messrs. E. Schwenk Nachf. published in 1929 - today Ulmer Weisskalk GmbH.

ly been ruined: Success only returned after the introduction of lime."

The reason for this was quite simple. Next to all the principal nutrients, calcium is also a vital nutrient that cannot be dispensed with.

4.1.1 The influence of lime treatment on the soil

A soil of sustained fertility simultaneously forms the location and supplier of nutrients which can be mobilised from its own reserves. The soil must be able to store fertilizer nutrients, transform them and make them available to plants. Moreover, it must have good regenerative capabilities. The plants must be well rooted in the soil so that they can make optimal use of the supply of nutrients, water and oxygen contained in the soil. A fertile soil has an optimal pH-range for the given location, a favourable soil structure, a humus content of between 1.5 and 4 per cent. Moreover it is able to store most of the nutrients that are not immediately used, permanently fix harmful substances and maintain the correct ratio between the solid, liquid and gaseous phase.

Soil fertility

The productivity of the soil, i.e. its fertility, is influenced by various factors which, in turn, depend on a balanced lime condition. This includes the soil structure which mainly determines the exchange of air, water drainage and storage, and the bearing capacity of the soil. And then there is the biotic activity of the soil since the conversion of the different humus forms depends primarily on the existing bacteria, fungi and other organisms in the soil. Finally, there is the ability of the soil to store certain mineral nutrients, such as phosphate, and its influence on the reserve capacity regarding the continued supply of nutrients.

The soil structure influences, by way of the calcium magnesium carbonate content, the pH value – and thus the structural properties – as well as the biotic activities and the ability to store and convert the nutrients contained in the soil. The pH-value is therefore the most important characteristic of all soils, thus making lime treatment the most influential and important factor of soil fertilizing.

The soil reactions of calcium

The most important calcium minerals in the soil are calcite ($CaCO_3$) and dolomite [$CaMg(CO_3)_2$]. They are released into the soil by the weathering of carbonate rocks. The speed of the weathering process depends mainly on the content of carbon dioxide formed in the soil during the decomposition of organic substances. The individual reac-

Soil type group[1]				Target pH values Class C[2]	
				Humus content[3]	
No.	Designation	Symbol	Clay share	0 - 4%	4.1 – 8%
1	Sand	S	< 5%	5.4 – 5.8	5.0 – 5.4
2	Weak loamy sand	l'S	6 – 12%	5.8 – 6.3	5.4 – 5.9
3	Intensely loamy sand	lS	13 – 17%	6.1 – 6.7	5.6 – 6.2
4	Sandy/silty loam	sL/uL	18 – 25%	6.3 – 7.0	5.8 – 6.5
5	Weak clayish loam to clay	t'L, tL, lT, T	> 26%	6.4 – 7.2	5.9 – 6.7

1 Soil type group 6 "Moor" was not taken into account
2 The soil's supply of lime was subdivided into five pH classes: A – very low; B – low; C – target, optimal; D – high; E – very high
3 Five grades of humus content: 0-4%; 4.1%-8%; 8.1-15%; 15.1-30%; > 30%. Arable land usually has a humus content between 0 and 8%.

Definition of the soil types and target pH-values for arable land (Source: VDLUFA, 1999).

tions of carbonate weathering can be summarised in the following manner:

$$CaCO_3 + CO_2 + H_2O \rightleftharpoons Ca(HCO_3)_2$$

$$Ca(HCO_3)_2 \rightleftharpoons Ca^{2+} + 2\ HCO_3^-$$

Most of the free calcium ions (Ca^{2+}) are adsorbed by the clay minerals and humus constituents. However, this process is reversible. An exchange reaction takes place with an excess of other cations such as magnesium (Mg^{2+}) or hydrogen ions (H^+), thereby liberating calcium ions.

Calcium carbonate, or more precisely the calcium carbonate/ hydrogen carbonate system is the most important buffer which maintains a permanent, stable pH-value that is not affected by external influences:

$$HCO_3^- + H^+ \rightleftharpoons H_2O + CO_2 \quad (I)$$

$$HCO_3^- + OH^- \rightleftharpoons H_2O + CO_3^{2-} \quad (II)$$

The buffering capacity of the soil defines the quantity of acids and alkalines that enter the soil and can be neutralised so that the reaction processes can be maintained.

The pH-value is defined as the negative decadic logarithm of the H^+ ion concentration in water. Consequently, a reduction of the pH-value by 1 (e.g. from pH 7 to pH 6), equals an increase of the acid concentration by a factor of 10, and a reduction of the pH-value by 2 equals an increase of the acid concentration by a factor of 100. The amount of lime required for neutralisation must therefore be increased accordingly.

Different types of soil and their optimal pH values

The properties of the soil are determined primarily by its constituents and the ratio of the constituents in relation to each other. The dominating constituents define the soil as sand, weak loamy sand, intensely loamy sand, from sandy to silty loam and on right up to clay and moor. Each of these types of soil has a different composition, depending upon its content of clay, fine soil and humus and is therefore defined accordingly. The subdivision is by soil type groups (see figure).

The humus content also influences the pH-value of the soil. The normal content varies between 0 and 4 per cent. The pH-value declines the higher the humus content, i.e. lime must be continously added to humus soil to maintain the required optimal pH-value.

Meadows with their permanent plant rooting and the associated organic substances have a higher humus content than arable land. Consequently the humus content can vary between 0 and 15 per cent, and in humus-rich locations between 15.1 and 30 per cent. A lower pH-value is therefore desirable for greenland soil than for arable land with the result that fertilizer recommendations specify correspondingly lower lime treatment quantities.

Soil acidification and soil acid

Natural soil acidification is the result of enrichment with different acids. These can be organic acids formed as secretions from plant roots when organic substances are converted under reduction conditions. They can also be inorganic acids that are formed in the soil by microbiological transformation. Finally there is also an acidic reaction of the carbon dioxide which is present in greater quantities in the soil's air.

Hydrogen ions are formed when these acids come into contact with water. The latter are then dissolved out of the clay-humus complex by an ion exchange reaction, for instance into calcium ions.

The calcium, in turn, is dissolved in the soil and is eventually washed out by precipitation water as hydrogen carbonate [$Ca(HCO_3)_2$], sulphate ($CaSO_4$), chloride ($CaCl_2$) or nitrate [$Ca(NO_3)_2$], depending upon the existing anions.

Modern fertilizer systems take the actual nutrient requirements of the crop plants into account so that slow-release fertilizing by way of mineral fertilizers is only of secondary importance. The reduction of the application rates of mineral fertilizers has resulted in a decline in the supply of nutrients stored in the soil. Moreover, most of the nitrogen fertilizers are physio-logically acidic fertilizers which require additional lime treatment for neutralisation (see figure). Only calcium nitrate and lime nitrogen fertilizers produce excess lime.

The immission-induced acidification due to acid rain should not be underestimated. Normally rain has a pH-value of 5.6. However, as a result of the high concentration of nitrogen (NO_x) and sulphur oxides (SO_x), the current pH-value is lower by one to two pH-values so that the rain water causes additional neutralisation requirements that can amount to as much as 80 kg calcium oxide per hectare (kg CaO/ha)[2].

In addition to the above-described factors, the calcium supply in the soil under average European climatic conditions is additionally

Acidic fertilzing is at the expense of lime – Normal commercial nitrogen fertilzers and the lime loss as a result of their application.

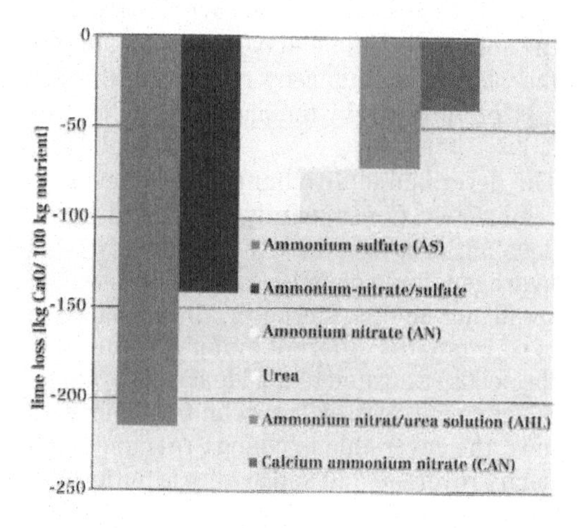

[2] Irrespective of the actual form in which calcium is present, calcium oxide is defined as a unit of measurement for the calcium content of lime fertilizers used in agriculture and forestry.

subject to intense exhaustion due to wash-out. The resulting lime losses can average between 250 and 350 kg CaO/ha per year. The actual quantity of washed-out calcium depends on a number of additional factors including precipitation rates, water storage and draining properties of the soil, the amount of clay contained in the soil, the humus content and the degree of coverage by plant growth. The soil's degree of calcium saturation is also important. Light sandy soils acidify more rapidly than heavy loam and clay soils because their calcium saturation level is significantly lower.

Some of these factors, among them the water regime and ground coverage, can be influenced by the manner of cultivation. This also applies to a certain extent to the saturation level since the calcium ions can be once again raised to the surface through more intensive ploughing down to a depth of 20-40 cm.

Soil acidification always requires a higher level of neutralisation, irrespective of whether the acidification is the result of natural, physiological or emmission-induced causes or due to wash-out. The more calcium ions are dissolved as a result of ion exchange reactions, the greater the acidification of the soil. The share of acidic hydrogen ions adsorbed by the clay-humus complex increases, while the share of calcium, magnesium, potassium and sodium ions within the overall level of adsorbed cations declines to the same extent. And at low pH-values even the aluminium ions are slowly washed out of the aluminium silicates, and manganese ions are dissolved in the soil. Both cations have a toxic effect on the plant roots, thereby restricting the intake of nutrients by the plant.

The described ion-exchange processes proceed slowly. Consequently, soil acidification does not progress rapidly or suddenly. The hydrogen ion concentration only rises in small increments. Moreover, since each type of soil contains different buffer systems and the soil is saturated in the ideal case by up to 80 per cent with calcium and magnesium ions, the reversible exchange reactions are partly reversed and soil acidity is buffered.

Acid damage to plants as an acidification indicator

The first signs of acidification are noticeable by a decline in yields which are unavoidable even with otherwise optimal tillage, sowing and fertilizer application. In spite of a measurable decline in yields, there are still no visible signs of plant damage. Only when the pH value continues to decline does the damage caused by lime deficiency and excessive acid become clearly visible. Blank patches appear where the crop has failed to grow because the plants were unable to survive the winter due to the poor soil structure (the fine roots were destroyed by the frost). Lime deficiency results in acidic stress due to the intensified absorption of aluminium, ammonium, hydrogen and potassium ions. External plant damage, similar to the symptoms of nitrogen deficiency, then becomes visible.

Whether or not a soil is acidic is indicated by so-called indicator plants. The typical flora of acidic soil include pansies (*Viola tricolor*), sheep's sorrel (*Rumex acetosella*), knawel (*Scleranthus annuus*), camomile (*Anthemis cotula*) and corn-spurry (*Spergula avensis*).

Soil structure and lime

The soil structure is determined by the density of the soil, the share of air pores, the useful water-holding capacity and the stability of the spatial arrangement of the solids in the soil. Any intervention in the overall structure automatically changes all the structure-dependent properties, but primarily the water, air and heat regime.

Such problematic interventions are often unavoidable in present-day agriculture. The soil is severely stressed by the operation of heavy machines and poor weather conditions during the harvesting season, as is often the case with sugarbeets and maize, with the result that the soil structure suffers severely. This makes it all the more necessary to stabilise, and if necessary, regenerate the soil structure through appropriate measures.

An adequate supply of lime with free calcium carbonate that is not bonded to soil particles plays a significant role in this context. With heavy soils this free lime (calcium and magnesium carbonate) is included in addition to the achievement of the desired liming target of pH 6.4 to 7.2. The following rule of thumb applies to marsh soils: pH 7 + 1 per cent free $CaCO_3$ independent of the lime form. The action of the free lime becomes apparent, among other things, by an increase in the soil's pore volume.

The total pore volume of a soil should be divided more or less equally by 50 per cent between water- and air-filled pores. An increase in air capacity promotes gas exchange which, in turn, reduces the CO_2 partial pressure in the root area so that the growth of the fine root system is promoted.

The rapid warmth increase in spring results in early vegetation. Excessive water is also more quickly drained off, i.e. the draining effect is significantly improved. The share of fixed water in the fine pore volume declines so that the soil water in the root area becomes more easily accessible for the plants. Moreover, soils containing adequate quantities of lime remain looser also during extended dry periods, and have a lower shear resistance so that less effort is required for tilling.

Small creatures and soil micro-organisms

Organisms and micro-organisms develop intensively in a balanced pH-value environment which also has a positive influence on the quality of the soil. Earthworms (Lumbricidae), in particular, contribute to an intensive stabilisation and loosening of the soil structure.

The number and effectiveness of soil micro-organisms depends upon the pH-value. For instance, micro-organisms that control nitrogen fixation in the soil are very active in all soils with a pH-value of 6.6 and higher. The number and activity of nitrate and nitrite bacteria declines rapidly when the pH-value drops below 5.0 and only revives when the value is once again raised to pH 6 and beyond. The same applies to nodule bacteria of leguminous plants.

Since the pH-value of the soil depends on the lime condition, lime treatment can significantly increase the share of microbial biomass and the activity of soil micro-organisms. Microbiological studies of the composition and anti-phytopathogenic potential of the soil microflora revealed that a neutral pH-value has a positive effect on bacteria and a negative effect on the fungi that are responsible for fungal diseases in plants.

Nutrients availability dependent on the pH-value

The availability of plant nutrients depends on the pH-value of the soil and, therefore, on the lime condition. To gain a more precise statement concerning the pH-value dependency it is necessary to consider the nutrients individually and to make a distinction between the nutrients nitrogen (N), phosphorous (P), potassium (K) and calcium /magnesium (Ca/Mg), and the micro-nutrients (trace elements) such as boron, copper, manganese and molybdenum (see figure, next page).

Elementary nitrogen is not available to the plant. The gas must first be integrated by so-called special ammonification bacteria into the ammonium ion (NH_4^+) which is then transformed into nitrate (NO_3^-) by other bacteria during nitrification. Both nitrate and ammonium are water-soluble so that the plant can absorb the two nitrogen compounds, except that nitrate is more easily absorbed than ammonium. The process of ammonification and nitrification is continuous and 98 per cent of the nitrogen supply in the soil is available in an organic-bonded form.

Nitrification depends greatly on the pH-value. The state of equilibrium between nitrogen storage (nitrification) and liberation (denitrification) is optimal at above pH 6.3. Little nitrogen is stored in the soil below pH 5.5.

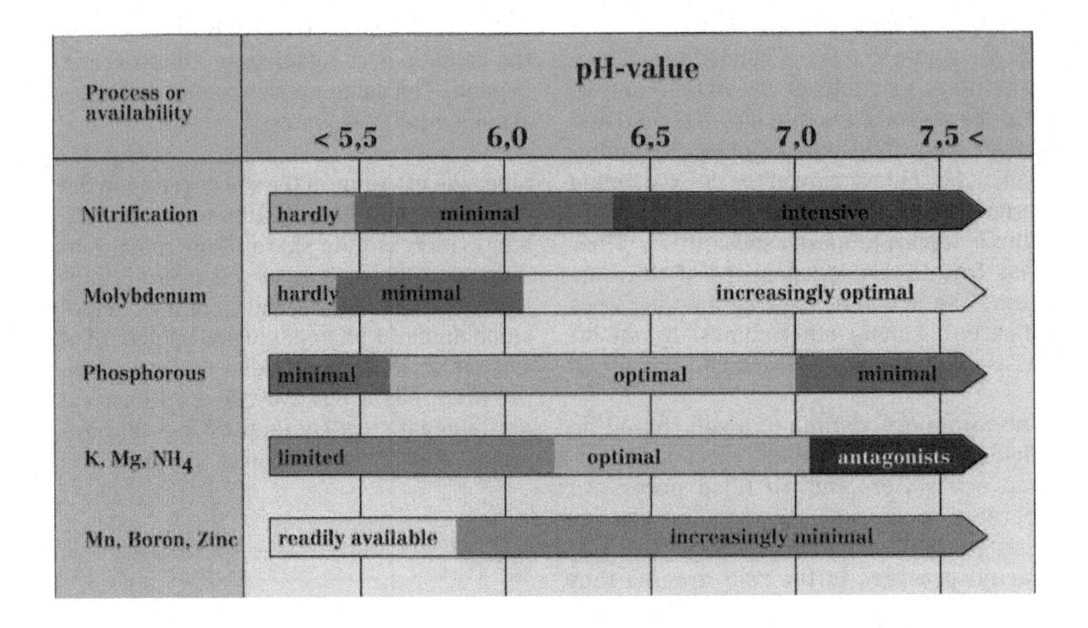

Influence of the pH-value on the soil and the nutrient availability (Source: LMS, 1998).

With phosphates the plant-available share is increased between pH 6 and pH 7. Consequently, the yield from an acidic soil can be significantly boosted merely by treating the soil with lime. This is because the resulting pH-value increase enables the soil to once again supply the plants with an adequate quantity of phosphate.

The absorption of potassium by the plants is less influenced by the pH value. It is optimal at pH 6.3. Antagonisms appear from pH 7.1 onwards, thus resulting in potassium deficiency symptoms.

Molybdenum plays an important role among the trace elements. Molybdenum deficiency impairs chlorophyll formation in the plant, thereby diminishing photosynthesis. In sugarbeet, for instance, this results in a reduction of the sugar content. Optimal molybdenum absorption by the plant commences at pH 6.1 and higher.

The micro-nutrients manganese and zinc control enzyme formation. Boron stabilises the cell membrane, while copper supports grain formation in cereals. Contrary to the other nutrients, the availability of these trace elements declines with a rising pH-value. Consequently, they are occasionally applied in suspensions so that they can be absorbed by the plant through its leaves.

4.1.2 The influence of lime treatment on the plant

Lime fertilizers are often referred to as soil fertilizers. Quite rightly, too, in view of the diverse effects they have on the fertility of the soil, especially by controlling the pH-value. Moreover, calcium and magnesium fulfil additional tasks since both are elementary constituents of plant nutrition.

Calcium and magnesium as plant nutrients

The plant roots take up calcium and magnesium from the soil solution and are conducted on to the vascular bundle by way of the plasma. The amount of calcium or magnesium that can be taken up by a plant does not only depend on the amount that is available in the soil, but also on the competing cations. For instance, calcium absorption declines when there is a lot of calcium and sodium in the soil solution whereas it increases if there are no competing ions in the solution.

Plants use calcium and magnesium primarily to stabilise the cell walls in which they are incrusted. They increase the elasticity of the cell walls and promote cell division and expansion.

Calcium and magnesium regulate, jointly with their antagonists potassium and sodium, the expansive state of the plasma. The plasma is responsible for all functions of metabolism including nutrient absorption, nutrient transport and transpiration. The expansion function of the plasma becomes disrupted if plant-accessible lime is lacking and the permeability of the root membrane is significantly increased. This, in turn, increases the absorption of aluminium and heavy metals which can result in metabolic disorders.

Calcium intensely promotes root formation. Cell division in the roots results in the formation of the first cell wall from the so-called pectinates which are created by the neutralisation of pectic acid. Lime deficiency slows down this reaction with the result that root and shoot development is disrupted, and the formation of mycelium hair roots and suckers is impaired.

Magnesium is the most important constituent of chlorophyll so that it holds a key function in photosynthesis and in the metabolism of carbohydrates, fat and albumin of plants. Magnesium deficiency symptoms are recognisable by the yellow colouration of the leaves during the vegetation period. The magnesium requirements of plants can be secured with dolomite [$CaMg(CO_3)_2$].

Lime requirements and lime extraction of different crop plants

The individual crop plants have widely different lime requirements. The reference magnitude in this instance is calcium oxide

Lime withdrawal rate of some typical crop plants related to cropped land (Source: BAD, BML, 1998).

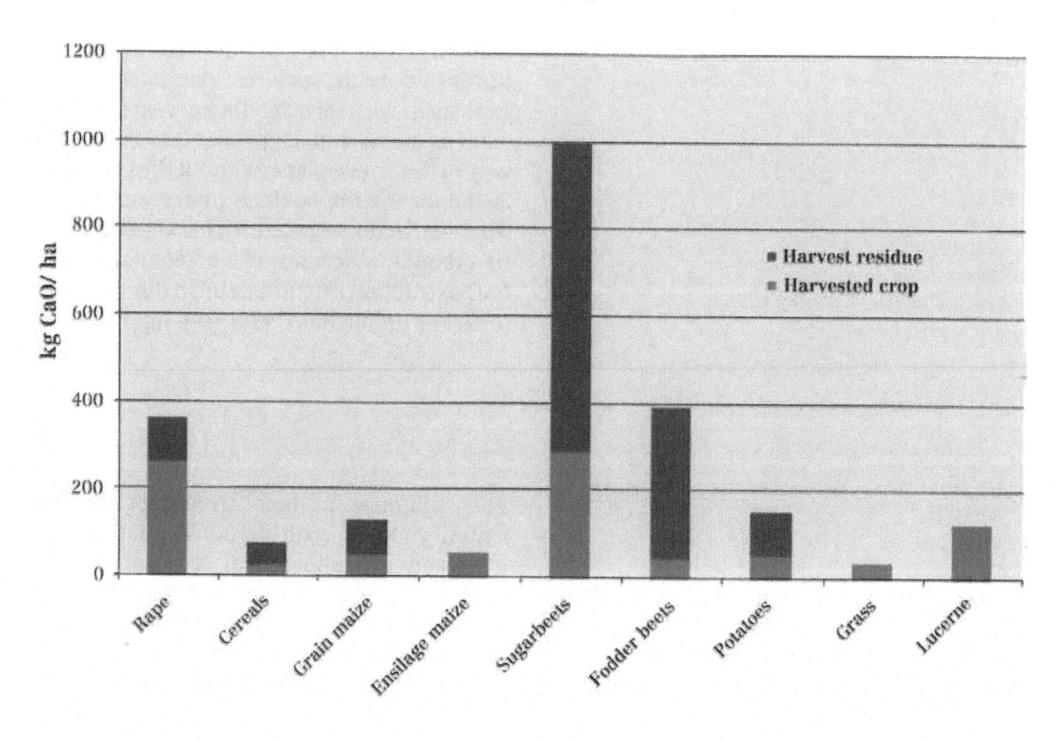

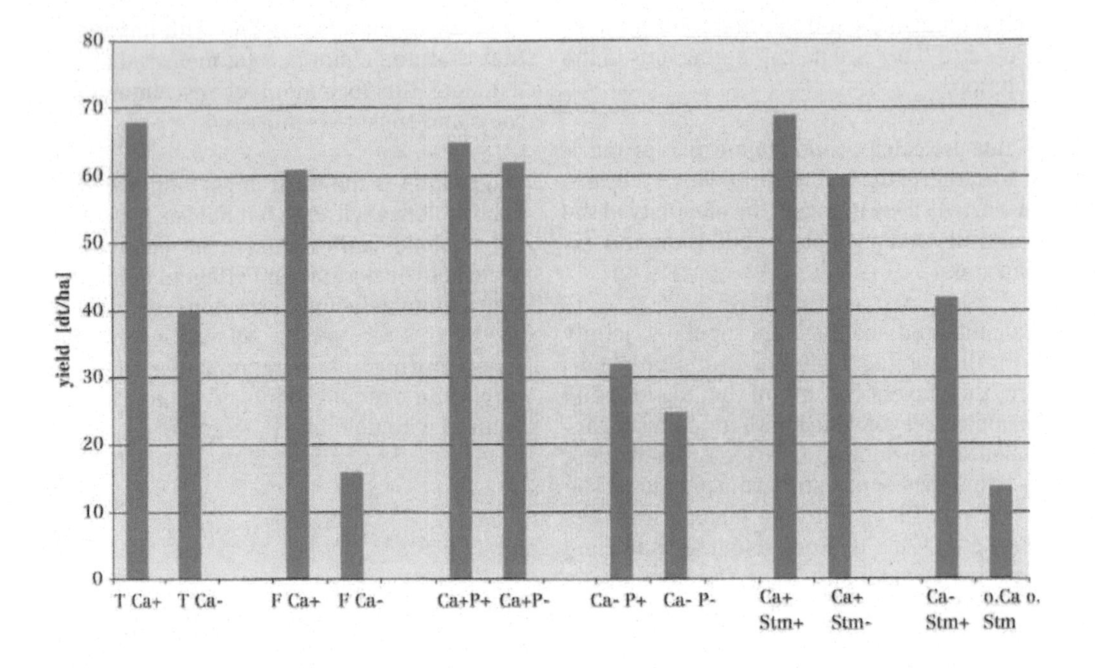

Dahlem static long-term test - Effect of long-term lime fertilzing on crop yields (acc. to Grimm, 1997).
The Dahlem test clearly proves that dispensing with calcium results in a significant decline in yields. This is independent of the crop, regular crop rotation and of the cereal crop rotation shown here.
The influence of deep (T, 28 cm) and shallow (F, 17 cm) tillage has a major influence, particularly if the soil is not limed: The calcium ions located in the deeper depths are regularly ploughed to the surface so that the calcium loss is relative low. This, in turn, means less decline in crop yields. Phosphate fertilzing (P) has little influence on the crop yield. However, if liming is simultaneously dispensed with, then crop yields decline by 55% to 70%.
Enrichment of the soil with manure (stm), combined with liming, produces the highest yields, whereas yields drop to nought with the untreated comparison variant.

(CaO). The lime requirements are calculated from the average amount of lime that is extracted per hectare of arable land and the achieved average yield in decitons (dt).

Plants continuously absorb nutrients from the soil during vegetation, and they use these nutrients for their growth. However, since crop plants are not left on the fields to rot, thereby once again releasing the nutrients, the soil looses some of its minerals with each harvest. The loss is caused by the harvested crop such as potatoes, grain, rapeseed, corncobs, by the harvest residues such as straw and sugarbeet leaves, and by way of the entire plant (without the roots), as is the case with ensilage maize and grass. This extraction caused by harvesting must be taken into account when calculating the CaO extraction rate to establish the required fertilizer application rates (see figure).

The effect of lime fertilizers on yields

The influence of lime treatment on the actual yield of crop plants cannot be established by short-term measurements. Long-term examinations are required for this purpose. One of the oldest well-known tests is the "Dahlem static long-term test" of the Berlin University in which the influence of soil utilisation on the long-term development of fertility and yields of light soils are examined (see figure). The result was definitive: The yield of winter wheat in 1995 after 64 years of continuous use of the test land was highest for all species in which lime fertilizer (Ca+) was used.

What should be noted is the evaluation of the results by the authors.

"Lime fertilizing during the first four decades did not result in any significant difference in yields. Since 1963 yields [in dt/ha] from the land that had not been treated [test variants] always diminished to an average of 40% [of the harvested quantity] of the land treated with lime. And an end to this development is not foreseeable. The yield reaction of crop plants due to a lack of lime is considerable. It was most significant with fodder beets, followed by winter wheat, potatoes and oats, and was lowest with winter rye. Almost all soil characteristics data deteriorated without lime. It was particularly evident that even with a normal content of heavy metals, lower pH-values of the soil drastically increased the solubility of heavy metals which were then either washed out or more easily absorbed by the plants. The threat to the ground water was diminished when the pH-value for light soils did not drop below approx. pH 5.5 to 6."

Determining the lime requirements for different soil groups with a humus content of < 4% (Source: VDLUFA, 1999).
The pH-value is measured in $CaCl_2$ to define the lime requirements. The lime requirements can be deducted with sufficient accuracy from the measured value so that the required pH-value of Class C (optimal) can be established. The recommended fertilzing rate is related to a period of 4-6 years. For very high quantities it is advisable to distribute the required application quantity over a period of several years. For instance a maximum of 15-20 dt CaO per hectare and year should be applied for sand, and 100 dt CaO for clay.

	Sand (BG 1)		**Weak loamy sand (BG 2)**		**Intensely loamy sand (BG 3)**		**Sandy/silty loam (BG 4)**		**Weak clayish loam to clay (BG 5)**	
pH-class[1]	**pH**	**RF[2]**	**pH**	**RF**	**pH**	**RF**	**pH**	**RF**	**pH**	**RF**
A	< 4.0	45	< 4.0	77	< 4.5	87	< 4.5	117	< 4.5	160
	4.1	42	4.1	73	4.6	82	4.6	111	4.6	152
	4.2	39	4.2	69	4.7	77	4.7	105	4.7	144
	4.3	36	4.3	65	4.8	73	4.8	100	4.8	136
	4.4	33	4.4	61	4.9	68	4.9	94	4.9	128
	4.5	30	4.5	57	5.0	63	5.0	88	5.0	121
			4.6	53			5.1	82	5.1	113
			4.7	49			5.2	76	5.2	105
			4.8	46					5.3	98
B	4.6	27	4.9	42	5.1	58	5.3	70	5.4	90
	4.7	24	5.0	38	5.2	53	5.4	65	5.5	82
	4.8	22	5.1	34	5.3	49	5.5	59	5.6	75
	4.9	19	5.2	30	5.4	44	5.6	53	5.7	67
	5.0	16	5.3	26	5.5	39	5.7	47	5.8	59
	5.1	13	5.4	22	5.6	34	5.8	41	5.9	52
	5.2	10	5.5	19	5.7	29	5.9	36	6.0	44
	5.3	7	5.6	15	5.8	25	6.0	30	6.1	36
			5.7	11	5.9	20	6.1	24	6.2	29
					6.0	15	6.2	18	6.3	21
C	5.4-5.8	6	5.8-6.3	10	6.1-6.7	14	6.3-7.0	17	6.4-7.2	20

[1] The lime supply in the soil is usually graded into five pH-classes. Since the pH-classes D and E have a high/very high lime content so that liming is not required, they have not been taken into account in this table.
[2] RF = recommended fertilzing rate in dt/ha CaO

Determining lime requirements in relation to the pH-value

The first indication of the lime requirements of a given soil is obtained from an examination of the soil. In Germany such tests are mandatory and must be carried out at intervals of no more than six years in keeping with the current Fertilizer Ordinance. The examination furnishes information on the current condition of the soil, its nutrients reserve and the pH-value. The binding examination methods to be applied are laid down in the Methods Manual of the 'Verband der landwirtschaftlichen Untersuchungs- und Forschungsanstalten' (VDLUFA). This guarantees comparable criteria in determining the nutrients, trace elements, pH-value, grain fraction and other aspects. Other countries use other methods. For instance in France the examinations must be based on the Methods Manual of the Institut National de la Recherche Agricole (INRA).

The application of an alkaline calcium fertilizer is necessary when the pH-value of the soil is below the optimal pH-range. This requirement is still receiving insufficient consideration worldwide with the result that, for many decades, calcium application rates have been too low.

Maintenance liming (latent lime requirements) defines the amount of lime that is needed to maintain the optimal pH range. The optimal pH-value of a soil depends on the soil type group, the composition of the grain fractions and the humus content.

Healthy lime dressing (acute lime requirements) defines the application rates that are required for the regeneration of a soil with a pH-value that is clearly below the optimum level. To establish the acute requirements it

Liming.

Country	Nutrient content	Designation	Coarse fraction	Middle fraction	Fine fraction
Denmark	> 70% $CaCO_3$	Fertilzer lime as powder	100% < 4 mm/ 90% < 2 mm		70% < 0.25 mm 50% < 0.1 mm
Germany	> 75% $CaCO_3$	Calcium carbonate	97% < 3 mm	70% < 1 mm	–
England	Neutralizations must be specified	Cl.1 limestone	100% < 5 mm	–	40% < 0.15 mm
		Cl.4 dolomite	95% < 3.35 mm		
France	> 45% CaO Additional neutral value	Cl.1 chalk as powder	–	80% < 0.315 mm	–
	> 43% CaO	Cl.2 dolomite as powder	–	80% < 0.315 mm	–
Italy	35% CaO	Cl.2 dolomite as powder	–	80% < 0.3 mm	
Norway	42% CaO	Limestone as	–	98% < 1 mm/ 80% < 0.4 mm	–
	50% CaO	Fine dolomite	–	98% < 1 mm	80% < 0.2 mm
Austria	> 90% $CaCO_3$	Limestone + dolomite	–	80% < 0.3 mm	
Sweden	42% CaO	Limestone	–	98% < 1 mm/	70% < 0.25 mm 50% < 0.125 mm
	46% CaO	Dolomite	–	(see limestone)	(see limestone)
	42% CaO	Chalk lime	95% < 3 mm	70% < 1mm/ 40% < 0.5 mm	20% < 0.125 mm
Switzerland	90% $CaCO_3$	Calcium carbonate	–	80% < 0.5 mm	–
	17% Ca	Dolomite	–	80% < 0.5 mm	–
USA	80% $CaCO_3$	CaMg-carbonate	95% < 2.36 mm		35% < 0.25 mm

Norms for calcium carbonate (Source: ILA 1994).

is first necessary to establish the actual pH-value of a mineral soil in a laboratory. A table is then used to determine the quantity of lime that is needed to achieve the required pH-value (see figure).

The establishment of the lime requirements under laboratory conditions is disputed because the results are often approximate values. For instance the examination of marsh soils reveals that the actual lime requirements are often significantly higher. This difference is attributable to the fact that laboratory examinations do not sufficiently consider the proton production that actually proceeds in the soil under the influence of use, fertilizing, soil tillage, microbial conversion and acidic deposition.

Periods for the application of fertilizer lime

The age-old principal that lime can be applied at any time remains true. However, liming should be carried out before intensive soil cultivation measures to achieve an intense mix and homogeneous distribution of the lime in the soil as this significantly boosts the initial effect. Particularly favourable are stubble liming in autumn after the harvest and lime application before seed sowing in spring. It is essential to ensure that no unhydrated lime is used on fields that are to be used for crops that are sensitive to high pH-values, e.g. oats.

Greenland liming should either be completed in spring, if possible when the soil is still frozen, either after the grass has been cut or at the end of the grazing period.

4.1.3 Lime fertilizer and its conversion

Lime fertilizers are produced from calcium carbonate rocks that are either free of magnesium carbonate or contain small to large quantities of this mineral. The type of rocks used for lime fertilizer production ranges from chalk to lime and dolomite. Traces of aluminium, silicates and iron are often present as impurities, but this is a minor consideration for application in agriculture. The different types of lime are classified according to physical and/or chemical aspects.

Product distinctions

Normally distinctions are made between hard and crystalline lime. Chalk belongs to the soft and porous lime minerals that have a high specific reaction surface.

Unhydrated lime containing magnesium is almost exclusively used in forestry, whereas in agriculture both the hydrated and unhydrated products are used. Hydrated fertilizer lime is marketed in a ground or granular form, or mixed with calcium carbonate.

Fertilizer lime products are defined by the regulations of national law. They are usually graded by total carbonate or total oxide content. Other criteria can include grinding fineness, alkaline effectiveness, the neutralising value (NV) and the different calcium and magnesium contents. Valuation is primarily on the basis of the oxides (CaO/MgO), partly also on the basis of the carbonates ($CaCO_3$/$MgCO_3$) or the element form (Ca/Mg) (see figure on page 289).

Grinding fineness as a measure of product conversion

Unhydrated fertilizer lime products are obtained by sieving or grinding. The purpose of grinding is to adjust the specific reaction surface of the products to specific applications. Thus, the larger the surface, the faster the effect of the lime fertilizer.

The soft and porous chalk raw materials have the largest specific surface which declines with the increasing hardness of the limestone. Consequently, hard limestones rich in magnesium must be more finely grounded to achieve an availability rate that is comparable to that of coarser grades of chalk.

Finely ground fertilizer lime products are superior to the coarse ground grades. Thus, dolomitic lime ground to a grain fineness of 0-1 millimeter can be used as a fertilizer on land suffering from magnesium deficiency. With such a degree of fineness, dolomitic lime has a better magnesium fertilizer effect than the traditional combination of kieserite ($MgSO_4 \cdot 2\ H_2O$) and calcium carbonate.

Coarse chalk granulate - the customary commer-c-ial application form for lime fertilzer.

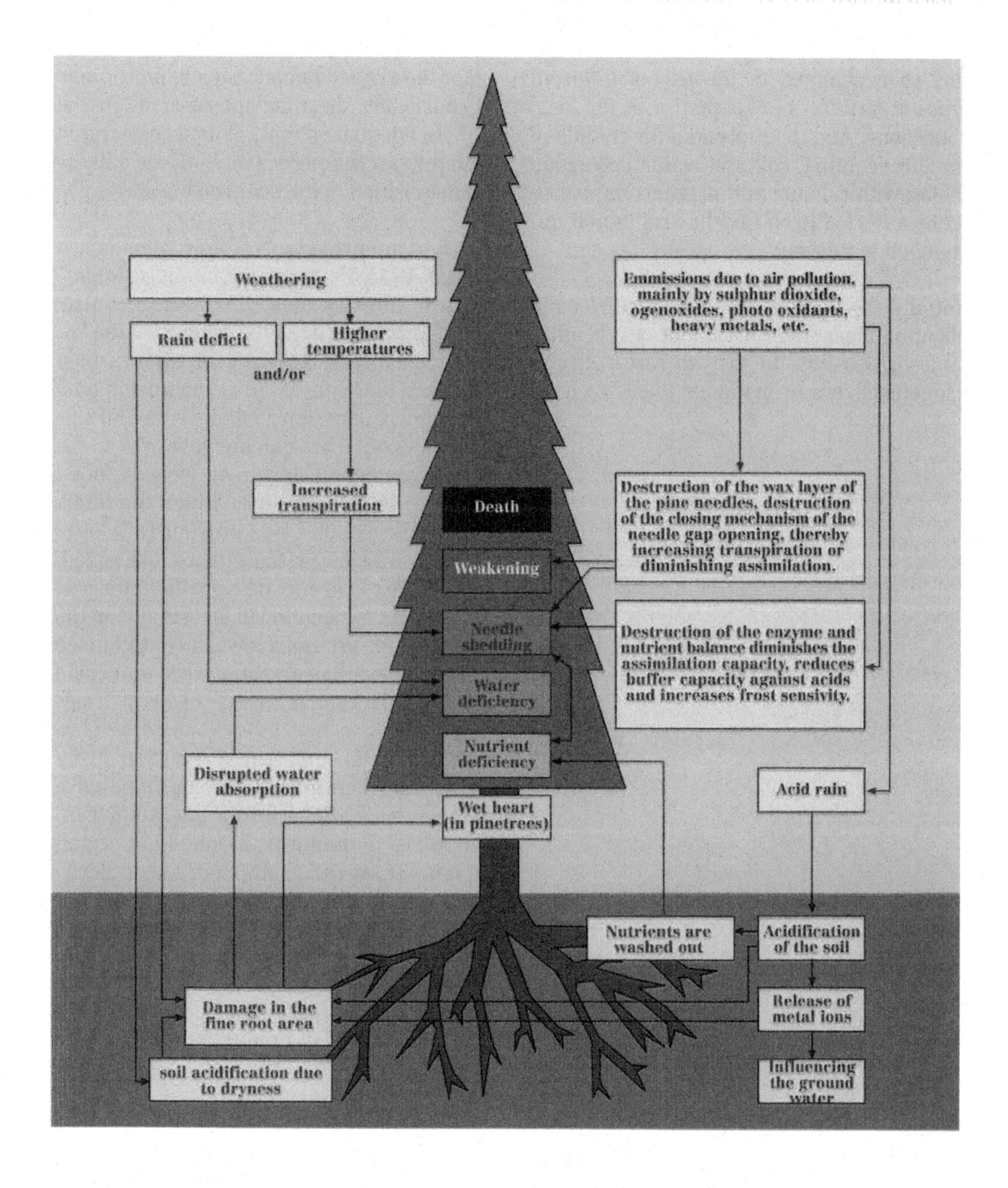

A simplified diagram illustrates the chain of circumstances causing the forests to die. The individual influences can act independently of each other or mutually intensify their effect.

4.2 The use of calcium and magnesium carbonate for forestry

The history of fertilizer application in forestry goes right back to the beginning of the 19th century when organic substances and ashes were spread on forest soils. Since then the purpose of fertilizer and lime application in forestry has continuously changed. Initially, the systematic application of fertilizers was focused on Liebig's minimum law, but this soon changed to the use of different types of fertilizers in keeping with the local conditions. And then towards the middle of the 20th century, soil and water protection liming with calcium and magnesium carbonates asserted itself and has remained unchanged to this day.

Initially, the primary aim of all fertilizer application measures was to increase profitability by boosting the growth rate with nutrient fertilizers, even though these were not specific or targeted in their application. Later the targeted application of lime aimed at mobilising the nutrient reserves in the soil. An adequate supply of lime accelerated both the raw humus conversion rate and the decomposition of the scattered layer.

The first intensified use of lime commenced around 1950. However, the immense importance of forestry lime dressings to control forestry damage was only realised at the beginning of the 1970s. Dying forests on a massive scale were observed at the time and no accurate explanation could be found for this phenomenon. This quickly gave rise to intensive studies of this "new form of forest damage" in the course of which the interdependencies in the previously closed "forest" nutrient circuit were examined. The studies concentrated on the effects of dry and wet application in forests and on the quality of the ground and surface water (see figure). It soon became apparent that "acid rain" was the principal cause of forest decay.

European-wide emission protection legislation was then enacted which imposed a number of conditions for the operation of incinerators to diminish the emission of pollutants. However, continued forest decay was inevitable because the acid contamination had resulted in irreversible damage. Lasting damage had been caused to the buffering effect of the forest soil as a result of long-term blockage of the ion-exchange capacity of the clay-humus complex.

The acidification front under the forest soil (acc. to Veerhoff, 1996). The acidification front has progressed to deeper soil depths due to continued introduction of acid. Whereas the state of equilibrium between the hydrogen and aluminium ions on the one side and magnesium ions on the other, was originally at 20 cm depth, the profile depth today is at 60-70 cm. This prevents the trees from forming an adequate fine root system within the main root area. Moreover the acidified soil horizon lacks organisms and micro-organisms. The scattered fertilzer cannot be disintegrated so that the nutrient circuit is interrupted.

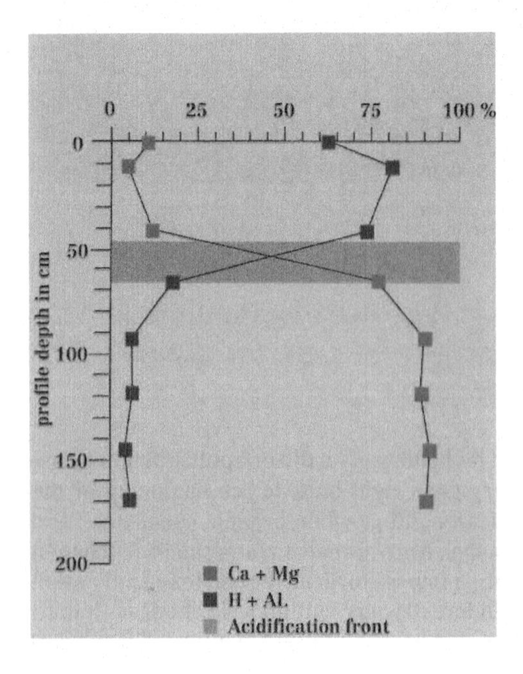

Condition of the forest soil

Increasing acidification was the principal reason for the damage caused to the forests and forest soil. The consequences of acidification are similar to those of arable land and greenland except that they are more severe since the forest soil has a much lower pH-value than healthy soil so that the supply of alkaline calcium and magnesium is correspondingly low.

Developments in recent years have resulted in a dramatic shift of the "acidification front" down to lower soil depths (see figure). Intense long-term acidification of the forest soil persists from the leaf-mould surface right down to a depth of 50-70 cm. Aluminium ions and protons are the dominating cations in the exchangers above the acidification front, while underneath there has been a dramatic increase in calcium and magnesium saturation.

Forest and water

Progressive acidification has damaged the ion-exchange function of the forest soil to such an extent that rain is reaching the ground water in an unfiltered state. This has had an adverse effect on the water quality. The amount of aluminium contained in the ground water is increasing continuously due to the declining pH-value. In many locations it has exceeded by a multiple the limit value of 200 microgram per litre (μg/l) of water specified for drinking water. Analyses of crude water in acidification endangered locations in the Rhineland Palatinate area in Germany revealed that valid limit values for aluminium had been exceeded in 26 per cent of the water samples examined in 1991/92.

Forest liming by helicopter.

The increasing discharge of heavy-metal ions and the low pH-values have resulted in a lasting deterioration of the drinking water quality. The acid buffering capacity of the ground water dropped from pH 5.3 to 0. Since the buffering system always endeavours to achieve a state of equilibrium, the discharge of pollutants increases with the declining buffering effect of the soil and rock formations. In the long run this action mechanism can only be overcome by a “concept for forest soil restoration”.

Forest soil restoration

The restoration of the forest soil holds a key position in re-establishing a stable, resiliently reacting forest ecosystem. The renewal of the “Forest Soil Buffering System” requires long-term intervention in the system under consideration of ecological compatibility: Chemically by repeated soil and water-protection liming; biologically by extensive reforestation that includes a large share of deciduous trees and, where necessary, the incorporation of larger quantities of lime in the soil before it is replanted.

In Germany some 100,000 to 150,000 hectares of forest land, i.e. less than one per cent of the total forest area, have been given a lime dressing each year since 1984. Three tons of carbonic lime with a high magnesium content are applied per hectare of forest land. The application of these 400,000 tons of calcium carbonate pursues the following aims:

- To neutralise the acidic air damage in the tree crowns and in the forest soil
- To improve the supply of magnesium for the trees and the plant-available magnesi-

um content in the soil
- To increase the biological activity of the organically influenced soil horizons
- To contribute to the formation of new and stable clay-humus complexes
- To intensify the nitrification process in the humus layer
- To stabilise the metal-organic complexes in order to prevent heavy-metal pollution of seepage water
- To promote soil vegetation

This is a truly Herculean task which cannot be satisfactorily resolved merely by lime dressing. The regeneration of the forest areas has only just started and the end of the measures is not yet in sight. The annual cost of the current programmes in Germany, alone, are in the order of 22.5 to 25 million Euro.

4.3 Lime fertilizers and their application in Europe

Lime fertilizer marketing is regulated worldwide by corresponding national laws. The uniform legislation that is in place within the European Union (EU) for nitrogen, potash and phosphate fertilizers does not apply to lime fertilizers.

The following details are therefore related to the respective statutory requirements of the EU member countries which have been summarised within the framework of the Committee for European Norming (CEN). In addition to the hydrated and unhydrated lime derived from chalk, limestone or dolomite – the so-called "natural limes" – there is also "industrial lime" resulting from iron and steel production, sugar production and numerous other industrial processes.

Chalk limes are separately classified in France, Great Britain and the Scandinavian Countries on account of their special geological peculiarities. Since these limes have a high specific reaction surface already in their crude state, these countries do not specify such degrees of grinding fineness for chalk as it is the case for crystalline lime rocks and devonian mass lime.

A distinction can be made between the pure calcium carbonates and those containing a high amount of magnesium by way of the so-called neutralising value (NV) since this value (related to CaO) rises super-proportionally with a rising magnesium content. Thus, carbonic lime with 90 per cent $CaCO_3$ has an NV of 50, whereas carbonate dolomite with 50 per cent $CaCO_3$ and 40 per cent $MgCO_3$ has an NV of 55. This means that, at an identical total carbonate content, the acid neutralising capacity of dolomite is 10 per cent higher.

On account of its specific properties, chalk has a high reactivity which makes it possible to shift the pH-value more quickly into an alkaline milieu.

The CEN was founded in preparation of uniform norms within the European Union. The main task of this Committee is to harmonise on a Europe-wide scale the individual national specifications. The implementation of the norms drawn up for lime fertilizers is expected for the year 2005.

In addition to fertilizer legislation, Germany also has an RAL quality certificate for fertilizer lime. Company participation in this system is voluntary. Accordingly, the companies undertake to submit their products to external, independent branch-specific scrutiny. The standard guaranteed by this certification system clearly exceeds the statutory norms.

Fertilizer sales in Europe

Each year more than 10 million tons of lime fertilizers are applied to Europe's agricultural and forestry land. At first sight these figures may well appear to be high, but if this figure is seen in relative terms, i.e. consumption is related to the total area used by agriculture in these countries, then this figure is soon seen to be very moderate.

If the annual extraction rate from the soil were estimated to be in the order of 450 kg

Country	Farming land [1000 ha]	Lime sales [1000 to CaO]	Average consumption [kg CaO/ha]
Belgium	1 337	299	223
Denmark	2 727	451	166
Germany	17 157	1 866	109
France	28 267	1 080	38
Netherlands	1 999	187	94
Finland	2 192	330	150
Norway	1 031	170	165
Sweden	3 060	127	42
Total	57 770	4 510	78

Lime sales in different European countries (Source: CEN, 1993). Although the data relate to 1992 they have remained valid to this day.

CaO/ha, then this would result in an annual lime deficiency of 370 kg CaO/ha at an average annual application rate of 70 to 80 kg CaO/ha. To offset this deficiency an additional 21.5 million tons of CaO would have to be applied as lime fertilizer.

4.4 Calcium carbonate in the feeding of livestock

Calcium carbonate forms a major part of an animal's body. This mineral regulates cell activity, including the compensation of lime loss. Moreover, calcium carbonate is involved especially in the development of the bones and the skeleton as a whole.

Calcium ranks among the vital minerals that also plays an important role in the metabolic process. It regulates the swelling function of the protoplasma and, thus, the intake, retention and conversion of nutrients. Typical calcium deficiency symptoms are metabolic disorders, poor fodder utilisation, softening of the bones and bone brittleness, particularly in young livestock.

Calcium deficiency in mother animals can result in the animals not being able to stand up (paresis), or only to a limited extent (cows must be standing while calving otherwise birth will result in an abortion). Calcium from the bones can be mobilised at short notice in acute deficiency situations, but this is only possible to a limited extent. In the long term this results in a typical softening of the bones and of the skeleton as a whole.

Since animals require differing quantities of calcium at different stages of their development, these requirements must be taken into account when formulating the correct fodder ration. The amount of calcium supplied by way of animal fodder is based on the amount of calcium in milk (breeding and dairy cattle), the inevitable calcium losses by way of the excrements (faeces and urine) and the net requirements for the performance of the animals (growth, meat, eggs). The form of chemical bonding of the calcium has only a minor influence on its bio-availability.

Calcium ranks among the major elements and is therefore one of the many mineral constituents contained in commercial fodders. It can be added as "carbonic fodder lime" from limestone or chalk with a calcium content exceeding 36 per cent. This carbonic fodder lime is processed into a ground or granular form and added in conformity with the corresponding fodder laws which define the chemical composition, purity and permissible dosing rate of the major elements. This carbonic lime is a natural product obtained from ground limestone, grained chalk, whiting or ground mussel and oyster shells. Carbonic fodder lime is permissible on its own or as an ingredient in mixed fodder preparations.

4.5 Calcium carbonate for environmental protection

There are three principal areas of environmental protection in which calcium carbonate plays an important role as a chemical reagent. It is used in large power stations for flue gas desulphurisation, for the preparation of drinking water in water works and, last but not least, for the neutralisation of over-acidified lakes and rivers.

4.5.1 Flue-gas cleaning

Air pollutants are held primarily responsible for most of the decades of pollution caused to forests, soil and ground and surface waters. The operators of combustion installations – be they power stations, industrial and other large-scale furnaces or domestic fireplaces and also motor vehicles – are the principal contributors to this pollution with sulphur dioxide (SO_2), nitrogen oxide (NO_x) and ozone (O_3).

The German Government reacted to this air pollution in 1983 by amending the Clean Air Act and introducing a separate administrative regulation governing the purity of the air. Although, in 1994, this regulation was amended and supplemented for the 22nd time in eleven years, the problems have still not disappeared.

The flue-gas cleaning installations are primarily based on a wet chemical process. Since there are a multitude of air pollutants with widely differing chemical properties, the reactions that take place during flue-gas cleaning are equally diverse. The leading cleaning process is flue-gas desulphurisation which is usually carried out with calcium compounds consisting of lime, limestone or chalk. There are various processes, among them the dry process with lime, the spray absorption process with lime or lime hydrate, and the wet process based on lime, limestone or chalk with gypsum hydrate. The following reactions take place when calcium carbonate is used:

$$CaCO_3 + \tfrac{1}{2} H_2O + SO_2 \rightleftharpoons CaSO_3 \cdot \tfrac{1}{2} H_2O + CO_2$$

$$CaCO_3 + 2\, H_2O + SO_3 \rightleftharpoons CaSO_4 \cdot 2\, H_2O + CO_2$$

The requirements expected of lime and limestone products are primarily determined by the demands expected of the end product: A high degree of whiteness, fine grinding (90 per cent of the particles have a grain size of less than 90 micrometers) and a low residual moisture for optimal control of the flowing and conveying properties of the limestone flour.

The pH-values in the scrubber drop down to pH 4 which is a good guarantee for the transformation of the sulphur dioxide into gypsum dihydrate. Air is often blown into the scrubber to oxidise the arising sulphite into sulphate:

$$CaSO_3 \cdot \tfrac{1}{2} H_2O + \tfrac{1}{2} O_2 + 1\tfrac{1}{2} H_2O \rightarrow CaSO_4 \cdot 2\, H_2O$$

The degree of desulphurisation depends on a number of process parameters, among them the crude gas concentration, liquid distribution, the dwell time of the gas and washing liquid in the scrubber, the pH-value and particularly the liquid-over-gas ratio (L/G) that is measured in litres of washing liquid per cubic meter of gas [l/m^3].

Desulphurisation levels in the order of 90 per cent can be achieved at an L/G ratio of 8 l/m^3 and an SO_2 crude gas concentration of 3,500 milligram per cubic meter (mg/m^3). To increase the desulphurisation level to 95 per cent, it is necessary to increase the L/G ratio to 14 l/m^3; and for 97% even to 20 l/m^3.

Rostock power station using Rügen chalk for flue-gas desulphurisation.

The gypsum produced as a reaction end product in power stations fired with hard coal fulfils current quality criteria of the gypsum industry and can be marketed as a high-quality building material.

4.5.2 Drinking water processing

Drinking water has become scarce. Even in a country like Germany that is so rich in water, not all sources and reservoirs are suitable for the production of drinking water, and many of the sources used today are threatened by a dramatic deterioration of the water quality. It is therefore essential to counteract this development trend.

Stage 2 filter of the Rieblich water treatment plant: Carbon dioxide is introduced into the water: For optimal adjustment of the total hardness of the drinking water carbon dioxide is introduced into the water where it reacts with the filter's Jura lime to form hydrogen carbonate.

Since it is hardly possible to rectify the water that is stored in the soil, the extracted water for drinking must be subsequently treated by environment-friendly methods in such a manner that it is perfectly suitable for human consumption. To prevent the dissolution of undesirable substances in pipeline systems and to neutralise acidic crude water, it is necessary to adjust the lime/carbon dioxide equilibrium with calcium carbonate or lime products.

The German Drinking Water Regulation governs the treatment of crude water with calcium carbonate and defines the requirements that drinking water producers have to fulfil. The pH-value is of special importance and it must be between 6.5 and 9.5.

The pH-value of natural water sources (moor water, ground water, springs and bank filtration) has dropped significantly as a result of anthropogenic influences. Moreover, the quality of crude water continues to decline as a result of a number of other factors, among them denitrification, an increase of carbon dioxide in the soil, nitrate washout, and the increased inclusion of nitrogen and sulphur oxides in rain water.

Defining size	Dense calcium carbonate			Porous calcium carbonate	
	Type 1	Type 2	Type 3	Type 1	Type 2
$CaCO_3$	> 98 %	> 94 %	> 80 %	> 97 %	> 85 %
$CaCO_3 + MgCO_3$	> 98 %	> 94 %	> 90 %	> 99 %	> 95 %
HCl-undissolved residues	< 2 %	< 6 %	< 12 %	< 1%	< 5 %

Products for treating water for human consumption (EN 1018).

The use of calcium carbonate in this area is very effective. The German Drinking Water Regulation and the European Norms (EN) explicitly define calcium carbonate as a suitable additive for water treatment. The quality of the individual additives does not only depend on the actual active ingredient content but mainly on the content of potentially harmful constituents such as lead or cadmium. That is why pH-value correction is achieved with calcium carbonate that is used as filter material in open or closed rapid filters because it is not contaminated with heavy metals.

The requirements expected of products used for drinking water treatment have been formulated in a corresponding norm by the Committee for European Norming (CEN). Accordingly, national norms such as the German Industrial Norm (DIN) will be gradually replaced with this European Norm (EN).

The EN 1018 "Products for the Treatment of Water for Human Consumption" distinguishes between dense and porous types of limestone and defines the minimum carbon content (see figure).

EN 1018 specifies A- and B-types for lime and dolomite products, and defines the per-

Limit values for the content of toxic substances in calcium carbonates for drinking water treatment (EN 1018).

Parameters	Calcium carbonate, limit values [mg/kg product]	
	Typ A	Typ B
Antimony (Sb)	3	5
Arsenic (As)	3	5
Lead (Pb)	10	20
Cadmium (Cd)	2	2
Chrome (Cr)	10	20
Nickel (Ni)	10	20
Mercury (Hg)	0.5	1
Selenium (Se)	3	5

Lake liming in Canada.

missible limit values for a total of eight heavy metals (see figure). This represents a marked increase in the stringency of the guidelines compared with past demands, particularly since the operators of treatment plants are expecting a lower content of toxic substances to preclude a possible potential hazard.

As a result of its mineralogical, crystallographic properties, the Jura calcite of the White Jura Epsilon (a formation of the Swabian Jura lime) is used as an example of an appropriate calcium carbonate of high purity and density that is suitable for drinking water treatment in conformity with the new European requirements.

4.5.3 Neutralisation of over-acidified water

The acidification of numerous open waters has become a major problem, particularly in the Scandinavian Countries, the USA and Canada. The effects on the ecological diversity of inland waters are quite dramatic. Fish is reacting sensitively to these changes and some species, such as the Canadian Aurora Trout, were or are still threatened by extinction.

There is only one way to counteract this growing acidification. The most effective and frequently used form is lake liming with calcium carbonate. Alone in Sweden some 7500 lakes and more than 10,000 river kilometers were repeatedly treated with lime during the 1990s. The cost of this programme in 1994 exceeded 25 million US dollars.

The calcium carbonate was either applied as a slurry or as a finely ground powder. Calcium carbonate slurry is usually sprayed onto the water surface from boats, whereas the powder is dispensed over the lakes by helicopters or fixed-wing planes.

The Canadian Bowland Lake is an example of how effective liming can be. The pH-value of the lake water before liming was 5.0. The application of dry calcium carbonate powder raised the pH-value to 6.8. At the same time the aluminium concentration declined from 130-150 mg/l to approx. 65 mg/l, thereby assuring the survival of the Aurora Trout.

4.6 Calcium carbonate - Everyday products

Calcium carbonate is a mineral raw material of enormous economic importance. The paper industry, alone, consumes several million tons of the mineral each year. Moreover, calcium carbonate is contained in numerous household products. In fact the few hundred tons of cleaning and blackboard chalk and the small amount of calcium carbonate used for toothpaste and medicines are irrelevant by comparison. These markets are merely small niches and the last relics of past decades and centuries, but they nevertheless continue to exist and have their importance.

In bygone days it was natural chalk - be it from Champagne or Rügen - that played the principal role in household products. However, "artificial chalk", known as precipitated calcium carbonate (PCC), gradually gained its share of the market, and in some

In the beginning of the 1950s natural chalk was used in a number of products.

areas it eventually ousted its "natural" competitor. Contrary to natural chalk the synthetic product is free of contaminants, nor does its composition fluctuate, i.e. the quality always remains the same. However, chalk and PCC as everyday products both feature the same characteristics: Both are white, can be finely reduced, and they have a low hardness value and high volume.

Blackboard chalk

"Chalk is white, rubs off and it can be used for writing" - these are the generally known properties which the geologist Wilhelm Deecke formulated well over one hundred years ago. It is therefore hardly surprising that chalk was widely used in schools, pubs and for writing that could be easily wiped away. And nothing has changed since then, except for the actual manufacture of blackboard chalk.

In the early days, individual chalk crayons and rods were sawn directly out of crude chalk blocks. Towards the middle of the 20th century a new process was introduced in France by which super-fine chalk powder was mixed with a water-based binder to produce a dough-like mass that was pressed through finger-thick nozzles. The resulting extruded rods were then cut to length while the chalk was still moist. Initially this was a manual process but nowadays everything is automated, including drying and assembling packs of 10 or 100 chalk crayons. Such an automatic machine can produce up to 1 million chalk crayons day after day.

The decisive advantage of the modern process is that blackboard chalk is guaranteed not to contain any small stones or sand which could damage the blackboard. However, the same applies to sulphate chalk produced from natural gypsum which, at first sight, is indistinguishable from natural chalk. This type has asserted itself in Germany, Austria and Switzerland over natural chalk which no longer enjoys its monopoly status in most of the other countries in Europe and the world. Only 20% of all blackboard chalk crayons, worldwide, are made of calcium carbonate.

Cosmetics

Whiteness that rubs off, yet still has excellent adhesion properties, are chalk properties that are also useful in cosmetics, for instance as a powder and in make-up. Nevertheless, chalk has never played an important role in cosmetics on account of its

Mixing a batch of green chalk.

Chalk rods are continuously formed by more than 100 parallel nozzles.

More than 1.5 million chalk crayons leave the factory each day.

low hiding power. Women in ancient Greece and Roman even preferred to use poisonous lead white rather than compromise when it was a matter of beauty. And leafy talcum was preferred to fine-grain chalk as a base for powder and coloured make-up, particularly since the ground talcum had a pleasant silky-like feel on the skin. Subsequent generations used mainly vegetable substances such as rice starch as a base for powder.

With the beginning of industrial manufacture of PCC in the USA in the 1920s and the need to find suitable markets for the new product, journals such as "Modern Cosmetics" praised PCC as the ideal constituent of powders:

"Rice powder and other starch types produce an excellent visual effect properly applied. Precipitated calcium carbonate can achieve the same effect, but it is not harmful. It is only the less valuable types that cannot be applied so easily and do not adhere so well. However, there are chalk types that assure significant sliding and distributing properties and tenacity so that they can be applied as a face powder, even in large quantities, without the slightest risk."

And "Perfumes, Cosmetics and Soaps" reported:

"The pleasant calcium carbonate is an excellent ingredient of facial powder. It is light and voluminous, but for reasons of convenience it should not amount to more than 30% by total weight."

Whether or not these advertising efforts were very successful is hard to judge today. However, natural chalk is no longer being used by the cosmetics industry, and even PCC now ranks as only one of many among the pigments as it has long been eclipsed by titanium dioxide and talcum. Precipitated calcium carbonate in significant quantities is now only found in hair-removing creams.

However, PCC has retained its importance in another area of cosmetics, namely in toothpaste.

Advertising in the magazine "Illustrierte Zeitung", volume 143, 1914. Advertising for pharmaceutical products was very popular at the beginning of the 20th century. Some companies assigned as much as 20% of their total turnover to their advertising budget.

Dental care products

Dental care products are first mentioned in Indian and Chinese scripts that are more than four thousand years old. However, tooth care in these civilisations was more a ritual than anything else, particularly when one considers the strange ingredients that were used for this purpose. Mouse excrements, urine and the ashes of wolf and hare skulls are listed alongside plant extracts. Such mundane products as ground egg and oyster shells were first mentioned by Pliny when clean, white teeth gradually became more than a mere ritual, thus making it an expression of beauty and health.

Since the 18th century the range of tooth powders has been extended by inorganic minerals. Chalk, pumice stone and detergents were first listed as tooth care agents towards the middle of the 19th century. From then on the ever-growing demands expected of hygiene and cleanliness resulted in the development and spread of new tooth care agents which were boosted by industrial production methods and targeted advertising for the new products.

In the period between 1860 and 1870 the range was extended by liquid mouth care products and tooth powder, soon to be followed by toothpaste in a tube in the USA in 1895 where it was marketed under the Colgate trade name. The famous toothpaste brands in Germany at the time were Chlorodont and Kalodont. The basic composition of these products was the same everywhere and it has remained almost the same ever since.

The basic substance of all toothpastes are hardly soluble inorganic minerals suspended in water and thickened into a stable paste with appropriate binders. Sweetening and aromatic substances, among them menthol and peppermint oil, merely improved the taste, while detergents or tensides ensured that the paste produced a great deal of foam.

Good taste and foaming were merely psychological factors with a sales promoting effect, while the inorganic minerals performed the cleaning function. They were formulated as fine polishing agents that removed plaque more quickly and thoroughly than would be the case merely with a brush and water.

However, not every mineral is suitable for cleaning and polishing teeth. The hardness of the mineral must be less than that of tooth enamel which has a hardness of 5 according to Mohs's scale. Moreover, the cleaning agent must have a low density and the highest possible volume so that it can bind large quantities of water and aroma substances, but also germs and odours in the mouth. Few substances fulfilled these demands at the outset of the 20th century, particularly since for sales psychological reasons the substances had to be white. For instance, black active carbon would be equally effective, but advertising can hardly establish a plausible and acceptable link between a black cleaning agent and shining white teeth. Consequently, apart from some exotic substances such as marble dust, ground oyster shells and coral powder, all that remained was chalk.

The formulation for Kalodont was typical for toothpastes at the time:

"Equal parts of detergent powder from good neutral soap, the finest possible chalk slurry and glycerin of 28° Bé are all carefully mixed together, and water is added until a semi-fluid mass is obtained. This is then coloured red with a water-soluble detergent or carmine solution, and perfumed with peppermint oil. The mixture is then constantly stirred while it is heated in a porcelain dish placed in a bath of hot water. After it has cooled down, the semi-fluid mass is finally filled into tubes."

However, not all grades of chalk were suitable. It had to be pure white, absolutely free of all sand and have a very small particle size. These demands were fulfilled in an ideal manner by the modified form of Aragonitic crystal because its hardness is lower than that of the calcitic modification, except that it is very rarely encountered in nature. It is therefore hardly surprising that PCC was soon being used for toothpastes, particularly since the crystal modification of the product could be quickly determined by selecting the corresponding precipitation conditions. Moreover, this precipitated chalk was guaranteed to be absolutely pure. By the 1930s PCC had entirely displaced natural chalk, and it remained the most used cleaning agent in tooth care right up to the 1960s. Since several tens of thousands of tons of toothpaste containing between 35 and 55 per cent PCC are consumed each year in the German Federal Republic, this market was extremely lucrative for PCC manufacturers.

However, the PCC star quickly waned when fluorine was added to toothpastes for caries prophylaxis in the 1960s. Although PCC is difficult to dissolve, it is nevertheless not insoluble. Small traces of calcium ions always enter into a solution and these minute quantities are sufficient to instantly precipitate fluoride into insoluble calcium fluoride and thus render it ineffective.

Other disadvantages became apparent during the subsequent period. It was not possible to use synthetic tensides because they formed insoluble lime detergents, while the high pH-value of a calcium carbonate solution of pH 9 not only corroded the cheap, unlacquered aluminium tubes but also had a detrimental effect on the mouth mucosa. The addition of different pyrogenic silicic acids may have offset some of the disadvantages yet, at least in western Europe and the USA, other cleaning agents such as di- and tri-calcium phosphate, sodium metaphosphate, magnesium carbonate and particularly the synthetic silicic acids, were all increasingly replacing PCC.

Today, the amorphic silicic acids dominate in the toothpaste market. At a content of only 10 to maximum 20 per cent, it achieves the same cleaning effect as a conventional toothpaste containing up to 50 per cent cleaning agent. Moreover, amorphic silicic acids do not cause such serious tooth abrasion. Its comparatively high abrasion rate is only one of the numerous reasons why crystalline PCC is rarely used in toothpaste nowadays. There are also certain economic factors which disfavour the use of PCC.

Although PCC is cheaper to purchase than amorphic silicic acids, the price advantage is nevertheless lost in North America and Europe because the raw material is purchased as a bulk mass, yet the sale of toothpaste is based on volume. Consequently, toothpaste manufacturers prefer cleaning agents of a low density such as amorphic silicic acids. However, in many countries of eastern Europe, Africa, Asia and South America, both the purchasing and selling price are related to bulk mass with the result that the comparatively heavier and particularly cheaper PCC is still preferred over silicic acids. The higher abrasion rate is not taken into account, particularly since only few people in these countries can afford to pay DM 4.- to 5.- for a tube of toothpaste based on silicic acids.

It is estimated that the entire German market for dental products consumes only 1800 tons of calcium carbonate per year, and this includes the amount used to produce impressions for dentures.

"Putzi" is one of the few toothpastes that still contains PCC as a cleaning agent. Natural chalk has not yet totally disappeared from the toothpaste market.

Polishing and cleaning agents

The same properties which make chalk and PCC suitable cleaning agents in toothpaste are likewise required for different household polishing agents.

There are two basic ways to remove dirt particles or the rust and oxide layers formed by chemical processes: either chemical or mechanical. And a combination of the two is also possible and forms the basis of most polishing agents. Most chemical cleaning agents are based on detergents and tensides which occasionally also include organic solvents such as turpentine. Mechanical cleaning agents consist of various substances that have a mild abrasive action.

Chalk is used as a polishing agent for metals, tiles and even glass panes since its hardness of 2.8 according to Mohs's scale is significantly lower than the hardness of any of these materials. The only precondition for use is careful processing of the crude chalk - as described by Robert Scherer in his 1922 publication "Chalk, its deposits, quarrying and uses":

"Chalk must feel like the finest flour and it must not contain any hard or sharp particles or grains of sand, no matter how small

they may be. Consequently it must be ground and prepared with the utmost care."

The first time that chalk is mentioned as a cleaning agent dates back to Pliny The Elder who said: "Another chalk is referred to as silver chalk (*creta argentaria*) because it can revive silver's lustre." The use of chalk was widespread during the subsequent centuries. In the 12th century the monk Theophilus Presbyter described in his painter's hand book "Schedula diversarum artium" scraped chalk as a customary cleaning agent. Some 600 years later the German poet Johann Wolfgang von Goethe, when accompanying the Duke of Weimar on his campaign in France in 1792, remarked:

"A soldier merely had to dig a small hole into the ground to find the cleanest white chalk which he required to polish his equipment. The army actually issued an order instructing the soldiers to provide themselves with as much chalk as possible since it cost nothing. This invariably gave rise to ridicule - in the midst of the most terrible filth, the soldiers were supposed to think of cleanliness and provide themselves with cleaning material at a time when they groaned for bread and had to be content with dust."

The heyday of chalk as a cleaning and polishing agent emerged towards the end of the 19th century and the beginning of the 20th century when the hygiene wave also affected households. Scherer mentions in his publication more than 50 different cleaning powders, pastes, detergents, liquid metal cleaning agents, cleaning stones and other "cleaning formulations" which all contained varying amounts of chalk. Even Schleich's face soap containing marble dust was much in demand at the time.

However, after the Second World War, this market steadily declined and, just as was the case with toothpaste, chalk was being rapidly superseded by synthetic silicic acid. Nowadays, in addition to silicic acids, cleaning and polishing agents usually contain amorphic aluminium oxides since their polishing properties are superior to those of chalk and they can be optimally adapted to the given requirements. Nowadays chalk has become a rarity on the shelves of supermarkets and drug stores.

Medicine

The scientific examination of the human body, and with it the study of the physiological effects of individual substances, only commenced in the second half of the 19th century. Over the past millenniums and centuries physicians and healers were guided by their experience in the treatment of diseases and injuries with the result that calcium carbonate formed an integral part of the art of healing throughout this period.

In pre-Christian days the Chinese and Japanese were already using calcium carbonate to stop wounds from bleeding by applying this finely ground mineral. The Greek physician Galen praised the astringent effect of calcium carbonate and calcium sulphate, and even recommended both minerals when patients were coughing blood as a result of pulmonary haemorrhage. Paracelsus extended the use of calcium carbonate to uterus bleeding for which he prescribed a preparation consisting of coral lime.

The store of knowledge of the 18th century was summarized by Johann Georg Krünitz in his "Economic-Technological Encyclopaedia":

"Chalk is recommended as a desiccating and acid neutralising agent, or as an alkaline and adsorbing earth. It can skilfully improve the acidic lymphs of the stomach and is therefore useful for disorders induced by this fault, for instance burning sensations at the entry to the stomach caused by foul acid in the stomach or the so-called heartburn. It helps with severe coughs caused by a sharp mucus. It often stops bleeding and is supposed to kill worms. It is also praised for use against calculus."

For the administration of chalk one should produce a watery suspension of the mineral and it can also be used "externally against shingles and other inflammations, to desiccate wounds and ulcers and for cracked breasts."

Towards the middle of the 19th century calcium carbonate - or more precisely calcium - gained its scientific ordination. This was the time when medicine obtained the first insights into the physiology of calcium metabolism. Gradually it was realised that calcium was present in the entire organism. The share of the element by body weight is just over 2 per cent of which 95 per cent are incorporated in the skeletal system and only 5 per cent in the blood which, on average, contains 10 milligrams of calcium per 100 millilitres of fluid. From the physiological point of view, however, these 5 per cent are far more significant since they regulate both cell activity and metabolism and influence the speed of blood coagulation. This is because the calcium ions are responsible for the formation of fibrin which agglutinates wounds.

Calcium is especially important during pregnancy as the growth and development of the foetus depends upon an adequate supply of calcium. Women require three times the normal amount of calcium during pregnancy. If, therefore, the calcium intake from food or medicine is insufficient then a massive transfer of calcium to the foetus takes place with the result that pregnant women suffer a very considerable loss of calcium, primarily from the skeleton and the teeth. This is because bone calcium is mobilised in connection with increased calcium requirements. Calcium deficiency in babies results in skeletal deformations and growth disorders of the teeth.

All these findings were reflected in the therapeutic measures with calcium preparations. The scientifically substantiated calcium therapy, first drawn up by the American doctor Almroth Edward Wright in 1896, was increasingly refined over the years. Today, calcium is used in medical therapy primarily for two different purposes: For *remineralisation* in connection with calcium metabolic disorders due to rickets or tetany, increased calcium requirements during pregnancy, the lactation period or growth, and for *transmineralisation* in connection with inflammatory processes, allergy disorders, bleeding and disorders of the vegetative nervous system.

Although the aims of therapy have remained the same over the decades, there have been continuous discussions concerning the most appropriate form of calcium administration. Since the organism synthesizes the body's own calcium compounds, such as the apatite of the bones, the decisive criteria of a calcium preparation are the ease with which the calcium can be resorbed and a pleasant – or at least not unpleasant – taste.

After the previously used forms of calcium carbonate – derived from oyster and mussel shells (*Calcarea carbonica, Calcarea ostrearum* and *Testa Ostrya*), crab's eyes (*Lapides Cancrorum or Oculi Cancrorum*), pincers (*Chelae Cancrorum*) and corals (*Corallium Rubrum)* – disappeared from the market, the carbonates available at the beginning of the 20th century were mainly in the form of natural chalk and PCC. In pharmaceutics these were known as *Creta praeparata* and *Calcis carbonis praeparatus.*

The calcium chloride that had dominated in oral therapy until then was increasingly being rejected and encountering intense criticism as the watery solution not only had a very unpleasant, salty, bitter and sharp taste but it also caused indigestion.

Yet calcium carbonate was still unable to win this market. Although it was neutral in taste and could be easily resorbed as a constituent of mineral water - in Austria in the 1920s there was a mineral water known as "Biokalk" ['biological lime'] which contained 3.97 g of calcium hydrogen carbonate per litre and was salt-free - the real breakthrough as a calcium preparation was achieved by calcium gluconate developed by the Swiss Sandoz pharmaceutical concern in 1927. And this has remained the most widespread application form of calcium until today.

Calcium carbonate has still retained a certain measure of importance for treating hyperacidity of the stomach, but a pure calcium carbonate preparation, such as 'Renocal', is nevertheless an exception. This contrasts with 'Antacidum' which unites a phosphate binder and a calcium supplement in a single preparation. Normally, the pharma-

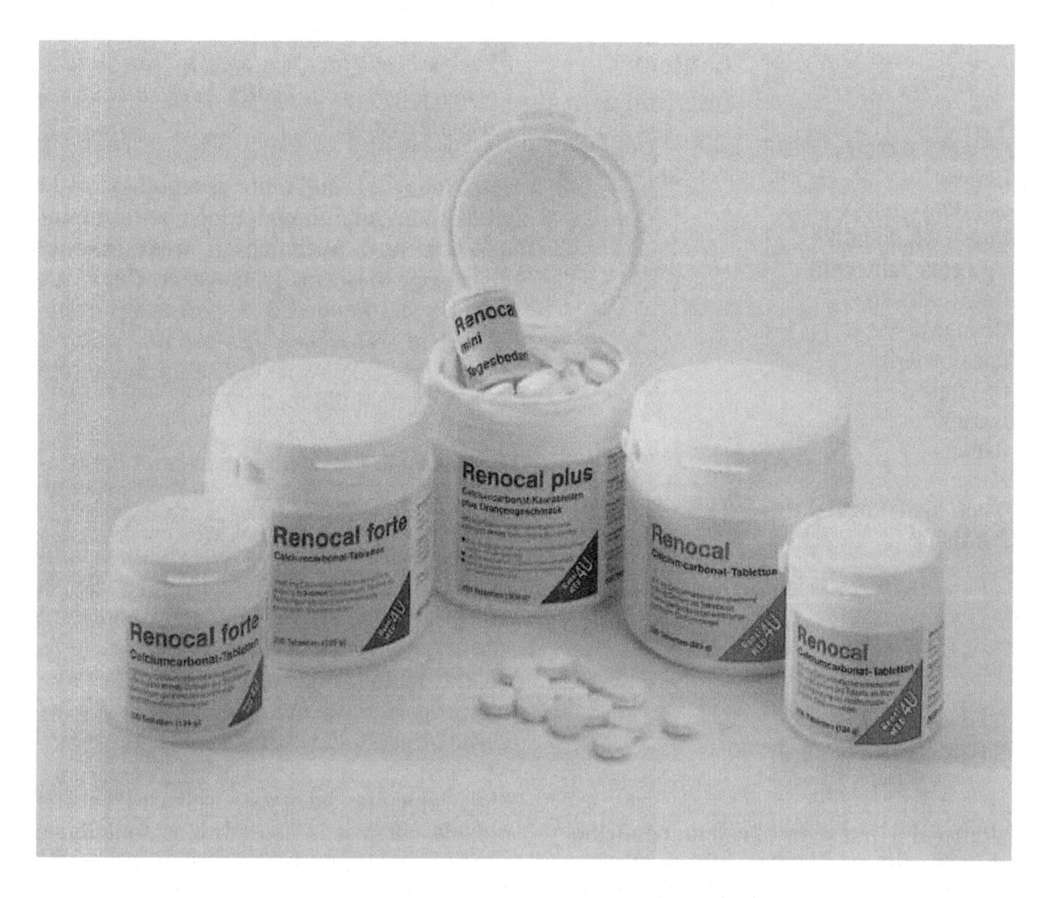

Calcium carbonate chewing tablets.

ceutical industry is only using this white mineral as a carrier material or as an auxiliary in tablets.

Nutrition

Attempts to incorporate calcium carbonate in foods to prevent the classic deficiency symptoms such as osteoporosis, decalcification of the teeth and growth disorders in babies and infants, have likewise failed – at least in Germany. The efforts of the German limestone industry to promote the importance of calcium in nutrition in the 1920s did not change this situation, nor did the book "The Cultural History of Limestone", published in 1923 with the support of the German Lime Association, in which the author Hans Urbach referred to 'lime hunger':

"Lime hunger also arises in people in the event of severe calcium deficiency, for instance in children who eat chalk."

He also described in vivid terms what an adequate supply of calcium can supposedly achieve:

"The limestone-rich state of Kentucky not only supplied the sturdiest and most capable soldiers in the American civil war, but also the best cattle."

However, there were also serious arguments favouring the addition of calcium carbonate in foods. Although normal nutrition can easily cover the requirements of just under 1 gram of calcium per day, poor nutrition can easily result in calcium deficiency. For

Food	Calcium Content [mg/100 g]
Cheese	860
Condensed milk	285
Cow's whole milk	123
Egg yoke (chicken)	141
Beans (dried)	125
Oranges	86
Peas (dried)	82
Oat flakes	53
Barley	44
Bread	32
Meat (veal, beef, pork)	12
Potatoes	6

Nutrients table.

instance the first signs of calcium deficiency arose in the post-war period in Germany when the daily calcium intake was barely 0.2 to 0.4 gram. This was due to the fact that arable land had become totally depleted of lime. Just under one quarter of the entire arable land had optimal lime levels while all the remaining land had become drained of lime due to over-cropping during the war years. This meant that crop plants were also lime deficient and this continued through the food chain with the result that the population was suffering from calcium deficiency.

Since good experiences had been gained in the USA and Great Britain with the addition of calcium carbonate to bread, the addition of 0.28 per cent chalk in flour was ordered in the British Occupation Zone of Germany in spring 1947. And it was only natural that chalk, i.e. calcium carbonate, was used for this purpose since it was available in large quantities and was cheap. Moreover, it did not influence the taste of the bread nor did it represent a health hazard as opposed to the claim made by Krünitz in his encyclopaedia:

"Others added chalk or slaked lime [...] to poor flour [...] to make the bread whiter. But all these bread falsifications are certainly to the detriment of our health, even causing serious poisoning."

Even today it is still quite acceptable to add chalk to bread although much better methods are now available to treat calcium deficiency. In Germany, however, chalk was withdrawn from bread as soon as the arable land had recovered. Nowadays, natural chalk and PCC play a very subordinate role as a constituent in food.

According to the German Food and Requisites Law (LMGB) the number E 170 (chalk, calcium carbonate) is a "technical auxiliary" that is a "natural colourant (white), acidification and free flowing agent" that is used in the production of chewing gum and curd cheese and for the decoration of foods, or as an anti-lumping agent to maintain the flowing properties of table salt and other powdered or crystalline substances.

Since calcium carbonate releases carbon dioxide when it is heated, it is sometimes used as a raising agent in baking powders. Pure precipitated grades are used to reduce the acid content of wine. The neutralisation of wine's acidity is based on a very simple reaction:

$$CaCO_3 + C_4H_6O_6 \rightarrow CaC_4H_4O_6 + CO_2 + H_2O$$

The resulting calcium tartrate precipitates as a hardly soluble salt which can be easily filtered out.

All told, the use of calcium carbonate in foods could be easily discounted if it were not for the sugar industry. Less than 1000 tons of calcium carbonate are consumed annually in Germany for baking powder, as a free-flowing agent or for rice hulling in contrast to the 400,000 tons that are used by the sugar industry.

For more than 100 years calcium carbonate has been used in sugar production to purify "crude sugar juice" by removing all non-sugar substances such as albumens and oxalic acid. Strictly speaking calcium carbonate is not actually used but rather calcium oxide is added to the "crude juice", followed by carbon dioxide, to precipitate all non-sugar substances as insoluble slurry which is then filtered out and processed into high-quality lime fertilisers. The purified sugar is then transformed into the final product by a succession of individual processing stages.

Since calcium oxide and carbon dioxide are formed when calcium carbonate is fired, many sugar mills have set up their own firing furnaces to cover their own requirements. These are quite considerable since between 230 and 300 kg of calcium carbonate are required per ton of sugar. And as in the case throughout the food industry, the demands expected of the purity of the used limestone are very high.

Precipitated calcium carbonate to deacidify wine and fruit juices.

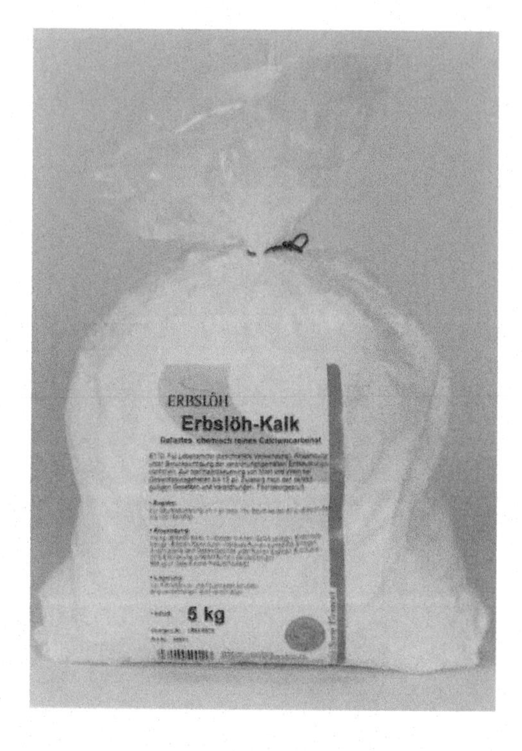

ANNEX

Bibliography

General

K. Werner Barthel, "Solnhofen – Ein Blick in die Erdgeschichte", Thun, 1978.

Robert S. Boynton, "Chemistry and Technology of Lime and Limestone", 2nd ed., New York/Chichester/Brisbane/Toronto, 1980.

DIN-Taschenbuch 49, Farbmittel 1 (Pigmente, Füllstoffe, Farbstoffe) [contains DIN 5033-1 through DIN 55929], 4th ed., Berlin, 1993.

DIN-Taschenbuch 157, Farbmittel 2 (Pigmente, Füllstoffe, Farbstoffe) [contains DIN 55943 through DIN 66131, DIN-EN and DIN-EN-ISO standards], 3rd ed., Berlin, 1993.

Walter Döbling, "Chemisches vom Kalk", Berlin, 1924.

Gmelins Handbuch der Anorganischen Chemie – Calcium, 8th ed., Weinheim, 1956/1957.

Rudolf Gotthardt and Werner Kasig, "Karbonatgesteine in Deutschland – Rohstoff, Nutzung, Umwelt", Düsseldorf, 1996.

Anton Herbeck, "Der Marmor – Entstehung, Arten, Gewinnung, Vorkommen", München, 1953.

ISO Manuals, volume 2: "Raw materials (pigments, extenders, binders, solvents)", Berlin, 1994.

Johann Georg Krünitz, "Oekonomisch-technologische Encyklopädie oder allgemeines System der Stats-, Stadt-, Haus- und Land-Wirthschaft und der Kunstgeschichte", volumes 1-242, Berlin, 1782-1856.

Olaf Lückert, "Pigment + Füllstoff: Tabellen", 5th ed., Laatzen, 1994.

Joseph A. H. Oates, "Lime and Limestone – Chemistry and Technology, Production and Uses", Weinheim, 1998.

Plinius C. Secundus the Elder, "Naturalis Historiae Libri XXXVII [Natural History in 37 volumes]", edited and translated by Roderich König, Darmstadt, 1978 (35), 1984 (33), 1992 (36), 1994 (17).

Plüss-Staufer AG Oftringen (ed.), "OMYA Kontakte".

Eberhart Schiele and Leo W. Berens, "Kalk – Herstellung, Eigenschaften, Verwendung", Düsseldorf, 1972.

Egon Trümperer, "Mineralogisches vom Kalk", Berlin, 1927.

Ullmanns Encyklopädie der technischen Chemie, 4th ed., Weinheim/New York, 1977 (13/14), 1978 (15), 1979 (18).

Hans Urbach, "Der Kalk in Kulturgeschichte und Sprache", Berlin, 1923.

Hans Urbach, "Die Verwendung des Kalkes", Berlin, 1931.

Hans Vogel, "Der Kalk und seine Bedeutung für die Volkswirtschaft", Stuttgart, 1941.

Winnacker, Karl (init.), "Chemische Technologie", volumes 1 - 6, Munich, 1981-1986.

Karl M. Zittel, "Die Kreide", Berlin, 1876.

I. Geology of calcium carbonate

Anthony E. Adams et al., "Atlas of sedimentary rocks under the microscope", London, 1984.

Calcit, Lapis extra, no. 14, Munich, 1998.

Peter W. Harben, "The Industrial Minerals Handy Book", 2nd ed., London, 1995.

Peter W. Harben and Milos Kuzvart, "Industrial Minerals – a Global Geology", London, 1997.

Jean Jung, "Precis de pétrographie", Paris, 1963.

Rolf Langbein et al., "Karbonat- und Sulfatgesteine. Kalkstein – Dolomit – Magnesit – Gips – Anhydrit", Leipzig, 1982.

Werner Lieber, "Calcit – Baustein des Lebens", Munich, 1990.

Maurice Mattauer, "Ce qui disent les pierres", Paris, 1998.

Paolo Orlandi and Marco Franzini, "Minerali del marmo di Carrara", Milan, 1994.

P. W. Scott and A. C. Dunham, "Problems in the evaluation of limestone for diverse markets", Proceedings of the Sixth Industrial Minerals International Congress, pp. 1-21, Toronto, 1984.

Helmut G. F. Winkler, "Petrogenesis of metamorphic rocks", 5th ed., Berlin/New York/ Heidelberg, 1979.

Bruce W. D. Yardley et al., "Atlas of metamorphic rocks and their textures", London, Stuttgart, 1990.

Journals

Industrial Minerals, London (UK).

Mines & Carrières – Revue de l'industrie minérale, Paris (France).

Zeitschrift für Angewandte Geologie, Hanover (Germany).

II. The Cultural History of Limestone

Anonymous, "Il marmo ... ieri e oggi", Carrara, 1970.

Robert Bedon, "Les Carrieres et les Carrieres de la Gaule Romaine", Paris, 1984.

Friedrich Behn, "Steinindustrie des Altertums", Mainz, 1926.

Norman Davey, "A History of Building Materials", London, 1961.

Deutscher Naturwerkstein-Verband DNV (ed.), "Naturstein und Architektur. Materialkunde – Anwendung – Steintechnik", Munich, 1992.

Hazel Dodge and Bryan Ward-Perkins (eds.), "Marble in Antiquity. Collected Papers of J. B. Ward-Perkins", Archaeological Monographs of the British School at Rome, no. 6, London, 1992.

J. Clayton Fant (ed.), "Ancient Marble Qarrying and Trade", British Archaeological Reports International Series, 453, Oxford, 1988.

Jan Gympel, "Geschichte der Architektur von der Antike bis heute", Cologne, 1996.

Norman Herz and Marc Waelkens (eds.), "Classical Marble – Geochemistry, Technology, Trade", NATO Advanced Science Institute Series E (Applied Sciences), vol. 153, Dordrecht/Boston/London, 1988.

Werner Kasig and Benno Weiskorn, "Zur Geschichte der deutschen Kalkindustrie und ihrer Organisationen", Forschungsbericht, Düsseldorf, 1992.

Luciana Mannoni, "Marmor – Material und Kultur", Munich, 1980.

Reclams Handbuch der künstlerischen Techniken, vols. 1-3, Stuttgart, 1984-1990.

Giorgio Vasari, "Künstler der Renaissance", Berlin, 1948.

Marcus Pollio Vitruvius, "De architectura libri decem [Ten Books on Architecture]", translated by Dr. Curt Fensterbusch, Darmstadt, 1964.

III. Calcium Carbonate – A Modern Resource

Carl Breuer, "Kitte und Klebstoffe", Bibliothek der gesamten Technik, vol. 33, Hanover, 1907.

Walter H. Duda, "Cement Data Book", vols. 1-3, Wiesbaden/Berlin, 1985.

Karl Höffl, "Zerkleinerungs- und Klassiermaschinen", 2nd ed., Hanover, 1993.

Hans Kellerwessel, "Aufbereitung disperser Feststoffe", Düsseldorf, 1991.

Otto Labahn, "Ratgeber für Zementingenieure", Wiesbaden/Berlin, 1982.

André Moussy, "La craie et l'industrie du blanc dans le département de la Marne", Châlons-en-Champagne, 1928.

Paul Ney, "Zetapotentiale und Flotierbarkeit von Mineralien", Vienna/New York, 1973.

Rheinische Kalkwerke (eds.), "Wülfrather Taschenbuch für Kalk und Dolomit", Wülfrath, 1974.

Robert Scherer, "Die Kreide – Deren Vorkommen, Gewinnung und Verwertung", Chemisch-technische Bibliothek, vol. 372, Vienna/Leipzig, 1922.

Gert Schubert, "Aufbereitung fester Stoffe", vols. 1-3, Leipzig, 1989.

Gert Schubert, "Mechanische Verfahrenstechnik", Leipzig, 1990.

Alfred Peter Wilson, "Precipitated Chalk. History, Manufacture and Standardization", 2nd ed., Birmingham, 1935.

Journals

Aufbereitungstechnik, Wiesbaden (Germany).

Industrial Minerals, London (UK).

Mines & Carrières – Revue de l'industrie minérale, Paris (France).

ZKG-International, Walluf (Germany).

IV. Industrial use of calcium carbonate

4.1 Paper

Team of authors, "Lehrbuch der Papier- und Kartonerzeugung", Leipzig, 1987.

Werner Baumann and Bettina Herberg-Liedtke, "Papierchemikalien", Berlin, 1993.

Dan Eklund and Tom Lindström, "Paper Chemistry – an Introduction", Grankulla, 1991.

Lothar Göttsching and Casimir Katz, "Papier-Lexikon", Gernsbach, 1999.

Robert W. Hagemeyer, "Pigmets for Paper", Atlanta, 1997.

Hans Kotte, "Welches Papier ist das?", Heusenstamm, 1972.

Wilhelm Sandermann, "Papier – Eine Kulturgeschichte", Berlin, 1997.

William E. Scott and James C. Abbott, "Properties of Paper – an Introduction", Atlanta, 1995.

Wolfgang Walenski, "Das Papier. Herstellung, Verwendung, Bedruckbarkeit", Itzehoe, 1999.

Jan C. Walter, "Coating Processes", Atlanta, 1995.

Journals

American Papermaker Magazine, Chicago (USA).

Das Papier, Heidelberg (Germany).

Demand – Supply Report / Newsprint and Magazine Paper Grades, Zurich (Switzerland).

European Papermaker Magazine, Surrey (UK).

JPPS – Journal of Pulp and Paper Science, Ontario (Canada).

ipw Internationale Papierwirtschaft – International Paperworld, Heusenstamm (Germany).

Papermaker Asia Pacific, Bondi Junction (Australia).

PIMA's North American Papermaker, Mount Prospect (USA).

PPI Fact & Price Book, San Francisco (USA).

Pulp and Paper International, Brussels (Belgium).

TAPPI Journal, Atlanta (USA).

Wochenblatt für Papierfabrikation, Frankfurt (Germany).

4.2 Plastics

Gerhard W. Becker et al. (eds.), "Kunststoff-Handbuch", 2nd ed., Munich, 1986-1998.

"Fine Carbonate Fillers", Industrial Minerals, April 1995, p. 11.

Reinhard Gächter and Helmut Müller, "Plastics Additives Handbook", 4th ed., Munich, 1993.

Harry S. Katz and John V. Milewski, "Handbook of Fillers for Plastics", New York, 1987.

Georg Menges, "Werkstoffkunde Kunststoffe", Munich/Vienna, 1990.

"Minerals and Polymers – High performance, high value", Industrial Minerals, June 1998, pp. 73 ff.

Hansjürgen Saechtling, "Kunststoff Taschenbuch", Munich/Vienna 1996.

Georg Wypych, "Handbook of Fillers", Toronto/New York, 1995.

Journals

Adhäsion – Fachzeitschrift für Kleben und Dichten, Munich (Germany).

GAK – Gummi, Fasern, Kunststoffe, Ratingen (Germany).

European Plastics News, Croydon (UK).

Modern Plastics International, Lausanne (Switzerland).

Kunststoffe, Munich (Germany).

Plaste und Kautschuk, Leipzig/Stuttgart (Germany).

Plastverarbeiter, Heidelberg (Germany).

4.3 Paints and coatings

Thomas Brock et al., "Lehrbuch der Lacktechnologie", Hanover, 1998.

Heinz Dörr and Franz Holzinger, "Kronos Titandioxid in Dispersionsfarben", Kronos Titan GmbH.

Artur Goldschmidt et al., "Glasurit-Handbuch der Lacke und Farben", 11th ed., Hanover, 1984.

Hans Kittel (ed.), "Lehrbuch der Lacke und Beschichtungen, volume II: Pigmente, Füllstoffe, Farbstoffe", Berlin, 1974.

Wilfried Morley Morgans, "Outlines of Paint Technology", 3rd ed., London, 1990.

Paolo Nanetti, "Lackrohstoffkunde", Hanover, 1997.

Gerald Patrick Anthony Turner, "Introduction to Paint Chemistry and Principles of Paint Technology", 3rd. ed., London, 1988.

Ulrich Zorll (ed.), "Römpp Lexikon Lacke und Druckfarben", Stuttgart/New York, 1998.

Journals

American Paint & Coatings Journal, St. Louis (USA).

Applica, Wallisellen (Switzerland).

European Coatings Journal, Hanover (Germany).

Farbe und Lack, Hanover (Germany).

Journal of Coating Technology, Blue Bell (USA).

Modern Paints and Coatings, New York (USA).

Phänomen Farbe, Düsseldorf (Germany).

Pigment and Resin Technology, London (UK).

Surface Coatings International, London (UK).

Welt der Farben, Cologne (Germany).

4.4 Calcium Carbonate – A Versatile Mineral

Rüdiger Bartels et al., "Kalkbedarf von Marschböden", VDLUFA Schriftenreihe, vol. 16, pp. 295-311, Darmstadt, 1985.

Bundesarbeitskreis Düngung (BAD) im Industrieverband Agrar e.V. (eds.), "Grundlagen der Düngung", Frankfurt, 1998.

Franz Greiter, "Aktuelle Technologie in der Kosmetik", Heidelberg, 1987.

Johannes Grimm and Knut Caesar, "Einfluß der Bodennutzung auf die langfristige Entwicklung von Fruchtbarkeit und Ertragsfähigkeit sandiger Böden", Ökologische Hefte 7, p. 35 ff, 1997.

Reinhold Gutser et al., "Kalk- und Magnesiumwirkung kohlensaurer Kalke mit unterschiedlichem Vermahlungsgrad", VDLUFA Schriftenreihe, vol. 33, pp. 323-328, Darmstadt, 1991.

International Lime Association (ILA), "Comparison of Different Liming Materials for Agricultural Use", Cologne, 1994.

Kalkdienst (ed.), "Der Wald braucht Kalk", 3rd ed., Cologne, 1959.

Kalkdienst (ed.), "Düngekalk – Leitfaden für Wirtschaftsberater", 4th ed., Efferen, 1965.

Manfred Kerschberger et al., "Beziehungen zwischen Kalkdüngung, Pflanzenertrag und Pflanzenqualität", VDLUFA Schriftenreihe, vol. 37, p. 591 ff., 1993.

Arno Mönkemeyer, "Der Markt für Kalkdüngemittel", Emsteten, 1928.

Albert Orth, "Kalk- und Mergeldüngung", Anleitungen für den praktischen Landwirt (edited by Deutsche Landwirtschaftsgesellschaft), pp. 1-50, 1896.

Norbert Peschen, "Reaktive Kalkprodukte für die Trinkwasseraufbereitung – Herstellung und Qualitätskriterien", bbr wasser und Rohrbau, no. 2, p. 3 ff., 1998.

Karlheinz Schrader, "Grundlagen und Rezepturen der Kosmetika", Heidelberg, 1979.

Gebhard Schüler, "Bodenschutzkalkung und deren Auswirkung auf Sickerwasser, Boden und Bodenbiozönose", Waldschäden, Boden- und Wasserversauerung durch Luftschadstoffe in Rheinland-Pfalz (edited jointly by the Rhine-Palatinate Ministry of Agriculture, Viticulture and Forestry and the Ministry of the Environment), pp. 117-131, Mainz, 1993.

Gebhard Schüler, "Stabilitätserhöhung im Ökosystem Wald durch Bodenschutz, Kompensation von Nährstoffverlusten und naturnahe Waldbewirtschaftung", Waldschäden, Boden- und Wasserversauerung durch Luftschadstoffe in Rheinland-Pfalz (edited by the Rhine-Palatinate Ministry of Forestry and the Environment), pp. 74-96, Mainz, 1997.

Manfred Schütz, "Stand der Rauchgasentschwefelungstechnik", VGB Kraftwerkstechnik, p. 943 ff., 1997.

Wilfried Umbach, "Kosmetik – Entwicklung, Herstellung und Anwendung kosmetischer Mittel", Stuttgart, 1988.

Wilhelm Windisch et al., "Calcium – Bioverfügbarkeit verschiedener organischer und anorganischer Calcium-Quellen", Journal of Animal Physiology and Animal Nutrition 77 (1997), p. 189 ff.

Wissenschaftlicher Beirat der Sandoz AG (ed.), "Calcium – Physiologie, Pharmakologie, Klinik", Basel, 1952.

Definitions and Measurement Methods

Identification of successful processing

It is not always possible for the seperating process to achieve a 100% distinction between the classification and grading. Thus it might happen that the crushing behaviour of two minerals in a raw material may differ. Let us look at a brittle, weathered limestone which contains quartz. If it is ground the softer mineral limestone is size-reduced better than the hard quartz. During subsequent screening the less reduced quartz becomes concentrated in the coarse fraction on the screen while the finer material that has passed through the screen is depleted of quartz.

During classifying, therefore, a sorting process has also taken place. By analogy, sorting can also lead to classifying. Such effects are undesirable if classifying alone is the objective. They can, however, also be utilised to support sorting with classifying.

The success of separation, either by classifying or by sorting, is described by the yield. A distinction is drawn between two types of yield:

- mass yield
- valuable product yield

The mass yield indicates how big the relative cut produced by a separation is. Thus, from the feed mass flow ma a screening results in two products: the coarse oversize mg and the fine undersize mf. The following formula applies:

$$\dot{m}_a = \dot{m}_g + \dot{m}_f \qquad (1)$$

or if the products for the feed are set in ratio

$$\frac{\dot{m}_g + \dot{m}_f}{\dot{m}_a} = 1 \qquad (2)$$

or

$$v_g + v_f = 1 \qquad (3)$$

where v_g is the mass yield in the coarse material and v_f the mass yield in the fine material.

This says nothing about the quality of separation, but only something about the quantity ratios. The quality of separation describes which proportion of desired valuable material enters the two product flows. This is explained below with reference to a screening.

After size-reduction of a limestone to 0-10 millimetres it is to be separated by screening at 4 millimetres in order to then feed the fines < 4 mm to a ball mill. To optimise screening, i.e. to obtain a large throughput on as small a screen area as possible, a screen deck is selected with a mesh width just bigger than 4 millimetres. Oversize > 4 mm will not be obtained which is completely free of < 4 mm material, and in the same way undersize < 4 mm will not be completely free of > 4 mm material. If the portion of the 0-4 mm fraction in the feed is designated as "a", that in the coarse material as "b" and that in the fine material as "c", the following applies by analogy to (1)

$$\dot{m}_a * a = \dot{m}_g * b + \dot{m}_f * c \qquad (4)$$

or

$$a = \frac{\dot{m}_g * b}{\dot{m}_a} + \frac{\dot{m}_f * c}{\dot{m}_a} \qquad (5)$$

or

$$a = v_g * b + v_f * c \qquad (6)$$

and

$$1 = v_g * b/a + v_f * c/a \qquad (7)$$

or

$$1 = f_g + f_f \qquad (8)$$

where f_g is the valuable product yield in the coarse material and f_f the valuable product yield in the fine material.

From equation 3 and 6 the mass yield can be calculated as

$$v_g = \frac{a-c}{b-c} \qquad (9)$$

or

$$v_f = \frac{a-b}{c-b} \qquad (10)$$

and then from (7) the valuable product yield in the cuts as

$$f_g = v_g * b/a \qquad (11)$$

in the example presented the 0-4 mm portion which has remained in the coarse material, and

$$f_f = v_f * c/a \qquad (12)$$

the portion of the fraction 0-4 mm which has entered the fine material as desired.

These equations can be used for classifying as well as for sorting, because it is all the same whether a grain in certain limits or a mineral being won is designated as "valuable product".

The terms mass yield and valuable product yield as well as the contents of the valuable material in the product flows do not adequately describe separation. The products considered here always consist of a particle size distribution (see 2.1). It is evident that during classifying on a screen not all particle classes behave the same. To stay with the example presented above, the particles which are only slightly smaller than the mesh width are discharged much more easily in the coarse material during screening than the particles which are very small compared with the mesh width. This means that the individual fractions in a grain collective behave differently from each other during separation.

To better indicate the quality of separation, the mass yield in the individual fractions is calculated and plotted against the particle size. This shows the probability of a certain grain class in the feed grain entering the desired product (see figure). The separation particle size of this classifying is then defined as the grain class which enters both product flows in equal parts. This separation curve can naturally be calculated not only for a separation according to particle size but also for any other separation feature such as density. An ideal separation is indicated by a vertical line for the separation particle size as all particles which are bigger than the separation particle size enter the coarse material and all

Typical separation curve in practice. Separation takes place at X3.

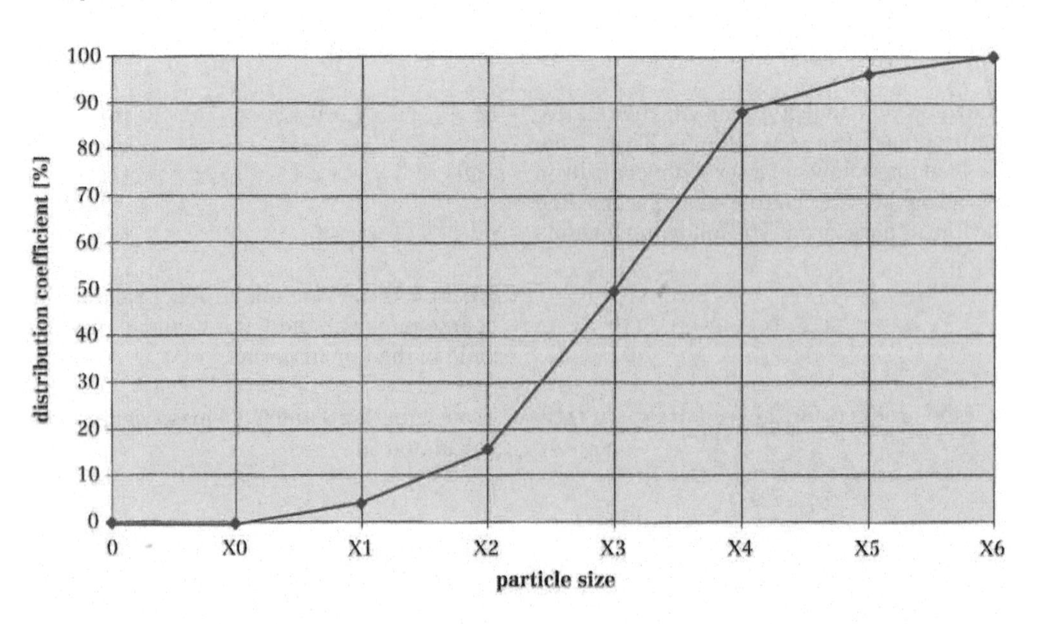

particles which are smaller enter the fine material. The steepness of the separation curve and its shape are applied for assessing a separation.

Determination of particle size distribution: Screening (a), density curve (b) and cumulative curve (c).

Screen tower	Screen No.
00	X n
0000	...
00000	X i+1
000000	X i
0000000	...
000000	X 1
:::::::::::::	X 0

a)

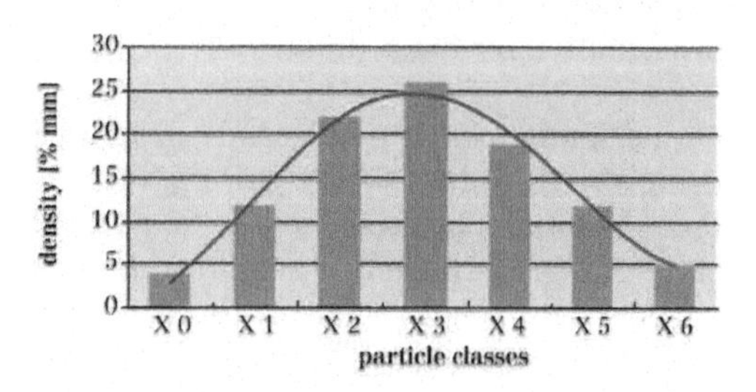

b)

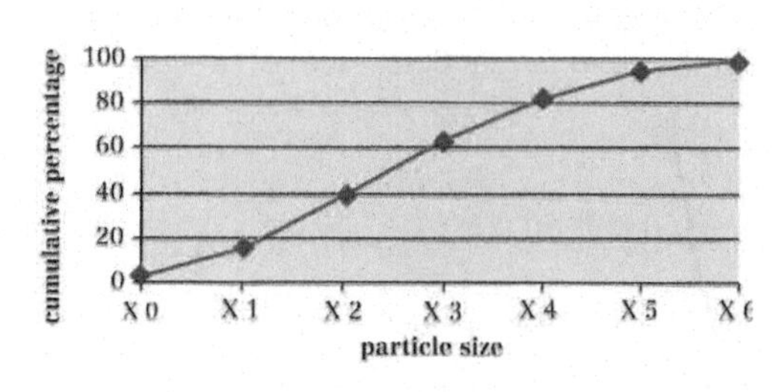

c)

Measurement methods

To be able to describe the properties of fillers the most important methods should be described with which these properties can be indicated.

Particle size distribution

A grain collective is described by its distribution of particle sizes. For this purpose the particle size range is divided into various classes. This is most simply described by imagining a series of screens on which a material is being screened. On the screen in each case with the smaller opening the grains remain lying which have passed through the screen above but which cannot pass through the screen with the smaller opening. The mass caught between the two screens is weighed and registered. The material which passes through the smaller screen is screened on the next-smaller screen, and so on (see figure a).

Finally, for the distribution the mass is obtained which is retained on each screen as well as that which falls through the smallest screen x_7 and is collected on the screen deck (xo). For each screen class between $x = x_{i+1}$, and $x = x_i$ of $x_o<x<x_n$, a value of m_o to m_n is obtained. As it is not appropriate to apply the absolute weight values, a comparability of results would not be possible, the relative masses $q_i = m_i/\Sigma m_i$ are formed.

As other variables than the mass can be envisaged (number, length, surface area or volume) these variables are given different indices in order to identify the dimensions of the variables in this way:

$q_{0.i}$ = number distribution
$q_{1.i}$ = length distribution
$q_{2.i}$ = surface area distribution
$q_{3.i}$ = mass distribution (volume or density included)

If in our example the relative masses $q_{3.i}$ are plotted against the respective class mean $(x_{i-1} + x_i)/2$ (see figure b), a component bar chart is obtained whose compensating curve represents the density distribution of the particle sizes.

A further representation is the cumulative curve of a distribution, in which the quantity portions are totalled up and plotted against the respective particle size value: $Q_{3.I}=\Sigma q_{3.i}$. This obtains the cumulative or passing distribution (see figure c).

The representation in the passing distribution is the most common form. To appropriately represent a wide particle size range, the particle sizes are frequently presented logarithmically. There are also other types of representation for the ordinates.

A distribution is in most cases described using several values. The x_{50} or d_{50} value is thus the **mean particle size**, i.e. 50 percent of all grains are bigger or smaller than this particle size. The d_{98} value or **top cut** is the particle size at which 98 percent of all grains are smaller than the stated diameter.

The calculation of the particle size distribution was represented here on the basis of a screen analysis. This can only be effectively used up to a particle size of around 20 micrometres. Fillers frequently exhibit very much smaller particle sizes, so that other measurement methods have to be applied. These include sedimentation techniques such as Sedigraph or laser diffraction methods. Using these techniques particle sizes down to approx. 0.5 micrometres can be measured. Finer particle sizes require scanning electron microscopy or other techniques.

Whiteness

Apart from particle size distribution, the whiteness of a product is decisive for its use (see figure). To determine this a tablet of the corresponding material is produced on a specially designed press and in a measuring device the intensity of the reflection of red, green and blue light of precisely defined wavelengths is then measured (Elrepho). The result is stated as a percentage of a standard value (the $BaSO_4$-Normal).

For the dry fillers the green value R_y is given and for the wet fillers the blue value R_z or TAPPI value R457. For the definition of the

Whiteness of different fillers.

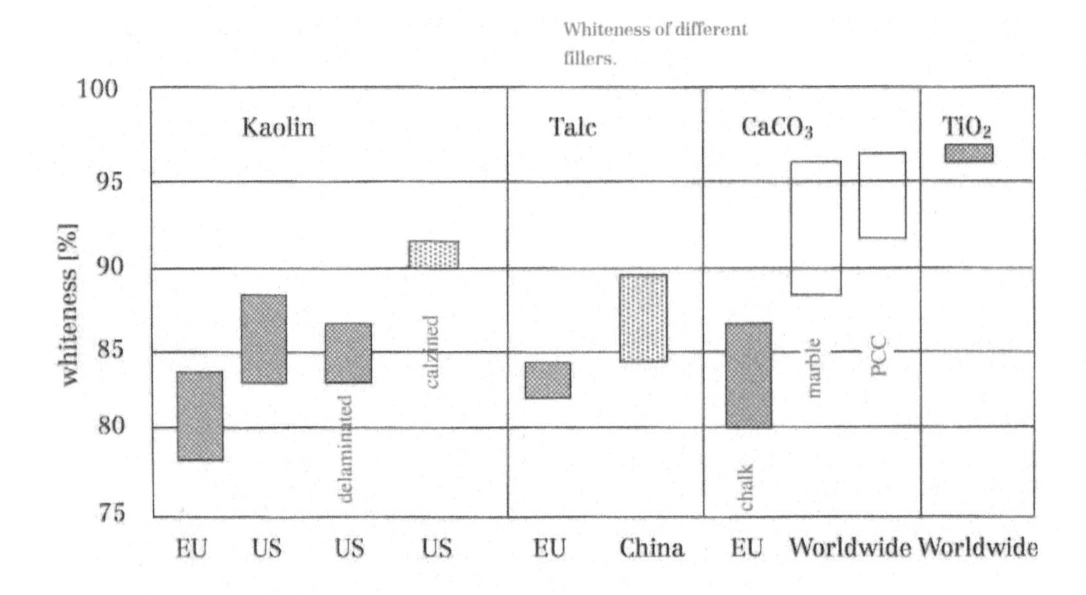

Product type	Raw material	Fineness [% < 2 µm]	Whiteness [% Tappi]	[% Ry]
Slurry	Chalk	60	89 - 91	
	Limestone	60	90 - 93	
	Marble	60	93 - 96	
	Marble	90	93 - 96	
	Marble	95	93 - 96	
	Marble	90 % < 1 µm	93 - 96	
Flour	Chalk	70		85 - 87
	Chalk	40		82 - 84
	Limestone	40		88 - 90
	Limestone	20		83 - 85
	Limestone	15		84 - 86
	Marble	80		90 - 92
	Marble	60		91 - 93
	Marble	40		93 - 95
	Marble	25		92 - 94
	Marble	15		90 - 92

Raw materials, grain sizes and whiteness of various calcium carbonate products.

yellow hue of the product the yellow value is determined:

$$W = \frac{R_x - R_z}{R_y} \cdot 100$$

The greater this value is, the more yellow is the hue of the material.

Opacity

If a crystalline body like marble is illuminated by light, three phenomena occur:

- **Reflection** – is the amount of light reflected by the crystal to the light source. Complete reflection is equivalent to a mirror.
- **Absorption** – is the amount of light which penetrates a crystal and is converted into heat. Complete absorption makes a body look black.
- **Transmission** – is the amount of light which passes through the crystal and is perhaps refracted but not otherwise influenced. Complete transmission makes the body appear invisible.

The total of the three phenomena is 1 because the nothing can be lost from or added to a ray of light:

$$r + a + t = 1$$

The transmitted light can be reflected by an object behind the crystal and penetrate back through the crystal. Thus we can see what is behind the crystal. A filler used either in paper or paint is required to have high hiding power, i.e. it should reflect as high a proportion of the incident light and only absorb a small part of it, while as far as possible not transmitting any of it as this reduces opacity (see figure).

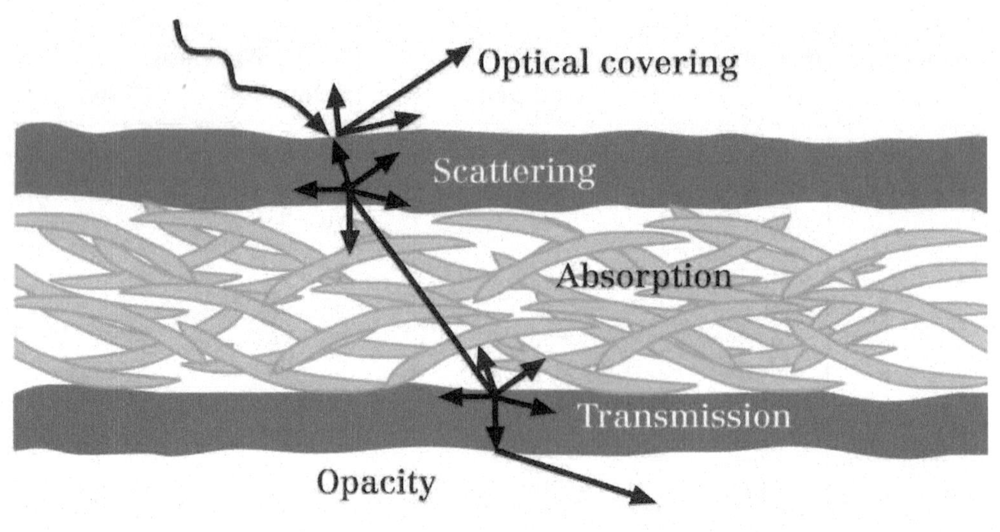

Interaction of light with paper.

To determine the hiding power the opacity is measured. To this end, a coating of the filler is applied in a specific thickness to one black and one white surface. The two samples are then compared. As this measurement can be easily influenced by subjective factors when making and assessing the samples, it is necessary to prepare the samples very carefully in order to obtain as objective a conclusion as possible.

Chalk (a), GCC (b) and PCC (c) under the scanning electron microscope.

Miscellaneous

In the paper industry the abrasivity of the filler is important. This variable, which depends on the particle size and on the chemical composition and the origin of the calcium carbonate, can only be determined indirectly. Thus, chalk is less abrasive than limestone and limestone is in turn less abrasive than marble, for the same particle size and chemical purity. In addition, quartz and also dolomite constituents increase abrasivity owing to the high hardness of these materials.

Information on the causes of the different behaviour of various calcium carbonate varieties is provided by scanning electron micrographs of the individual product particles. The scanning electron microscope renders the grain shapes and even the particle size differences in the individual products visible.

Glossar

Abrasion
Mechanical wear to a surface when subjected to rubbing, scraping or other mechanical means.

Agglomerates
An assemblage of primary particles, aggregates or a mixture of the two which may be broken down during normal paint-making processes.

Amorphous
Solids whose molecules are not arranged regularly in the crystal lattice.

Anisotropy
(from the Greek *anisos* = unequal and *tropos* = direction) Dependence of the properties of a medium on the direction in which they are determined.

Aspect ratio
Ratio of the maximum length of an individual particle to its thickness.

atro
(= German 'absolut trocken' = absolutely dry) Material without any residual moisture.

Calcination
Decomposition of a chemical compound or driving off water of crystallisation from minerals by heating.

Calcrete
Lime or dolomite encrustations which form on the surface as a result of evaporation of the water in dry areas.

Chalking
The appearance of a loosely adherent fine powder on the surface of a paint coating, arising from the degradation of one or more of its constituents.

Coccoliths
(from the Greek *kokkos* = berry and *lithos* = stone) Disc-shaped plates of chalk covering the spherical cell (coccosphere) of coccolithophorids (unicellular sea algae). Through accumulation they form calcareous minerals such as chalk.

Contrast ratio
The ratio of the tristimulus values Y of identical films applied over black and white substrates of defined reflectance.

Corona treatment
Important process for the surface treatment of film. It is often required before film can be coated, laminated or printed. In most cases the surface is activated by bombardment with high-energy ions in a high-voltage field.

Dadmac
(Diallyl dimethyl ammonium chloride) Monomer unit for the production of polydadmac. Polydadmac is a frequently used commercial cationic polymer in paper manufacture.

Delamination
Grinding process for foliated minerals such as kaolin and talc with the aim of releasing individual platelets or thin layers.

Deposit
Natural accumulation of useable minerals and rocks which because of their size can be considered for economic extraction. If the accumulations are too small they are designated as occurrences.

Diagenesis
Totality of the physical and chemical processes which impact on sedimentary deposition and convert this gradually into solid sedimentary rock.

Diapir
(from the Greek *diapeirein* = break through) Mushroom-shaped, anticlinal fold whose core (consisting of salt) has broken through the layers lying above it.

Diatomite
(Diatomoceous earth) Earthy sediment of silicic acid structures of expired diatoms. Used as a filler for gloss/sheen control and thixotroping.

Dilatant
Suspensions which react to increased agitation (= increase in shear stress) with a disproportional rise in viscosity. Examples of dilatant suspensions are: starch in water, wet sand, sediment of oil-based paint.

Dispersibility
Ease with which pigments and fillers finely disperse.

Dolomite
(after Déodat de Dolomieu, French mineralogist, 1750-1801)
1) Sedimentary carbonate rock
2) Mineral of the orthorhombic system, calcium/magnesium/double-carbonate $CaMg(CO_3)_2$

Dry blend
Mixture of individual formulation constituents of a powder coating or plastic composition (PVC powder, calcium carbonate, stabiliser, pigments, etc.) which is made in the extruder before processing.

Elrepho
Spectrophotometer for determining whiteness. The measurement is made in comparison with a standard of barium sulphate. The red (R_x), green (R_y) and blue (R_z) portion of the reflected light is measured. The wavelengths of the filters are precisely determined.

Facies
Totality of the features of a geological body permitting conclusions to be drawn on the conditions under which it was formed.

Fissure
A cleft, without dislocation of the separated parts, which can occur in almost any rocks.

Flint
Silicon-dioxide (SiO_2) mineral of biochemical origin which occurs embedded as an impurity in calcareous minerals, especially chalk. Flint mainly occurs in the form of nodules and rarely also in long bands.

Full colour
Coating which contains a high concentration of a colour pigment.

GCC (Ground Calcium Carbonate)
Ground, natural calcium carbonate

Geode
An internally hollow formation in rocks ranging from a few millimetres to a few centimetres in size. Its inner walls are covered with internally facing crystals (mainly quartz and calcite crystals).

Granule
Designation for coarse solids (e.g. fillers, plastics) with a grain size in the millimetre range.

Grid
Particles which during slurrying of chalk are not suspended and sink to the bottom.

Heat-set-offset printing technique
Offset printing technique with hot-drying printing inks. This printing method is mainly used in web printing. In sheet-fed offset printing oxidative or UV-drying printing inks are used.

Hiding power
Ability of a coating material to cover the colour or colour differences of a background.

High-gradient magnetic separators (HGMS)
Separators which create magnetic fields with high gradients and thus high field strength changes.

Impact strength/Notch impact strength
Impact toughness is designated as the energy required for fracture or deformation. It is referred to the critical cross-section and stated in mJ/mm^2. For notched specimens the term notch toughness is used. The energy is referred in this case to the residual cross-section in the root of the notch.

Igneous rock
Rock formed by the solidification of magma.

Immobilisation point
Limit value for the rise in the solids content of a coating at which the freshly coated, still moist paper surface changes from gloss to matt.

Karst
(from *Karst*, a region in Yugoslavia) Morphology type of the generally plateau-like chalk regions. A karst forms when the rock is dissolved by water with a high content of carbon dioxide.

Metamorphite
Rock formed from another rock as a result of changes in temperature and pressure.

Modulus of elasticity
Measurement of the strength of a plastic. The higher the modulus of elasticity, the less a material elongates under the same strain, and the higher its rigidity.

Muschelkalk
Middle department of the Germanic Trias (Secondary), generally represented by calcareous and dolomite rocks.

Oil absorption
(DIN-ISO 785-5). The content of refined linseed oil absorbed by a pigment or filler sample.

Opacity
(DIN 53146) Degree of light impermeability (of a paper). The opposite of opacity is transparency (see Appendix "Definitions and Measurement Methods").

Orogenesis
(from the Greek *oros* = rock and *gennan* = produce) Process of mountain formation.

Oversize particles
(DIN 66150) The portion of fine material which is above a fixed separation limit of the top cut.

Packing density
In coatings the packing density designates the proportion of the volume made up by solids (fillers, pigments, etc.).

PCC (Precipitated Calcium Carbonate)
Synthetic calcium carbonate.
Petrogenesis
Statement on the formation of a rock.

Petrography
Systematic description of rocks.

Pigment compatibility
Maximum volume concentration of pigments and fillers in a binder. If it is exceeded pigments and fillers combine to form flocculates.

Primer
First layer of a coating system applied directly onto the substrate. The main task of a primer or coat is to impart adhesion for the entire paint coating, as well as to protect against corrosion.

Refractive index
Measurement of the change in direction of a ray of light in the transition from one non-absorbing medium to another.

Resistance to saponification
Resistance of the binder to hydrolytic decomposition through the influence of alkalis.

Rheology
The scientific study of flow and de-composition behaviour of materials

Runnability
Qualitative statement on the running properties of a paper machine.

SC (supercalendered) paper
Class of papers which were finished by supercalendering

Sedigraphy
Method for determining particle size distributions in which the different sedimentation speed of particles is measured in water using X-rays.

Sediment
Rock formed by deposition of weathering and decomposition products.

Serum
Liquid after separation of all solids e.g. by filtering or centrifuging.

Shell marl/limestone
Non-solidified sedimentary rock consisting of numerous residues of shells and a sand or clay matrix

Silicosis
Disease of the lungs caused by the inhalation of particles of silica.

Simplex formation
Combined layering of macromolecules as a result e.g. of opposite charges.

SMC/BMC
Sheet moulding compounds are mats of textile glass pre-impregnated with thermosetting plastics which are produced on continuous units.
Bulk moulding compounds are pasty, in most cases polyester resin moulding compounds with long-glass-fibre reinforcement produced in mixers similarly to thermosetting moulding compounds.

Stock model
In the case of a laboratory stock formulation the paper maker uses the term stock model as the laboratory conditions can only be translated into operating practice as a model owing to the specific cyclic conditions in a paper machine.

Stopper
Collective term for highly filled coating materials used for levelling out surface irregularities which are too big for fillers or primers.

Synthetic resin bound plasters
Easy-to-process pasty materials for decorative coating of facades and walls.

Top cut
The upper grain size of a product is defined as the top cut or d98: 98% of the particles are smaller than this grain size.

Trochites
Petrified remains of the stems or arms of crinoids (sea lilies) recognisable by their symmetry of the 5th order and their shiny fracture surface which corresponds to the crystalline cleavage of a calcite monocrystal. Certain calcareous rocks consist exclusively of trochites (trochite chalk).

Whiteness
Perception of high lightness, high diffusion (scattering) and absence of hue generally applied to opaque or translucent solids (pigments, extenders) or liquids (paints).

White pitches
Sticky, white type of deposit on paper machines which can transfer to the paper. The softening or melting point is frequently between 40 and 120° C.

Whiskers
Designation for the very fine, hair-like inorganic (e.g. SiC or boron nitride) or organic monocrystals used for reinforcing plastic bodies. Their mechanical and physical properties are much better than those of normal crystals, but broader use of whiskers is hindered by processing problems.

Yellowness index
Any of several numerical indices of the degree of departure from a preferred white reflecting material or colourless transparent material towards yellow (see "Definitions and Measurements Methods").

Selection of major standards

DIN EN ISO 3262-1: 1998-08
Fillers for coating materials – requirements and testing methods – Part 1: Introduction and general testing methods

DIN EN ISO 3262-4: 1998-09
Fillers for coating materials – requirements and testing methods – Part 4: Chalk

DIN EN ISO 3262-5: 1998-09
Fillers for coating materials – requirements and testing methods – Part 5: Natural crystalline calcium carbonate

DIN EN ISO 3262-6: 1998-09
Fillers for coating materials – requirements and testing methods – Part 6: Precipitated calcium carbonate

DIN 55625-4: 1996-11 (Draft)
Fillers for plastics – requirements and testing – Part 4: Chalk

DIN 55625-5: 1997-01 (Draft)
Fillers for plastics – requirements and testing – Part 5: Natural crystalline calcium carbonate

DIN 55625-6: 1996-11
Fillers for plastics – requirements and testing – Part 6: Precipitated calcium carbonate

DIN 55943: 1993-11
Colouring agents – terms

DIN EN 971-1: 1996-09
Paints and coatings – Specialist terms and definitions for coating materials – Part 1: General terms

Register

Geography

People and Companies

Technical Terms

Index of important addresses and institutionens

Universities and research institutes

Abo Akademi University
Domkyrkotorget 3, FIN-20500 Abo (Turku), Finland
Tel.: +358-2-215 31, Internet: www.abo.fi

CTP Grenoble, Centre Technique du Papier,
B.P. 251, F-38044 Grenoble Cedex 9, France
Tel.: +33-4-76 15 25 40 15, Fax: +33-4-76 25 15 40 16
E-mail: ctpdoc@ctp.inpg.fr, Internet: www.ctp.inpg.fr

Deutsches Kunststoffinstitut, TU Darmstadt
Schlossgartenstr. 6, 64289 Darmstadt, Germany
Tel.: +49-61 51-16 21 04; Fax: +49-61 51-29 28 55

Fachhochschule Esslingen, Studiengang Farben, Lacke, Umwelt
Kanalstrasse 33, 73728 Esslingen, Germany
Tel.: +49-711-397 30 11, Fax: +49-711-397 30 12, Internet: www.fht-esslingen.de

Forschungsinstitut für Pigmente und Lacke (FPL)
Allmandring 37, 70569 Stuttgart, Germany
Tel.: +49-711-687 80-0, Fax: +49-711-68 80 79

Institut für Kunststoffverarbeitung, RWTH Aachen
Pontstr. 49, 52062 Aachen, Germany
Tel.: +49-241-80 38 06, Fax: +49-241-888 82 62, E-mail: zentrale@ikv.rwth-aachen.de

Institut für Papierfabrikation, Technische Universität Darmstadt
Alexanderstraße 8, 64283 Darmstadt, Germany
Tel.: +49-61 51-16 21 54, Fax: +49-61 51-16 24 54, E-mail: ifp@papier.tu-darmstadt.de

Institut für Verfahrenstechnik Papier e.V. (IVP), Fachhochschule München
Lothstr. 34, 80323 Munich, Germany
Tel.: +49-89-12 65-15 01, Fax: +49-89-12 65-15 02

KCL The Finnish Pulp and Paper Research Institute
P.O. Box 70, FIN-02151 Espoo, Finland
Tel.: +358-9-437 11, Fax: +358-9-46 43 05, Internet: www.kcl.fi

LULEÅ University of Technology, Division of Mineral Processing,
S-97187 Luleå, Sweden
Tel.: +46-920-910 00, Fax +46-920-973 64, Internet: www.km.luth.se

Montanuniversität Leoben
Peter-Tunner-Strasse 5, A-8700 Leoben, Austria
Tel.: +43-384-24 02 (03), Internet: www.unileoben.ac.at

Papiermacherzentrum Gernsbach
Scheffelstr. 29a, 76593 Gernsbach, Germany
Tel.: +49-72 24-64 01-0, Fax: +49-72 24-64 01-114

Papiertechnische Stiftung (PTS)
Heßstr. 134, 80797 Munich, Germany
Tel.: +49-89-121 46-0, Fax: +49-89-123 65 92

PTS, Institut für Zellstoff und Papier
Pirnaer Straße 37, 01809 Heidenau, Germany
Tel.: +49-35 29-54 35-00, Fax: +49-35 29-54 35 74

STFi, Swedish Pulp and Paper Research Institute
Tel.: +46-8-676 70 00, Fax: +46-8-411 55 18, E-mail: info@stfi.se, Internet: www.stfi.se

Süddeusches Kunststoff-Zentrum (SKZ)
Frankfurter Str. 15-17, 97082 Würzburg, Germany
Tel.: +49-931-41 04 164 (184), Fax: +49-931-41 04 2 74(227), E-mail: info@skz.de

Technische Universität Braunschweig, Institut für mechanische Verfahrenstechnik
Volkmaroder Straße 4/5, 38104 Brunswick, Germany
Tel.: +49-531-391 96 11, Fax: +49-531-391 96 33
E-mail fb-mb@tu-bs.de, Internet: www.tu-bs.de

Technische Universität Clausthal, Institut für Aufbereitung und Deponietechnik
Walther-Nernst-Str. 9, 38678 Clausthal-Zellerfeld, Germany
Tel.: +49-53 23-72 20 37 (38), Fax: +49-53 23-72 23 53
E-mail: info@tu-clausthal.de, Internet: www.tu-clausthal.de

Technische Universität Bergakademie Freiberg
Institut für Mechanische Verfahrenstechnik und Aufbereitungstechnik
Agricolastraße 1, 09599 Freiberg, Germany
Tel.: +49-37 31-39 27 95, Fax: +49-37 31-39 29 47
E-mail: hoeschle@mvtat.tu-freiberg.de, Internet: www.tu-freiberg.de

Technische Universität Graz
Rechbauerstrasse 12, A-8010 Graz, Austria
Tel.: +43-316-873 0, Fax: +43-316-873 65 62
E-mail: info@tu-graz.ac.at, Internet: www.tu-graz.ac.at

University of Turku
FIN-20014 Turku, Finland
Tel.: +358-2-333 51, Internet: www.utu.fi

Associations, institutions

American Forest & Paper Association
1111 19th. Street, NW, Suite 800, Washington, DC 20036, USA
Tel.: +1-202-46 27 00, Fax: +1-202-463 24 71

APME – Association of Plastics Manufacturers in Europe
Avenue E van Nieuwenhuyse 4, Box 3,B-1160 Brussels, Belgium
Tel.: +32-2-676 82 59, Fax: +32-2-675 39 35, E-mail: info.apme@apme.org

ASC – The Adhesive and Sealant Council, Inc.
7979 Old Georgetown Road, Suite 500, Bethesda, MD 20814, USA
Tel.: +1-301-968 97 00, Fax: +1-301-968 97 95, E-mail: alies.muskin@ascouncil.org

British Lime Association
156 Buckingham Palace Road, London SW1W9TR, UK
Tel.: +44-171-730 81 94, Fax +44-171-730 43 55,
E-mail: quarry_products.association@virgin.net

Bundesverband der deutschen Kalkindustrie e.V.
Annastr. 67-71; 50968 Cologne, Germany
Tel.: +49-221-93 46 74-0, Fax: +49-221-93 46 74 10 (14)
E-mail: info@kalk.de, Internet: www.kalk.de

CEN – European Committee for Standardization, (für 2000 vergleiche DIN)

CEPE – European Council of the Paint, Printing Inks and Artists' Colours Industry
Avenue E Van Nieuwenhuyse 4, B-1160 Brussels, Belgium
Tel.: +32-2-676 74 80, Fax: +32-2-676 74 90, E-mail: secretariat@cepe.org

CEPI – Confederation of European Paper Industries
250, Avenue Louise, 1050 Brussels, B-Belgium
Tel.: +32-2-627 49 11, Fax: +32-2-646 81 37

Chambre Nationale Syndicale des Fabricants des Chaux Grasses et Magnésiennes
30, avenue de Messine, F-75008 Paris, France
Tel.: +33-1-45 63 02 66, Fax: +33-1-53 75 02 13

Deutsche Landwirtschafts-Gesellschaft e.V. (DLG)
Eschborner Landstr. 122; 60489 Frankfurt on the Main, Germany
Tel.: +49-69-2 47 88-0, Fax: +49-69-247 88 10, E-mail: info@dlg-frankfurt.de

Deutsche Kautschuk-Gesellschaft e.V.
Zeppelinallee 69, 60487 Frankfurt on the Main, Germany
Tel.: +49-69-79 36-153, Fax: +49-69-79 36-155

Deutsches Institut für Normung e.V. (DIN)
Burggrafenstraße 6, 10787 Berlin
Tel.: +49-30-26 01-0, Internet: www.din.de

European Agricultural Societies' Partnership (EASP)
c/o Deutsche Landwirtschafts-Gesellschaft

European Fertilizer Manufacturers' Association (EFMA)
Avenue E. Van Nieuwenhuyse 4; B-1160 Brussels, Belgium
Tel.: +32-2-675 35 50, Fax: +32-2-675 39 61, E-mail: main@efma.be

European Lime Association (EuLA)
c/o Bundesverband der Deutschen Kalkindustrie
E-mail: eula@kalk.de

Fédération d'Associations de Techniciens des Industries des Peintures, Vernis, Emaux et
Encres d'Inprimerie de l'Europe Continentale (FATIPEC), Secretariat General,
Maison de la Chimie, 28 rue St. Dominique, F-75007 Paris, France

Federation of Societies for Coatings Technology
492 Norristown Road, Blue Bell, PA 19422-2350, USA
Tel.: +1-610-940 07 77, Fax: +1-610-940 02 92, E-mail: fsct@coatingstech.org

FEICA – Association of European Adhesives Manufacturers, Sekretariat Düsseldorf
P.O. Box 230169, 40087 Düsseldorf, Germany
Tel.: +49-211-679 31 10, E-mail: mats.hagwall@feica.com

GKV – Gesamtverband Kunststoffverarbeitende Industrie e.V.
Am Hauptbahnhof 12, 60329 Frankfurt on the Main, Germany
Tel.: +49-69-271 05-0, Fax: +49-69-23 27 99

IK – Industrieverband Kunststoffverpackungen e.V.
Kaiser-Friedrich-Promenade 89, 61348 Bad Homburg, Germany
Tel.: +49-61 72-92 66 67, Fax: +49-61 72-92 66 69

International Lime Association (ILA)
c/o National Lime Association

International Paint and Printing Ink Council (IPPIC)
c/o Steve Sides, NPCA, 1500 Rhode Island Avenue, NW, Washington, DC 20005-5503, USA
Tel.: +1-202-462 62 72, Fax: +1-202-462 85 49, E-mail: ippic@paint.org

ISO – International Organization for Standardization, Central Secretariat
1 rue de Varembé, Case postale 56, CH-1211 Genève 20, Switzerland
Tel.: +41-22-749 01 11, Fax: +41-22-733 34 30, E-mail: central@iso.ch

KRV – Kunststoffrohrverband e.V.
Dyroffstr. 2, 53113 Bonn, Germany
Tel.: +49-228-22 35 71, Fax: +49-228-21 13 09

National Lime Association (NLA)
200 North Glebe Road, Suite 800; Arlington, VA 22203-3728, USA
Tel.: +1-703-243 54 63, Fax: +1-703-243 54 89, E-mail: natlime@aol.com

Normenausschuss Pigmente und Füllstoffe (NPF) im DIN,
10772 Berlin,
Tel.: +49-30-26 01 29 30, Fax +49-30-26 01 12 31, E-mail: fritzsche@fa.din.de

PRA – Paint Research Association
8 Waldegrave Road, Teddington, Middlesex, TW11 8LD, UK
Tel.: +44-181-614 48 00, Fax: +44-181-943 47 05, E-mail: coatings@pra.org.uk

Technical Association of the Pulp and Paper Industry (TAPPI),
P.O. Box 10 51 13, Atlanta, GA 303 48-51 13, USA
Tel.: +1-800-333 86 86, Internet: www.tappi.org

Verband der Chemischen Industrie (VCI)
Karlstr. 21, 60329 Frankfurt on the Main, Germany
VdL – Verband der Lackindustrie, Tel.: +49-69-25 56-14 11
VKE – Verband Kunststofferzeugende Industrie, Tel.: +49-69-25 56-13 00
Industrieverband Klebstoffindustrie, P.O Box 230169, 40087 Düsseldorf, Germany
Tel.: +49-211-679 31 10, Fax: +49-211-679 31 88
E-mail: ansgar.v.halteren@klebstoffe.com

Verband Deutscher Papierfabriken e.V.,
Adenauerallee 55, 53113 Bonn, Germany;
Tel.: +49-228-267 05-0, Fax: +49-228-267 05-62, E-mail: info@vdp-online.de

Verband der landwirtschaftlichen Untersuchungs- und Forschungsanstalten (VDLUFA)
Bismarckstr. 41 A, 64289 Darmstadt, Germany
Tel.: +49-61 51-264 85, Fax: +49-61 51-933 70, E-mail: info@vdlufa.de

Verein der Zellstoff- und Papier-Chemiker und -Ingenieure (ZELLCHEMING)
Berliner Allee 56, 64295 Darmstadt, Germany
Tel.: +49-61 51-332 64

Wirtschaftsverband der deutschen Kautschukindustrie (WdK)
Zeppelinallee 69, 60487 Frankfurt on the Main, Germany
Tel.: +49-69-79 36-0

Interesting Web-Sites

www.apme.org

www.coatings.de

www.coatings.site.de

www.coatingsworld.com

www.kunststoffweb.de

www.paintsandcoatings.com

www.paperonline.org

www.rubberstudy.com

Index of illustrations

Museum für Lackkunst, Münster: 63
Naturhistorisches Museum Vienna (Austria), Photo: Alice Schumacher: 69
Naturkunde-Museum Coburg, Coburger Landesstiftung: 134
Anna Neumann/laif: 135
Friedrich Rakob, Rome (Italy): 88
Rausche, BML: 293
Reifenhäuser GmbH, Troisdorf: 261
RenaCare, Hüttenberg: 163 (3. v.l.), 309
Friedrich Rinne, Gesteinskunde, Leipzig 1928: 41 top
Robert Scherer, Die Kreide, Vienna and Leipzig 1922: 145
SLUB, Dresden: 75, 92
SMPK, Berlin: 82 (Kupferstichkabinett), 101 bottom, 109 bottom
SPSG, Berlin-Brandenburg: 124
Staatliche Museen Kassel: 108
Staatliches Mexikanisches Verkehrsamt, Frankfurt on the Main: 100
Taittinger, Reims (Frankreich): 68
Tierpark & Fossilium Bochum, Sammlung Helmut Leich: 133 top
Philippe Tourtebatte, Courville (France): 67
Türkisches Fremdenverkehrsamt, Frankfurt on the Main: 7, 22
Ulmer Weißkalk GmbH: 280
Vinnolit, Ismaning: 241 bottom, 253 both
Gerd Weisgerber, Bochum: 84, 131
F. Zanetti/laif: 114, 127 bottom
Stefan Zenzmaier, Krastal (Austria): 129

Printed by Printforce, United Kingdom